LANDSCAPE INTERFACES

Landscape series

VOLUME 1

Aims & Scope:

The Landscape Series publishes manuscripts approaching landscape from a broad perspective. Landscapes are home and livelihood for people, house historic artefacts, and comprise systems of physical, chemical and biological processes. Landscapes are shaped and governed by human societies, who base their existence on the use of the natural resources. People enjoy the aesthetic qualities of landscapes and their recreational facilities, and design new landscapes. The Landscape Series aims to add new and innovative insights into landscapes. It encourages contributions on theory development as well as applied studies, which may act as best practice. Problem-solving approaches and contributions to planning and management of landscape are most welcome. The Landscape Series wishes to attract outstanding studies from the natural sciences, social sciences, humanities as well as the arts and does especially provide a forum for publications resulting from interdisciplinary and transdisciplinary acting teams. Ideally, the contributions help the application of findings from landscape research to practice, and to feed back again from practice into research.

LANDSCAPE INTERFACES

Cultural Heritage in Changing Landscapes

Edited by

Hannes Palang
University of Tartu, Estonia

and

Gary Fry
Norwegian Agricultural University, Norway

KLUWER ACADEMIC PUBLISHERS
DORDRECHT / BOSTON / LONDON

A C.I.P. Catalogue record for this book is available from the Library of Congress.

ISBN 1-4020-1437-6

Published by Kluwer Academic Publishers,
P.O. Box 17, 3300 AA Dordrecht, The Netherlands.

Sold and distributed in North, Central and South America
by Kluwer Academic Publishers,
101 Philip Drive, Norwell, MA 02061, U.S.A.

In all other countries, sold and distributed
by Kluwer Academic Publishers,
P.O. Box 322, 3300 AH Dordrecht, The Netherlands.

Printed on acid-free paper

Cover photograph by Bärbel Tress and Gunther Tress

Printed in the Netherlands.

Foreword by the series editors

When initiating this book series, it was our most prominent motivation to provide a forum for dealing with the complex and challenging variety of landscapes. The series should bring to the fore the positive and connective aspects of dealing with this variety instead of seeing them as barriers and separating elements. Yet there is not only the variety of the landscapes as such, but also the multiplicity of academic disciplines and approaches that characterize the study of landscapes. We also intended to provide examples of integration of academic knowledge cultures on one topic, rather than having different volumes presenting knowledge from different disciplines.

'Landscape Interfaces', the first volume of the series, mirrors this intention excellently. It is a book that embraces the variety in landscape research. Instead of understanding 'interfaces' as areas where contrasting, opposing or diverging elements clash, Palang and Fry understood and interpreted 'interfaces' as the area where two elements meet and can be integrated for their common benefit. Interfaces are many in landscapes and landscape research: between different academic disciplines, experts and laypersons, cultures, past and future, time and space. These interfaces are discussed in relation to cultural heritage management in changing landscapes in different geographical localities. We hope that this book gives all who are interested in the management of cultural heritage - students as well as professionals - the opportunity to be inspired by the multiplicity of approaches that are presented and the subjects that are discussed.

Toulouse and Wageningen, May 2003

Henri Décamps
Bärbel Tress
Gunther Tress

Contents

Contributing Authors

Chad W. Adams is an environmental planner and landscape architect with Andropogon Associates, Ltd. in Philadelphia, Pennsylvania, USA.

Helen Alumäe is PhD student at the Institute of Geography, University of Tartu, Estonia. She also works as an editor for the journal *Eesti Loodus/Estonian Nature.*

Teresa Andresen is associate professor in the School of Sciences at the Oporto University, Portugal, in the Landscape Architecture program. She is co-author of the Douro inscription in the World Heritage List.

Marc Antrop is professor of geography at the Ghent University, Belgium.

Invild Austad works as professor of landscape ecology at the Faculty of Science, Sogn og Fjordane University College in Sogndal, Norway. She is also a member of the Norwegian Academy of Sciences and Letters.

Margot Cantwell is the president of EDM – Environmental Design and Management, Ltd. in Halifax, Nova Scotia, Canada.

Denis Cosgrove is the Alexander von Humboldt professor in geography at the University of California Los Angeles, USA.

Maria José Curado is landscape architect and PhD candidate at the University of Aveiro, Portugal.

Stein Tage Domaas is PhD student at the Department of Landscape Planning at the Swedish Agricultural University in Alnarp, Sweden. He works as a research fellow at the Faculty of Science, Sogn og Fjordane University College in Sogndal, Norway.

Graham Fairclough is an archaeologist working in English Heritage, one of the UK's four national heritage boards. He is currently head of EH's Characterisation Team.

Robert Fish is lecturer in human geography at the University of Nottingham, England. Robert's research interests are in the culture and environment of the British countryside.

Gary Fry holds professorship in landscape planning at the Norwegian Agricultural University in Ås.

Katsue Fukamachi is a landscape architect working at the Kansai Research Center, Forestry and Forest Products Research Institute, Japan.

Roland Gustavsson is professor in landscape planning at the Department of Landscape Planning, Swedish University of Agricultural Sciences in Alnarp, Sweden, specialising in planting design and landscape management.

Roy Haines-Young is professor of physical geography at the University of Nottingham, England. The focus of Roy's research is on managing for sustainability at the landscape scale.

Jussi S. Jauhiainen works as professor of geography and regional planning at the University of Oulu, Finland, and associate professor in urban geography at the University of Tartu, Estonia.

Michael Jones works as professor of geography at the Norwegian University of Science and Technology in Trondheim, Norway. He is also a member of the Norwegian Academy of Science and Letters.

Egle Kaur is PhD student at the Institute of Geography, University of Tartu, Estonia.

Renata Krukowska is PhD student at the Department of Regional Geography, Institute of Earth Sciences, Maria Curie-Sklodowska University, Lublin, Poland.

Päivi Maaranen works as curator in the National Board of Antiquities in Finland. Her research focuses on landscape change and human impact on environment.

Jan Diek van Mansvelt is co-ordinator of the European Academy for the Culture of Landscape PETRARCA.

Ann Norderhaug was associate professor at the Faculty of Science, Sogn og Fjordane University College in Sogndal, Norway. She is currently working as a researcher at The Norwegian Crop Research Institute.

Hirokazu Oku is a landscape architect working at the Kansai Research Center, Forestry and Forest Products Research Institute, Japan.

Hannes Palang is senior researcher at the chair of human geography, Institute of Geography, University of Tartu, Estonia.

Bas Pedroli is senior researcher at Alterra Green World Research, Wageningen, The Netherlands.

Tiina Peil is research fellow at Estonian Institute for Sustainable Development, Stockholm Environment Institute Tallinn Centre, Estonia.

Anna Peterson is a landscape architect and works with her PhD concerning trees in cultural landscapes at the Department of Landscape Planning, Swedish University of Agricultural Sciences in Alnarp, Sweden.

Anu Printsmann is MSc student at the Institute of Geography, University of Tartu, Estonia.

Oliver Rackham carries out research in the Botany Department at Cambridge University, where he is a Fellow of Corpus Christi College, United Kingdom.

Jorge Rubiano has recently completed his PhD at the University of Nottingham, England. Jorge's research concerns itself with the role of GIS in stakeholder planning at the landscape scale.

Ewa Skowronek works as academic researcher at the Department of Regional Geography, Institute of Earth Sciences, Maria Curie-Sklodowska University, Lublin, Poland.

Helen Sooväli is PhD student at the Institute of Geography, University of Tartu, Estonia working parallelly as a lecturer in cultural geography at the same institute.

Grete Swensen is dr.art in ethnology and works as researcher in Norwegian Institute for Cultural Heritage Research, Oslo, Norway.

Andrzej Swieca works as professor at the Department of Regional Geography, Institute of Earth Sciences, Maria Curie-Sklodowska University, Lublin, Poland.

Jan Anders Timberlid works as associate professor at the Faculty of Social Science, Sogn og Fjordane University College in Sogndal, Norway.

Bärbel Tress is senior landscape researcher at the Landscape Centre, Alterra Green World Research in Wageningen, The Netherlands. She specializes on future studies and interdisciplinarity in landscape research.

Gunther Tress is senior landscape researcher at the Landscape Centre, Alterra Green World Research in Wageningen, The Netherlands. His main fields of interest are integrated approaches for research and management of landscapes and scenario studies.

Irmine Vermandere is MSc student at the Department of Geography, Ghent University, Belgium.

Preface

This book has been initiated by the workshop on *Cultural heritage in changing landscapes*, held during the IALE (International Association for Landscape Ecology) European Conference that started in Stockholm, Sweden, in June 2001 and continued across the Baltic to Tartu, Estonia, in July. The papers presented at the workshop have been supported by invited contributions that address a wider range of the cultural heritage management issues and research interfaces required to study cultural landscapes.

The book focuses on landscape interfaces. Both the ones we find out there in the landscape and the ones we face while doing research. We hope that this book helps if not to make use of these interfaces, then at least to map them and bridge some of the gaps between them.

The editors wish to thank those people helping us to assemble this collection. First of all our gratitude goes to the authors who contributed to the book. We would like to thank Marc Antrop, Mats Widgren, Roland Gustavsson, Marion Potschin, Bärbel Tress, Tiina Peil, Helen Sooväli and Anu Printsmann for their quick and helpful advice, opinions and comments during the different stages of editing. Helen Sooväli and Anu Printsmann together with Piret Pungas – thank you for technical help.

Tartu and Ås
April 2003

Chapter 1

LANDSCAPE INTERFACES

Introduction

Hannes Palang & Gary Fry
Institute of Geography, University of Tartu, Estonia
Institute of Landscape Planning, Norwegian Agricultural University, Norway

1. INTRODUCTION

This book is based largely on the outcome of a workshop on cultural heritage aspects of landscapes, which was held during the IALE European conference in Stockholm and Tartu in 2001. The workshop initially targeted the management of cultural heritage and identifying the links between landscape ecology and archaeology, but during the workshop the scope grew wider. As a result, the conference papers have been supported by invited contributions that address a wider range of related issues.

The workshop raised several questions for both researchers and managers of cultural heritage. First, can historic landscapes be characterized? Is it possible to determine a typology of historic landscapes and if so can this be used across regional and national boundaries? What about the non-visual aspects of landscape, how are they incorporated into the way we perceive and categorise landscapes? How continuous are the landscapes and for whom? Second, how far can one read the landscape? Is everything that one sees and perceives in the landscape equally understandable to everyone? Or does knowledge and experience determine our landscape literacy and the weighting we give to particular historic layers in the landscape? Third, when we speak of landscape restoration, what is the authentic landscape against which to measure success and what is legitimate to restore? How objective is that 'past authentic landscape'? And, finally, who is in position to make decisions about landscape change?

H. Palang and G. Fry (eds.), Landscape Interfaces, 1–13.

This brief introduction tries to map some of the intertwined interfaces researchers face in the landscape as well as in their research.

2. LANDSCAPE INTERFACES

There are four things to bear in mind when studying landscapes: forms, functions, processes, and context (Widgren 2002). These four attributes can all be studied separately, but then one is not studying landscape. Forms have meaning in a specific context. Landscape conservation decisions force one to choose between restoring the processes shaping that landscape or trying to restore the state, the appearance of the landscape. While it is possible to restore the state of a landscape (mostly its appearance), it is much more difficult to restore the context that gave rise to the processes and functions of the landscape.

Do we need living landscapes with forms, functions and processes in the right context, or should cultural heritage management be about creating museum landscapes that possess either forms or functions, but no context? Can we really conserve cultural landscapes in new contexts where the original reasons for their existence no longer exist? Is it possible to find new markets or other drivers of change to provide new contexts for new and valued landscapes?

Landscapes have a role to play in the shaping of local, regional and national identity. But, whose identity is it? Are we talking about those people who have lived in the landscape for generations, new minority groups or should we also consider those who come to see the landscape for a couple of days and then leave?

Landscape is an agreement, but an agreement between whom? Landscapes can be elitist in situations where a small group of powerful people is in a position to make decisions about future developments, but we also find democratic landscapes, where local people participate in planning and managing. Which of these do we value? And is insiders' subjective perspective on the landscape more or less important than the 'objective' analysis of an expert-outsider?

All of these questions and many more debates are, and perhaps always will be, part of the challenge facing all those who ask what landscape is. Similarly, there will always be interfaces between different approaches and 'readings' of the landscape, each valid in its specific social and historical context.

2.1 Humanities/natural sciences interface

The major conflict in landscape studies seems to be between the humanities and the natural sciences. Both lay claims to landscape studies and both have acquired 'ownership rights' over certain aspects of landscape knowledge. How then do we combine these and other knowledge cultures that may find landscape a challenging research topic? Cosgrove (2003) compares landscape concepts through reference to two influential texts representing each side of this divide, and points to an approach that could bring the two perspectives, ecological and semiotic, closer together. Another way would be to increase interdisciplinary approaches in landscape studies. Although landscape as a concept already implies interdisciplinarity, its study rarely achieves levels of theoretical integration (Tress et al. 2003).

The problems are connected to the interface between material and mental landscapes but go deeper into very basic ontological issues regarding what constitutes reliable knowledge and what can be accepted as data.

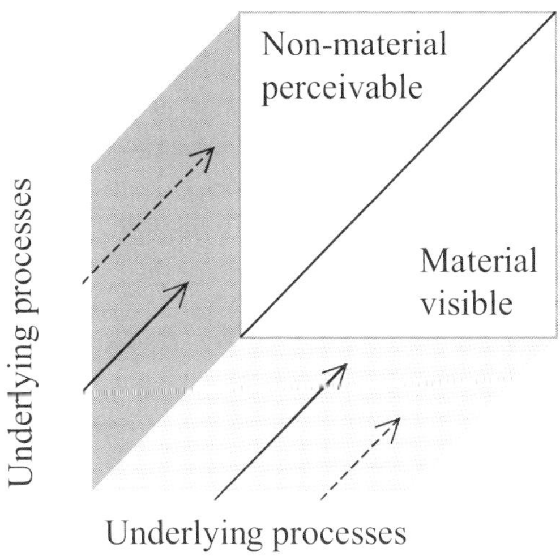

Figure 1-1. Landscape model used by Keisteri (1990)

The human landscape is formed in our minds and studied through a wide range of humanistic approaches, particularly phenomenological perspectives and techniques. The material landscape is the one we can touch and smell and see and measure. In addition to these two layers, there are different driving forces and underlying (spatial) processes involved in forming the character of landscapes. These factors also contribute to landscape studies. The study of visible elements falls into the scope of natural sciences. The non-material, perceivable layer interests social sciences and humanities. And the time dimension rising from the underlying processes broadens the study beyond the traditional limits of a landscape study, towards geology and political science for example. All this leads to understanding that in fact landscape could be handled as something indicating the quality of life of a society.

Linking the two approaches together has been and continues to be a difficult process. The main argument for combining several methods is the need for us to understand the relationships and interactions between mental and material landscapes. Similar physical settings often have different histories and hence evolved different visual character and meanings, yet they may also share many commonalities; neither aspect can be effectively studied with conventional disciplinary tools. Fortunately, there are several research initiatives making progress in this field and include phenomenological interpretations of physical landscape attributes or the use of computer-aided visual analysis techniques to provide insights into human cognition of the view (Fry et al. 2003). We fully support the place of both aspects in landscape research. The challenge lies at the interface between them and, especially, how to develop joint theory. Holism is claimed to offer the possibility of increasing collaboration between different approaches. Although, the detail of what this means at an operational level requires further clarification. The development of new theoretical insights in landscape studies remains a clear necessity and we believe that such developments are most likely to arise at the interfaces between existing knowledge cultures.

Part of the problem with the interface is that the extreme relativist approach to landscape has rather little to offer landscape management, which deals with very real problems in material landscapes. How can insights gathered from studying the mental landscape best be incorporated into the knowledge base needed for solving current landscape conflicts is a key question.

As demonstrated by Fish et al. (2003) and Cantwell & Adams (2003) in this book, local knowledge can often significantly add to, if not replace, the methods conventionally used in landscape planning and management. So, the challenge is in how to make 'relativistic' and 'subjective' research arguments valid for natural sciences. This will require careful, sensitive and committed research at the interface.

2.2 Culture/culture interface

This interface deals with the differences in landscape values resulting from different cultural perspectives. It is the way different cultures perceive and interpret landscape that is in focus, as well as the way cultures give landscape the symbolic meanings. This interface between cultures is closely associated with diverse cultural and political histories and is to be seen reflected in dissimilarities in material and mental landscapes (Figure 1). The interface is often observed along borders between countries trough patterns of land use that may reflect different national products or land tenure systems.

The study of the cultural interface in landscape research is extremely important for the understanding of how landscape evolved in the past and how conflicts may arise in the future. Societies are not homogeneous; all subcultures may have different understandings of the landscape. Usually these issues have been studied in towns and cities, while rural landscapes have been thought to be much more stable and similar in that sense. The interface also links with the past/future interface since cultures can be from various geographic, ethnic, socio-economic groups or the same group at different time periods. This leads to the issue of cultural heritage management.

Most landscape research has its roots in the European understanding of landscape, although, within this, several branches, approaches and trends exist. However, common history and similar religious background are the features that form the basis for European landscape approach (see Jones 2003 in this book; Claval 2002; Keisteri 1990 for more). There are numerous examples in the world showing how the invasion of populations with different cultural backgrounds have started to reshape the landscape so that it reminded them their needs (e.g., Meurk & Swaffield 2000 for New Zealand; Rechtman & Amir 2001 for Israel; and Skowronek et al. 2003 for the Polish-Ukrainian borderland in this book). Different cultures also attach different meanings to similar landscape elements – the example of hedgerows on England and Japan in this book (Fukamachi et al. 2003) is an example. Here is another link to the past/future interface – during time cultures change, something that was heritage for one culture could be handled by another as virgin wilderness that should be tamed and domesticated.

2.3 Past/future interface

The past/future interface in landscape studies focuses on temporal landscape dynamics. The temporal dynamics have often been seen as a fundamental aspect of landscape studies and this is also reflected in most landscape ecology definitions. The interface is in the present but at the junction where we both look forwards and backwards. The techniques used by futures studies and those of history have some similarities but they are not the same – can we find new ways of looking at the past or new ways of predicting the future through combining approaches?

Landscape can here be handled as a palimpsest, consisting of elements from different time periods. Every succeeding socio-economic formation (*sensu* Cosgrove 1984) has tried to erase the elements erected by the previous period and replace them by elements of its own. However, never all the elements are removed; the landscape we have today has remnants of several former periods (Figure 2).

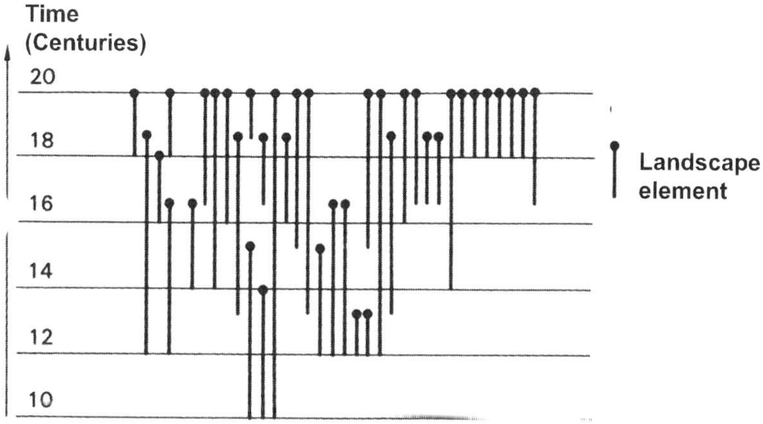

Figure 1-2. Landscape is a palimpsest of elements (after Vervloet 1986)

Cosgrove (1984) has shown how different socio-economic formations create their own landscapes with their characteristic features – value systems, etc. Antrop (2000) has distinguished between traditional and modern landscapes. The former are landscapes that have remained the same during several human generations, while modern landscapes have changed several times during one generation. The recent political changes in Eastern Europe have shown that the changes in socio-economic formations create time barriers in the landscape that are not transparent: younger people who have not lived in the previous formation are unable to understand those landscapes (Figure 3; Palang et al. 2002). For example, the elements created during the period of collective farms are still visible in the landscape, but their meaning and function remain incomprehensible for younger generations and people living in the cities. This once more confirms the claim of Jones (1991) and Widgren (2002) that landscapes can be read, interpreted and understood only in the historic and political context specific to the time the objects originated from.

How to avoid such incomprehension and alienation from 'own' landscapes in the future? Scenario testing is one of the important tools in future studies but how do we assess the reliability of both the scenarios and their consequences? Several studies (e.g. Emmelin 1996; Schoute et al.

1996; Tress & Tress 2003 etc.) have tried retrospective future studies by predicting the present from historical data. Combining a wide range of research approaches will be necessary to advance our ability to predict alternative landscape futures and their social and environmental consequences.

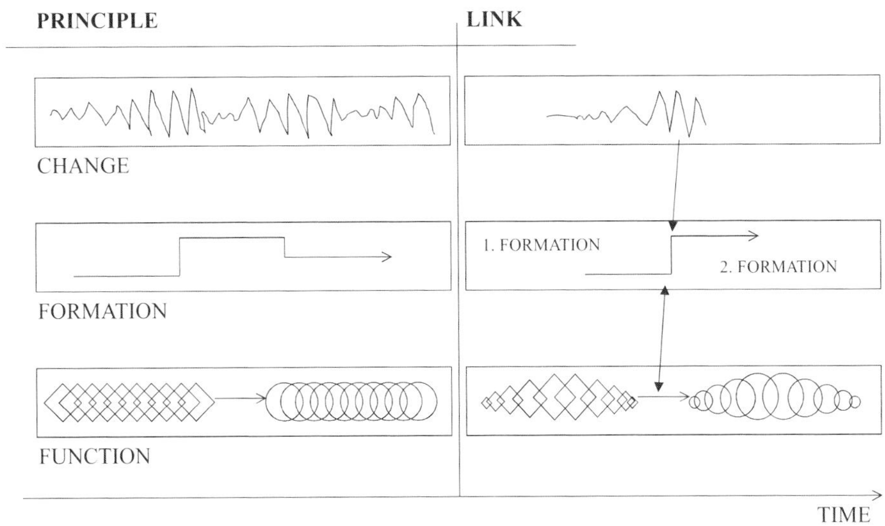

Figure 1-3. Time barriers in landscape (Palang et al. 2002)

2.4 Time/space interface

This interface represents the interaction of time-space continuum. It can also be viewed as underlying processes in Figure 1, which influences the design and planning of landscape research and management. Often we focus on just one of these two fundamental variables treating the other as fixed, such that we study spatial processes via a snapshot in time, and temporal processes at just one point in space as if this was not connected to other similar points.

Time/space is an important interface for the development of our understanding and hence theory development in the realm of landscape development and dynamics. For example, land use change transition matrices can provide important insights into the dynamics of landscape change through the study not only of mass change but also *where* that change took place and *how* each parcel of land became transformed to another use.

It is also an important interface for combining technical approaches with classical historical-geographical methodology. The web model developed in

1960s by a Swedish geographer Torsten Hägerstrand, later used in geography by e.g., Anne Buttimer (2002) and landscape studies by Lars Emmelin (1996) might provide a useful tool for further research. Old diffusion-theory actually combines both – time and space and with modern means of analysis – and the outcome is still unclear.

One of the most common shortcomings in landscape research is the treatment of landscapes as static units that can once be made finally ready. On the contrary, landscapes are dynamic (see, e.g. Antrop 2000), they are in permanent change, and we humans can only try to steer that change in directions we think suit us.

2.5 Expert/lay person interface

This interface has two main starting points. One is the old debate about whether local subjective knowledge is as valuable as the more objective scientific knowledge; another is derived from the previous one and studies how to involve lay people in planning and management exercises.

One of the current social trends is the involvement of lay people especially stakeholders, into the research process. This participatory research and the related action research are both popular social science approaches offering a way for the end users of research to have input to defining the research objectives, research methods, data collection, analysis and interpretation.

This interface also leads to the moral issues of landscape. Who is in position to decide what? Whose landscapes are the right ones? Are these the landscapes of the local people or are the outsiders/experts in position to decide? And how are related the issues of landscape as a common good and land underneath it as private property? In this book this interface has been one of the most important ones tackled by several authors. It seems that while human geographers are used to the 'lure of local', as Mitchell (2001) put it, and able to handle and analyze it, also landscape ecologists and planners are increasingly finding this issue appealing, be it participatory planning (*sensu* Luz 2000, or e.g., Swensen (2003), Andresen & Curado (2003), Alumäe et al. (2003) in this book) or other approaches. It becomes more and more apparent that landscapes are not primarily for scientists to study and planners and decision makers to manage, but more importantly, for local people to live in and for tourists to gaze at and roam.

This is maybe one of the most challenging interfaces in the landscape study at the moment, as it poses many unanswered questions. Does it remove landscape conflicts? How does the position of the researcher change and is this a good thing? How do we get the best form this interface, while at the same time avoiding the process of participatory research becoming an end in itself?

2.6 Preservation/use interface

This seems to be one of the major challenges in landscape planning and management. It addresses the basic questions we have to solve – do we need museum landscapes that preserve the appearance of a certain time or do we want landscapes that live the life it used to. Things that loose their functions tend to disappear. The same holds good for landscapes – change is part of the landscape, and new technologies, new ideas, new developments alter the traditional ones.

With preservation, authenticity is one of the major concerns. Things have to be old to be appreciated by locals, argue Alumäe et al. (2003) in this book. Old things could be described (Fairclough 2003) and then maybe included in the heritage lists (Andresen & Curado 2003), but are they really authentic? People's memory seem to go back three generations, sometimes the landscape of their childhood is the ideal against which to measure the changes. Authenticity is about subjectivity and ethics (Gustavsson & Peterson 2003) and should focus much more on places and landscapes, not just objects and habitats.

A link goes to the expert/lay people interface from here. Local people live in the landscape and use it. Mostly they also appreciate it, as shown by Cantwell & Adams (2003). Management changes can only be carried out by convincing the land owners/users in some way. Planners sometimes forget that they do not start from a white sheet – there may have been previous planning initiatives, some more successful and acceptable, others stepping on the toes of local people (Luz 2000). How to preserve spiritual, cultural, associative values? Insiders carry on this knowledge, outsiders can find it out, but it takes time to get into these layers of the landscape.

Identity is important. While globalization makes landscapes all over the world look the same, there is a tendency towards using landscape as a cornerstone of regional and national identities. Research is being carried out into Englishness (Palmer 2002), Finnishness (Häyrynen 2002) etc. of the landscape, traditions and values are being inventoried, mapped and characterized. Also, the *European Landscape Convention* stresses this interface.

And tourism brings in an extra dimension, on one hand as an income-generator, but on the other hand as the major threat to both landscape and identity.

3. INTERFACES IN THIS BOOK

The chapters in this collection address several interfaces landscape contains. These interfaces often indicate the division lines between the various academic disciplines, but also maybe knowledge cultures. The landscapes explored are manifold, as are the approaches and viewpoints. The book contains landscapes of cultural geographers, landscape ecologists, archaeologists, planners, landscape architects, but as well those of Norway, Estonia, the Netherlands, England, Japan, Belgium, Finland, Poland, seen from both insiders and outsiders perspective, having in mind research and management.

The introductory part of the book contains three chapters. After this one, Denis Cosgrove discusses the two sides of landscape, the ecology and the semiosis. While competing with each other, these two sides in fact are both needed in landscape discourse. Then Michael Jones studies the origin of the term *cultural landscapes* in European (English-speaking) research tradition. There are several developments, intertwined discourses and narratives, both language-wise and discipline-wise, that influence our understanding of the landscape

The two following chapters focus on the way different cultures understand and handle the landscape. Katsue Fukamachi, Hirokazu Oku and Oliver Rackham compare the role of hedgerows in two cultures, in England and in Japan. In one of them, trees and hedgerows have more significance as landmarks, while in the other the spiritual link between people and the nature is still there and the hedges have a different role to play. Ewa Skowronek, Renata Krukowska and Andrzej Swieca explore the changes the Polish-Ukrainian borderland has undergone during several centuries. They show how very heterogeneous cultural setting also creates differences in the landscape. And while the human population became more and more uniform due to political changes, the landscape did the same.

The third block of papers tries to explore how cultural values are interpreted and assessed in the landscape. The part starts with Marc Antrop's explanation of what are cultural values in the landscape and how the Flemish landscapes were assessed, based on this. Teresa Andresen and Maria José Curado follow that line, showing the importance of local heritage and how this has supported creating of a UNESCO World Heritage site in Portugal. Helen Alumäe, Anu Printsmann and Hannes Palang share the experience gained during a planning project with finding out what has been seen valuable in the landscape by local people. Finally, Robert Fish, Roy Haines-Young and Jorge Rubiano show how the local knowledge could be linked with more 'scientific' approach to landscape management, the one using GIS and other analytical tools.

The latter issue is more deeply discussed in the following part that studies the new tools that could be used in landscape studies. First, Margot Cantwell and Chad Adams show how technical approach to landscape can reach to very similar results to the local knowledge. A landscape suitability analysis using the state-of-the-art computer-based methodology was used to study the future options for development of an area in the Big Cove Reserve in Canada. Finally "when all the maps were ready, we approached the local community to show the results. They knew it all". Bärbel and Gunther Tress display the possibilities of explaining the future options of landscape development with computer visualization technique. This is useful for explaining the possibilities for future developments, but as well their impact. It also helps to gain a better contextual overview of future options and the results of people's choices. Stein Tage Domaas, Ingvild Austad, Jan Anders Timberlid and Ann Norderhaug show how historical cadastral maps can be of use while making landscape ecological decisions. Historical maps also show that elements that in the local knowledge are depicted as 'ancient' or at least 'old' might not be that in reality. Modern analytical tools enable to make the full use of old maps containing important information.

Several chapters explore cultural landscapes and authenticity, the interface between the present and the past. Gary Fry discusses the possible links between landscape ecology and landscape archaeology. Päivi Maaranen shares the experience gained in Finland in cultural heritage management, how management of archaeological sites is moving from site based approach to place and landscape based ones. Grete Swensen takes the issue of cultural landscape management in Norwegian small towns and show how the stakeholders' role-play works in planning. After that Graham Fairclough describes one of the methods used in England by English Heritage to assess the value of historical cultural landscapes, the Historical Landscape Characterization. We regret that another similar project (see Warnock et al. 2001) did not make it into this book. Finally, Roland Gustavsson and Anna Peterson pick up the discussion about what is authentic in landscape management and whether we should aim at historical authenticity – the object, or authentic illusion – the idea of the landscape.

The last three chapters outline some possibilities for future landscape research. Helen Sooväli, Hannes Palang, Egle Kaur, Tiina Peil and Irmine Vermandere show the power of combining different perspectives in landscape studies. They explore the visible changes and the shifting meanings, the ways landscape is presented in various languages, for 'our' people and for strangers, for locals and for tourists. None of these approaches separately taken can give a full understanding how the material landscape and the perception of landscape have changed during time; while combined, they allow getting a broader understanding of what is happening in the landscape. Jan Diek van Mansvelt and Bas Pedroli stress the need for insider view in landscape discourse. Landscape is not just about gathering the facts, it is rather a collection of places that have distinct identity, the sense of place. And there is a need to pass this sense over from one generation of people to another. Finally, Jussi Jauhiainen explores the landscapes of the post-postmodern era.

But there are more landscapes and more interfaces to discover.

REFERENCES

Alumäe, H., Printsmann, A., & Palang, H. (2003). Cultural and historical values in landscape planning: locals' perception. *This volume.*

Andresen, T., & Curado, M.J. (2003). Shaping the future of a cultural landscape: the Douro Valley wine region. *This volume.*

Antrop, M. (2000). Background concepts for integrated landscape analysis. *Agriculture, Ecosystems and Environment, 77*, 17-28.

Antrop, M. (2003). The role of cultural values in modern landscapes. *This Volume.*

Buttimer, A. (2002). *Sustainable Landscapes and Lifeways. Scale and appropriateness.* Cork: Cork University Press.

Cantwell, M., & Adams, C.W. (2003). An aboriginal planning initiative: sacred knowledge and landscape suitability analysis. *This volume.*

Claval, P. (2002). *The language of the rural landscape.* Paper presented at 20th Session of the Permanent European Conference for the Study of Rural Landscape, Tartu-Otepää, Estonia.

Cosgrove, D. (1984). *Social Formation and Symbolic Landscape.* Madison: University of Wisconsin Press.

Cosgrove, D. (2003). Landscape: ecology and semiosis. *This volume.*

Domaas, S.T., Austad, I., Timberlid, J.A., & Norderhaug, A. (2003). Historical cadastral maps as a tool for valuation of today's landscape elements. *This volume.*

Emmelin, L. (1996). Landscape impact analysis: a systematic approach to landscape impacts of policy. *Landscape Research, 21* (1), 13-35.

Fairclough, G. (2003). 'The Long Chain': archaeology, historical landscape characterization and time depth in the landscape. *This volume.*

Fish, R., Haines-Young, R., & Rubiano, J. (2003). Stakeholder landscapes and GIS. *This volume.*

Fry, G. (2003). From objects to landscapes in natural and cultural heritage management: a role for landscape interfaces. *This volume.*

Fry, G., Skar, B. G., Jerpåsen, B., & Bakkestuen, V. (2003). Predicting archaeological sites: a method based on landscape indicators. *Landscape and Urban Planning,* In press.

Fukamachi, K., Oku, H., & Rackham, O. (2003). A comparative study on trees and hedgerows in Japan and England. *This volume.*

Gustavsson, R., & Peterson, A. (2003). Authenticity in landscape conservation and management – the importance of the local context. *This volume.*

Häyrynen, M. (2002). *The Representation of the Countryside in Finnish National Landscape Imagery.* Paper presented at 20th Session of the Permanent European Conference for the Study of Rural Landscape, Tartu-Otepää, Estonia.

Jauhiainen, J.S. (2003). Learning from Tartu – towards post-postmodern landscapes. *This volume.*

Jones, M. (1991). The elusive reality of landscape. Concepts and approaches in landscape research. *Norsk Geografisk Tidsskrift, 45* (4), 153-169.

Jones, M. (2003). The concept of cultural landscape – discourse and narratives. *This volume.*

Keisteri, T. (1990). The study of changes in cultural landscapes. *Fennia, 168* (1), 31-115.

Maaranen, P. (2003). Landscape archaeology and management of ancient cultural heritage sites. *This volume.*

Meurk, C.D., & Swaffield S.R. (2000). A landscape ecological framework for indigenous regeneration in rural New Zealand-Aotearoa. *Landscape and Urban Planning, 50*, 129-144.

Mitchell D. (2001). The lure of the local: landscape studies at the end of a troubled century. *Progress in Human Geography, 25* (2), 269-281.

Luz, F. (2000). Participatory landscape ecology – a basis for acceptance and implementation. *Landscape and Urban Planning, 50*, 157-166.

Palang, H., Külvik, M., Printsmann, A., Kaur, E., & Alumäe H. (2002). Maastik, sidusus, identiteet. In Ü. Kaevats (Ed.) *Usaldus. Vastutus. Sidusus. Eesti sotsiaalteadlaste aastakonverents* (pp. 27-32).Tallinn: TTÜ Kirjastus.

Palmer, C. (2002). Christianity, Englishness and the southern English countryside: a study of the work of H.J. Massingham. *Social & Cultural Geography, 3* (1), 25-39.

Rechtman, O., & Amir, S. (2001). The impact of the Zionist development ideology on the traditional Mediterranean landscape. In Ü. Mander, A. Printsmann, & H. Palang (Eds.) *Development of European Landscapes. Vol.* 2. *Conference proceedings IALE European Conference 2001.* Publicationes Instituti Geographici Universitatis Tartuensis (pp.198-202). Tartu: Tartu University Press.

Schoute, J.F.T., Finke, P.A., Veeneklaas, F.R., & Wolfert, H.P. (Eds.) (1995). *Scenario studies for the rural environment.* Boston: Kluwer.

Skowronek, E. Krukowska, R., & Swieca A. (2003). Transformations of cultural landscape: the case of the Polish-Ukrainian borderland. *This volume.*

Sooväli, H., Palang, H., Kaur, E., Peil, T., & Vermandere, I. (2003). Combining approaches in landscape research. *This volume.*

Swensen, G. (2003). Pressure on the fringe of the cities. *This volume.*

Tress, B., & Tress, G. (2003). Communicating landscape development plans through scenario visualization techniques. *This volume.*

Tress, B., & Tress, G. (2003). Scenario visualisation for participatory landscape planning – a study from Denmark. *Landscape and Urban Planning*, In press.

Tress, B., Tress, G., van der Valk, A., & Fry, G. (Eds.) (2003). *Interdisciplinary and transdisciplinary landscapes studies: potential and limitations.* DELTA Series 2, Wageningen.

van Mansvelt, J.D., & Pedroli, B. (2003). Landscape – a matter of identity and integrity. *This volume.*

Vervloet, J. (1986). *Inleiding tot de historische geografie van de Nederlandse cultuurlandschappen.* Wageningen. Pudoc.

Warnock, S., Griffiths, G., Porter, J. & Simmons, E. (2001). The cultural component of landscpe character assessment: data sources and typology. In Ü. Mander, A. Printsmann, & H. Palang (Eds.) *Development of European Landscapes. Vol.* 2. *Conference proceedings IALE European Conference 2001.* Publicationes Instituti Geographici Universitatis Tartuensis (pp.229-230). Tartu: Tartu University Press.

Widgren, M. (2002). *Three perspectives on landscapes.* Paper presented at 20th Session of the Permanent European Conference for the Study of Rural Landscape, Tartu-Otepää, Estonia.

Chapter 2

LANDSCAPE: ECOLOGY AND SEMIOSIS

Denis Cosgrove
Department of Geography, University of California Los Angeles, USA

A principal purpose of this collection of essays is to bring together two approaches to landscape whose divergence has in some ways been exaggerated by trends within recent scholarship. For the sake of simplicity we might call one of these *ecological*, the other *semiotic*. I am conscious that my use here of these highly technical terms is loose, but I employ them to signify *discourses* rather than scientific and technical concepts and methods. An *ecological* landscape discourse focuses on the complex interactions of natural processes (geomorphological, climatic, biological, vegetational, etc.) shaping characteristic land areas, and extending its concerns to the ways that human activities interact with these natural processes. Frequently, but not invariably, there is an implicit (sometimes explicit) declensionist tone to ecological landscape study: human interactions – at least in the modern world – are seen as detrimental to balanced or stable systems and the landscapes that emerge from these. A *semiotic* approach to landscape is skeptical of scientific claims to represent mimetically real processes shaping the world around us. It lays scholarly emphasis more on the context and processes through which cultural meanings are invested into and shape a world whose 'nature' is known only through human cognition and representation, and is thus always symbolically mediated. The semiotic approach is less insistently pessimistic in its conclusions about the future of landscape; more commonly it is politically progressive, even activist, placing those landscapes within a broader critique of social form and process.

Anyone seriously concerned with understanding and perhaps regulating the changing appearance of Europe's landscapes and the natural and social processes that have shaped and sustain them needs to be attentive to both these discourses, and to hope for some kind of open dialogue between them. This is the goal of many of the papers in this collection, although any such

H. Palang and G. Fry (eds.), Landscape Interfaces, 15–20.

project faces the problems of communication that emerge from the lack of a shared ontology, epistemology and scientific language. In this brief introductory essay, my intention is not to offer a summary or judgement on the success of the papers in meeting the objective of communication. Rather, it is to comment upon the possibilities for moving forward in our appreciation of landscape today by drawing upon two recently published historical landscape studies that, while to some degree representing conceptually and methodologically the ecological and the semiotic approaches to landscape respectively, may demonstrate the possibilities for a convergent understanding of landscape. Such convergence may ultimately help shape practical responses to the complex questions posed by Europe's rural landscapes in the 21st century. The two books are Franciscus Wilhelmus Maria Vera's *Grazing ecology and forest history* (2002) and Kenneth Olwig's *Landscape, Nature and the Body Politic: from Britain's Renaissance to America's New World* (2002). Each is a highly specialized work, firmly set in its own scholarly and scientific tradition. Their principal arguments require summary before commenting on their potential joint contribution to current landscape study.

Vera's work is a closely argued study of the European temperate forest that scholars have long believed covered the continent from the time of the last glacial retreat until its aggressive clearance by medieval farmers. Stretching from the boulder clay lowlands of Britain, through upland Flanders and South Germany, Poland and Lithuania until it merged into the great boreal forests of Russia, this wooded landscape has played a powerful role in Europe's cultural myths and its geographical imaginations. The forest has been crucial in framing national identities in most countries north of the Alps. In *Landscape and memory* (1995), Simon Schama devotes one of his three sections to 'Wood'. He explores the meaning of various forest landscapes: the 'greenwood' of English imagination; the *Wald und Berg* that played so central role in shaping German identity from the time of Tacitus to the art of Albrecht Altdorfer and Caspar David Freidrich, Anselm Kiefer and Josef Beuys; the impenetrable mixed stands of Bielowieza Forest that still protect the last European bison in eastern Poland and Byelorussia. Associated with myths of liberty, but also with the dark terrors recorded by the brothers Grimm, the wild-wood has long been regarded as northern Europe's *Ur-Landschaft*, a gloomy world of sodden earth and tangled undergrowth with great trunks rising into the gloomy canopy. Conventional ecological understanding has it that medieval farmers, under the authority of secular and religious orders, pressed east with their grazing animals, thus opening out ever-larger clearances in the forest and eventually confining it to island remnants. 18th century foresters and 19th century ecologists gradually came to recognize these remnants as in need of protection in order to preserve them as a key biological resource and a vanishing heritage for Europeans.

Vera challenges this whole scientific picture on grounds of historical ecology. He argues that the European forest never existed except in relatively discreet pockets largely confined to upland regions. His primary appeal is to the existence, long before domesticated animals were introduced into northern Europe, of equally effective grazers whose effects on the ground species and lower story arboreal vegetation was similar to that of later introduced cattle, pigs and horses. The grazing habits of Holocene large mammals such as aurochs, bison, tarpan and elk occasioned the evolutionary response of undergrowth shrub and bush plants in developing thorns, spines and tough bark. These formed protective nursery zones for mixed stands of deciduous oak, ash, beech and conifers, which, together with the open grazed areas produced a mosaic of meadow and woodland over large areas of the continent. Human occupance by a mixed farming economy of cultivation, grazing and woodland panaging thus exploited and sustained a pre-existing landscape pattern rather than creating it. This mixed oak woodland has long been regarded as Europe's richest, most biodiverse ecology, but one whose origins are human rather than natural. For more than a century the aim of ecological landscape conservation policy has been to keep separate the elements making up this landscape in order to 'preserve' the remnant islands of primeval forest by excluding grazing animals.

Vera argues that his historical revision is critical for current landscape policy which, in an attempt to preserve the few remaining islands of the 'primeval oak forest' has sought to enclose them from grazing animals. The result is that remaining oak forests are being choked by trees and understory. The density of growth in the Bielowieza Forest has increased so much since bison have been provided with winter hay rather than stripping tree bark that their grazing meadows have to kept open by artificial mowing. If this argument is accepted, the consequences for landscape policy across much of Europe, which is based on the assumption that traditional agriculture sustains the continent's biodiversity, could be radically modified. Here, I am less concerned with the accuracy of Vera's scientific argument (which I am not in a position to evaluate) than its significance as a revisionist ecological statement that fully supports the semiotic or culturalist argument that 'nature' is as much an outcome of human discourse – in this case scientific ecology – as of 'natural' processes.

Kenneth Olwig's *Landscape, Nature and the Body Politic* also explores the origins of landscape in the German-speaking lands of northern Europe. But this is a philological and historical rather than a naturalistic exploration. From a rich study of etymology and legal history Olwig demonstrates that *landscape* is primarily a socio-legal term that refers to the areal extent of the customary law practiced by an agrarian community, or fellow*ship* (the suffixes *-scape* and *-ship* are cognate) in governing itself. Areas of local government in Schleswig-Holstein and surrounding regions still bear the term *Landschaft* and sustain a degree of autonomy from urban and state regulation that dates back to a pre-modern agrarian world in which custom and culture defined the *Land* and where “the physical environment was a reflection of the political landscape” (2002: 21).

But the focus and originality of Olwig’s book is not in identifying this landscape, but rather explaining its appropriation and transformation in the construction of the modern nation-state, first in England-Britain, and subsequently in Denmark, Germany and beyond. Working primarily with literary and legal documents, Olwig argues that in extending the power of the monarch over the territorial state, appeal had to be made to ‘natural’ law, valid for all times and places, rather than the customary laws of landscape. In a detailed analysis of the early Stuart court masques through which King James I of England sought to legitimate his authority over all of England and Scotland, Olwig shows that a new meaning of landscape emerged, one that emphasized the distanciated, perspectival and commanding gaze of the sovereign. This is *landscape* as *scenery*, a dominantly pictorial meaning that experienced historically first through drama and stage design, painting, photography and related art works and which was then applied to the design of actual, material landscapes – first in aristocratic estates and in more recent times in the picturesque appropriation of traditional rural landscapes. In reworking the meaning of landscape, this newer, more pictorial sense did not wholly erase the former collective and localized sense, especially in areas where it had been most deeply rooted. Rather, it was a question of the new term serving the ideological role of incorporating the sense of unity and harmony of a territory while enlarging the scale of its application (to the whole state in the case of iconic national landscapes – of which the primeval German forest might serve as an example), and rendering its harmony aesthetic rather than social.

Like Vera’s thesis, Olwig’s is not merely antiquarian in focus, it has implications for landscape policy today. To take just one example, the European Union has for a decade been involved in a long-running dispute with the USA and some other major trading partners over that part of the Trade Related Aspects of Intellectual Property (TRIPS) accord within the World Trade Organization. The dispute arises over EU Council Regulation 2081/92, which protects a large number of food and agricultural products by regulating use of specific regional names to products actually made in or deriving from defined regions. To use the terminology of the TRIPS agreement, these are “indications which identify a good as originating in the territory of a member, or a region or locality within that territory, where a given quality, reputation or other characteristic of the good is attributable essentially to its geographical origin". Examples of such designations are familiar to most consumers in both Europe and the USA: Parmigiano or Roquefort cheeses, Scotch whiskey, Champaign. In trade negotiations, the United States is opposed to such geographically immobile designations, arguing for an extension of trademark legislation that affixes to an individual or company and is inherently mobile. The European approach to the matter, often framed in anti-globalist terms, is closely bound up with a cultural commitment to protecting unique associations of land, people, craft and product in specific locations: in a word, landscapes. This association of scales between local and national, the central role of law negotiating universal principles and customary practices, all echo the complex meaning of landscape outlined by Olwig. Further, in many regions protected by this legislation, highly sophisticated, ultra-modern production and marketing systems which may prove not to be environmentally sustainable are hidden behind the picturesque façade of unchanging agrarian landscapes which Europeans feel morally and politically committed to protect.

Olwig's argument, like Vera's is much more complex and sophisticated that I am able to outline here. But they both offer points of convergence for any thoughtful discussion of landscape in Europe today. Over a decade after the end of the Cold War, with almost every state on the continent an actual or potential member of the European Union, and with a global agricultural market pressing ever more strongly on conventional modes of agrarian life and production across the continent, crucial decisions need to be made about how Europeans' lands are to be managed for production and recreation, and for community and national self-consciousness. For two centuries landscape in Europe has been conceived primarily in agricultural and nationalist terms. This reflected both the dominance of agriculture in national economies and the ideological thrust to 'naturalize' the nation as a distinct territorial unit, home to a culturally, even ethnically distinct 'people' or folk. The hesitant decline of nationalism has opened the landscape debate to new transnational perspectives, both ecological and semiotic, of which Vera's and Olwig's studies are examples. By exploring landscape at this meta-scale, these studies distance themselves from the deep local attachments that still govern much landscape policy. This may be seen as a weakness, especially in those parts of Europe where democratic control over the destiny of community life (both human and non-human) at a local and even state level is very recent. But in a world where the interdependency of territorial scales is increasingly apparent in all aspects of experience, landscape cannot be closed around any one scale, least of all the 'local'. This is not to say that locality – at the historical roots of landscape meaning – does not remain significant, only that it must not be the scale of closure, and that much work, both scientific and interpretative, remains to be done if genuinely 'European' policies of landscape protection and planning are to be achieved.

Beyond their specific conclusions, studies such as Vera's and Olwig's also lead to the common recognition that landscape policy cannot afford to ignore a constantly revising ecological science, but that such science is itself historically and geographically contingent. Nor can policy ignore the powerful shaping forces of culture that give meaning to the nature through which human life at all scales is sustained. Yet it must recognize too that meaning is always rooted in the material processes of life.

REFERENCES

Olwig, K. (2002). *Landscape, Nature and the Body Politic: from Britain's Renaissance to America's New World.* Madison: University of Wisconsin Press.

Schama, S. (1995). *Landscape and memory*. New York: Alfred A. Knopf.

Vera, F.W.M. (2002). *Grazing ecology and forest history*. CABI.

Chapter 3

THE CONCEPT OF CULTURAL LANDSCAPE: DISCOURSE AND NARRATIVES

Michael Jones
Centre for Advanced Study at the Norwegian Academy of Science and Letters, Norway

1. INTRODUCTION

In the 1990s the term *cultural landscape* was adopted by various international bodies as a conservation category. UNESCO's World Heritage Committee agreed in 1992 on revised operational guidelines specifying that cultural landscapes could be protected in accordance with the World Heritage Convention of 1972 (Eidsvik 1993; Rössler 1995; Aitchison 1996). In 1995 the Committee of Ministers of the Council of Europe adopted a Recommendation on the Integrated Conservation of Cultural Landscapes Areas as Part of Landscape Policies (Darvill 1996). As an academic term, *cultural landscape* goes back to Friedrich Ratzel (1895-1896), and was in frequent use among other German geographers in the early 20th century. The term was introduced to the English-speaking world by Carl O. Sauer (1925) and became central in the work of the Berkeley school of geography. Geographers and ethnologists used it in the Nordic countries in the interwar period (e.g. Tschudi 1934-1935; Smeds 1935; Campbell 1936). From the 1960s, the term *cultural landscape* became increasingly adopted in other disciplines and entered the terminology of environmental management. In Norway, for example, architects, landscape architects, ethnologists, archaeologists, historians, botanists, zoologists and agricultural economists have taken up the term, and it has become part of the rhetoric of agricultural policy-makers as well as the nature conservation and cultural heritage management agencies (Jones 1988; 1991; 1998a; Daugstad & Jones 1994; 1998; Jones & Daugstad 1997).

H. Palang and G. Fry (eds.), Landscape Interfaces, 21–51.

The use of the term has not been unproblematical. Referring to the use of the term *cultural landscape* in Norway, the present author has noted:

> "... it has come to be used in a number of related ways with different shades of meaning. The divergent concepts are not in all cases mutually irreconcilable and the use of the term cultural landscape may have several connotations at the same time" (Jones 1988: 153).

Different disciplines have used the term in different ways, dependent on the academic traditions and methods into which the practitioners of different disciplines have been schooled or socialized (Jones 1988). Also within the discipline of geography, the term has led to discussion. Lester B. Rowntree opens his essay on *The cultural landscape concept in American human geography* with the following words:

> "While the concept of the cultural landscape is frequently used in human geography, the term is an ambiguous one that carries a variety of meanings. This etymological elusiveness is both a liability and asset; to some, the notion of cultural landscape is an appropriate bridge between space and society, culture and environment, while to others its definitional fluidity weakens the concept and disqualifies it from serious analytical usage" (Rowntree 1996: 127).

In the fourth edition of *The Dictionary of Human Geography*, Denis Cosgrove (2000: 138) states that *cultural landscape* is: "Conventionally, a principal object of study in cultural geography and still the subject of intense debate among cultural geographers". He concludes:

> "In the aftermath of often heated debate over the definition and methods for studying cultural landscape within geography, the concept itself has been rejuvenated" (Cosgrove 2000: 140).

Others, on the contrary, have argued that the term *cultural landscape* serves little useful purpose, and that it is sufficient to speak simply of *landscape*. The Swedish geographer Mats Widgren states:

> "Concepts such as cultural landscape and cultural environment thus belong to the administrative sphere – they function in a unifying political programme declaration that wants to show that the cultural aspects of the landscape have greater importance than one until now has been prepared to see. ... The insight that human influence is found everywhere in the Nordic landscape has as a result laid the basis for doing away with the cultural landscape concept, since it has been shown that the opposite of cultural landscape – untouched nature – does not exist as an areal category. ... There is an increased understanding for the landscape of meaning – that landscape has a cultural content beyond its humanly

> created physical expressions. ... In keeping with this development in the concept of cultural landscape, cultural landscape researchers have gone over to talk of their field of study as landscape or landscape history" (Widgren 1997: 9-10).

Similarly, in his book on the Norwegian landscape, the ethnologist Arne Lie Christensen writes:

> "The concept of cultural landscape is newer than the concept of landscape, and shows itself in practice to be more problematical. ... It has become more and more clear how humans have physically influenced the landscape to such a degree and over such large areas that the whole earth is in the process of becoming a cultural landscape. ... Cultural historians have also increasingly focused on the total environment of human beings, regardless of whether it simply physically is influenced by humans. ... [They have] become steadily more aware of the landscape's cognitive sides. ... As a consequence many scholars within the cultural disciplines stopped talking of cultural landscape; instead they talk of landscape or of human surroundings" (Christensen 2002: 197, 205).

It is noteworthy that Rowntree's (1996) presentation of the concept of cultural landscape in American human geography ends by talking more of *landscape* without the adjective than of *cultural landscape*, while Cosgrove (1998), throughout a chapter entitled *Cultural landscapes* in a book on European geography, consistently talks of landscape rather than cultural landscape.

Rather than arguing in favor of one concept or the other, the intention of the present essay is to attempt to provide an overview of different ways in which the term cultural landscape has been used. A brief discussion of the role of concepts, discourse and narratives will provide a framework. The essay confines itself to discussing the concept of cultural landscape, rather than taking up a more extensive discussion of the concept of landscape. It will be suggested in conclusion that in recent usages of the concept of cultural landscape there can be identified a number of parallel types of narrative that reflect different methodologies and serve different research agendas and power interests.

2. THE POWER TO DEFINE A CONCEPT

David N. Livingstone, in his book *The Geographical Tradition* states:

> "To have command of definition is to have control of discourse. ... key words frequently have a certain plasticity about them which means that

their meaning can be expanded or contracted to serve the purposes of their users. Success in managing vocabulary – and thereby solidifying conceptual slipperiness – therefore brings considerable advantages in attempts to map out conceptual territory. ... The adjudication of definitions is, of course, an inherent boundary-marking, or boundary-making, enterprise designed to demarcate the true from the false, the legitimate from the illegitimate, the relevant from the irrelevant. Accordingly, the ownership of terminology is of enormous consequence in dialogue, for by it both ideas and people can be positioned on particular sides of debates. To dictate definition is to wield cultural power. Definitions of a discipline's cognitive domain are thus frequently less an ontological exercise about the piece of reality that the subject has a rightful claim upon, than a strategy for delimiting the scope of vocabulary that will be allowed as good currency within the professional division of labor" (Livingstone 1992: 304).

Geographers seem to be attracted to complex concepts expressed through terms such as *landscape*, *cultural landscape*, *nature*, *environment*, *region*, *place* and *space*. These are concepts that can be defined in many ways, and they lead to endless discussion. Such complex concepts, which are used in different contexts with varying, although generally overlapping, meanings, might be termed 'chaotic concepts'. In his *Methods in Social Science. A Realist Approach*, the geographer Andrew Sayer (1992 [1984]: 138-139) refers to the term 'chaotic conceptions' used by Karl Marx. Sayer, and others, argue that such concepts should be avoided because scientific thought requires precise and unambiguous concepts. On the other hand, it can be argued that 'chaotic conceptions' survive because they capture the complexity of the real world, and are thus needed and cherished by their users. The geographer John Eyles, in a lecture (unpublished) held in Trondheim in 1994, criticized the idea of 'chaotic conceptions' for assuming that certain concepts are 'correct' and others 'uncorrect', whereas he would claim that there is not one fundamental truth. In much the same way as there are 'cherished landscapes' that have a 'welcome complexity' (Jacques 1992), one might talk of 'cherished concepts' that capture the complexity of the world as it is experienced. The fact that these concepts and their associated terms are tenacious indicates that there is a need for such complex concepts (Jones 2001).

The analysis of concepts can be approached in various ways. One approach aims to arrive at a precise and correct definition and usage of terms, since the unambiguous use of terms contributes to, and is a necessary prerequisite of, logical argumentation and clarity of thought. This is undoubtedly true in many scientific contexts. However, to claim that a concept that has been precisely defined has a single, 'correct' definition is to conceal a value judgement. In not accepting other definitions and usages, this approach might be termed 'authoritarian'. An alternative approach aims to find out how and why certain terms are used differently in different contexts. This approach accepts that a term can have different meanings and usages for different users. The argument here is that a systematic presentation of a term's different meanings and usages can contribute to better understanding, communication and dialogue by making the users aware of their differences in their understanding of concepts. This approach has a disadvantage in that by not providing a single, 'correct' meaning it may pose problems for the young and inexperienced, or for people outside a particular discourse. This approach might be termed 'democratic' or 'pluralistic' (Jones 1993). This second approach is adopted in the present essay in discussing different usages of the concept of cultural landscape.

3. DISCOURSE AND NARRATIVES

3.1 Discourse

In a pluralistic understanding, concepts and terms can be expected to change in meaning over time. Concepts – mental notions and abstract ideas – and terms – words and expressions that express and help to define concepts – form constitutive parts of discourses. A discourse can be broadly understood as any discussion or exchange of ideas, expressed through conversation and dialogue, talks and lectures, and/or writings that treat a subject systematically and at some length. In particular academic disciplines or other fields of activity, such as administration and management, discourses follow certain accepted rules and conventions regarding methods of presentation and argumentation (Daugstad & Jones 1998). Referring to the theories of Michel Foucault, Livingstone (1992) explains discourses as the broad systems of thought, which inform how we conceive of the world and gain practical expression in regulative institutions; they are vehicles of social power and serve as strategies of moral manipulation.

In *The Dictionary of Human Geography*, Derek Gregory (2000: 180) defines discourse as: "A specific series of representations, practices and performances through which meanings are produced, connected into networks and legitimized". He further characterizes a discourse as:

- heterogeneous, i.e. it involves many authors, many types of text and statement, and contains multiple meanings and implications;
- regulated, i.e. it is coherent and systematic, although often containing contradictions, and has a "regime of truth" policing what is considered relevant and acceptable;

- embedded, i.e. it is implicated in social life and embedded in institutions;
- situated, i.e. it provides partial, situated knowledges, and is characterized by particular constellations of power that may be contested or negotiated;
- performative, i.e. it has variable meaning, force and effect, and itself constitutes the objects of which it speaks.

A discourse has its own specific history and geography. The history of a particular discipline, such as geography, can be conceived as the production of geographical knowledge through particular discourses (Gregory 2000). Discourses are "frameworks that embrace particular combinations of narratives, concepts, ideologies and signifying practices, each relevant to a particular realm of social action" (Barnes & Duncan 1992: 8).

In the following, the role of narratives for supporting particular definitions or interpretations of the concept of cultural landscape will be focused upon. Narratives lie within and contribute to discourses. Different types of narrative can be identified.

3.2 Narrative-descriptive approach

Yi-Fu Tuan (1991) has discussed *the narrative-descriptive approach* applied to understanding the role of language in 'the making of place'. The narrative-descriptive approach, says Tuan, is favored by cultural and historical geographers, historians and cultural anthropologists. While all narratives and descriptions contain interpretative and explanatory strategems, built into the language, the explicit formulation of theory is not their main purpose. "Theory, by its clarity and weight, tends to drive rival and complementary interpretations and explanatory sketches out of mind," making complex and ambiguous human experience schematic. "... in the narrative-descriptive approach, theories hover supportively in the background while the complex phenomena themselves occupy the front stage" (Tuan 1991: 686).

Tuan argues that words have the power to render objects, which were formerly invisible because unattended, visible. He is concerned with the moral dimension to speech – for example, "warm conversation between friends can make the place itself seem warm; by contrast, malicious speech has the power to destroy a place's reputation and thereby its visibility" (Tuan 1991: 684). Places are made, he argues, by means of place-names, informal conversation and written texts. Speech is thus an integral part of the construction of places.

> "... although speech alone cannot materially transform nature, it can direct attention, organize insignificant entities into significant composite wholes, and in so doing, make things formerly overlooked – and hence invisible and nonexistent – visible and real" (Tuan, 1991: 685).

The right to speak and to be heard, and the right to name, is empowerment, says Tuan. Naming makes a place familiar, and story-telling converts objects 'out there' into real presences. "Naming is power – the creative power to call something into being, to render the invisible visible, to impart a certain character to things" (Tuan 1991: 688). In modern societies, written text has largely supplanted story-telling, says Tuan. Words supply a temporal dimension to landscape that visual image alone cannot provide. Words – names, proper names, taxonomies, description, analyses – draw attention to things: "aspects of reality hitherto invisible, because unnoticed, become visible" (Tuan 1991: 693).

The power of words to render previously unattended objects visible applies not only to the way we structure reality 'out there' but also to historiography. In the same way that naming makes places and landscapes familiar and real, so can the same be said of a field of study when it is named and the story of its development is told.

3.3 Historical narrative

Historiography can be seen as one form of historical narrative. In his book on historical method, the Norwegian historian Knut Kjeldstadli (1999) discusses the use of narrative as a method in the discipline of history. The historical narrative provides a chronological presentation of events. The narrative has a beginning, middle and end, with events being placed in sequence and tied together. The narrative tells consecutively what has happened, and how one situation or state of affairs has led to another. Explanation lies in the narrative itself. This approach contrasts to consciously theory-informed approaches common in the social sciences, whereby events are structured according to whether and how they correspond with or diverge from theories, laws and concepts. Kjeldstadli points out a number of pitfalls in the narrative approach. While a narrative describes the course of actions that led to the event we are interested in, other lines of development are paid less attention to. By relating that things happened in a certain manner, the implication is that they had to happen in that way. A rich and nuanced narrative thus appears as 'natural'. A central question is how one determines what is to be included in a particular narrative or not. To present a narrative as if the facts are objective truths that speak for themselves conceals the role of the narrator in making the selection of facts to be presented. A narrative is one version of the events described, but not the only one possible. A subjective element lies the narrator's judgment as to which facts are considered the important ones to tell. In describing a particular development, the narrator writes his or her own meaning or moral into the events. The choice of topic and the way it is delimited implies an a priori standpoint, often unreflected, as to what is meaningful and coherent. The historical patterns that are described are real, but which of them that are seen as significant depend on the historian. Patterns and regularities can be identified in reality, and not everything is chaotic or fortuitous, but it is the historian who helps decide which patterns are made manifest through the questions asked and the way in which reality is divided up when it is described (Kjeldstadli 1999).

Contrasting with stories of places and narratives of historical events are personal narratives, discussed in relation to geography by Berit Helene Vandsemb (1995). These are biographical or autobiographical narratives based on the time- and place-specific life stories of individuals. By means of the personal narrative, order and coherence are imposed on experiences and activities, and meanings are ascribed to incidents and events. In narratives, events and causal patterns are interpreted and reinterpreted. Personal stories involve remembering, but also reconstructing the meaning of past events in the light of the present. Stories are subject to reformulation in the light of new stories. Narratives help give form to experience. However, they have clear limitations as a research strategy. One aspect is the narrator's subjectivity, biases and beliefs, influencing what the narrator remembers or chooses to remember. Narratives have an explicit or implicit message, which may be influenced by ideological or moralizing considerations. The personal narrative gives the illusion of telling a complete story, with a beginning, middle and end, when it is only one part of an unstructured and continuing life. Narratives tend to emphasize the temporal dimension at the expense of the synchronic. Personal narratives emphasize the variety of individual experience, but this may be at the expense of social context. In research, personal narratives are told by informants and retold by the researcher. The researcher's perspective is of necessity present. Thus the narrative is constructed by both the informant and the researcher (Vandsemb 1995).

This brief discussion of narratives has focused on three types of narrative: narrative-descriptive accounts of places and landscapes; narratives of historical events; and personal narratives. An important point is the role of the researcher in judging what facts are important to select and emphasize in presenting a narrative. An implication of this is that the researcher's personal narrative, focusing on the personal and academic influences that have influenced his or her choices of research topics and approaches (Jones 1985), may throw light on this process. In the following, a short personal narrative will attempt to provide some insights into the present author's interest in the cultural landscape concept. Thereafter, reference will be made to recent historiographical accounts of the concept of cultural landscape; these can be regarded as one variant of the historical narrative. Finally, different types of concrete landscape narratives (place narratives) will be described on the basis of different usages of the cultural landscape concept within the broader landscape discourse.

The essay also confines itself to a discussion of the cultural landscape concept and its usages in the Scandinavian, German and English language contexts, leaving aside a much broader discussion of equivalent concepts in other languages, such as the French *paysage* and similar terms in the Latin languages, and the Finnish *maisema* and related terms in the Finno-Ugric languages (see Keisteri, 1990).

4. *CULTURAL LANDSCAPE* – A PERSONAL RESEARCH NARRATIVE

If it is accepted that narratives, like other forms of knowledge, are always only partial representations of the truth, and are formulated in social interaction as part of a discourse, then it is relevant to focus on the role of the narrator in order to counter the tendency for scientists: "to write themselves out of their own stories" (Barnes 2000: 742). However, it should always be born in mind that "all situated knowledge is partial, including the situated knowledge we have of our own knowledge about ourselves" (Barnes 2000: 743). My own personal narrative as presented in this section is intended to throw some light on the way in which particular contexts and events in an academic career can lead to particular interests developing, particular choices being made and particular segments of knowledge being focused upon. Like all narratives, this one is necessarily selective (see also Jones 2001), and will summarize my work on the cultural landscape concept.

My awareness of the concept of cultural landscape arose in the initial lecture of the first-year undergraduate course in human geography at University College London, where I began studying in autumn 1963. The concept was defined in accordance with the classical definition of Ratzel (1895-1896) as "landscape modified by human activity", as opposed to the primeval natural landscape. The way in which humans had changed the landscape through the clearing of the forests, the draining of the marshes, the cultivation of the heaths and the growth of towns was the principle thrust of teaching in historical geography (Darby 1951). The way in which humans created a new landscape on emergent land resulting from shore displacement on the coast of western Finland later formed the topic of my doctoral thesis; to understand human responses to land uplift required study of land tenure and legal institutions, but the concept of cultural landscape seemed unproblematic (Jones 1972; 1977; 1987).

4.1 *Cultural landscape* in Norway

When I took up a research fellowship at the Agricultural University of Norway in 1973, I came in contact with landscape architects and architects who had other understandings of cultural landscape. Often the cultural landscape meant simply the agricultural or rural landscape, lying between the *natural landscape* of the mountains on one side and the urban landscape on the other (Langdalen 1965; Surnevik 1970; Austad 1983). Influenced by German terminology, others made a threefold division of the landscape into *primeval landscape*, untouched by human influence; *natural landscape*, where human impact is small and dominated by the natural conditions; and *cultural landscape*, where human impact dominates the landscape (Bruun 1967; Gabrielsen 1969). New directions in landscape architecture were widening the field from a focus on esthetics and landscape gardening to broad landscape history and ecology; the significance of vegetation was emphasized, and the cultural landscape was understood as the combined product of nature and human impact (Asheim 1975; 1978).

After I moved to the newly established Geography Department in Trondheim in 1975, I began teaching a course in historical geography with its principle emphasis on processes of change in the Norwegian cultural landscape, adapting the approach I had learnt in London, and in part inspired by the Berkeley school (Thomas 1956; Wagner & Mikesell 1962). The following years saw an increasing interest for cultural landscape in Norway, and the concept was taken up in physical planning and environmental management. The agricultural authorities and nature conservation authorities came together for a seminar at Sogndal in 1987 to discuss management problems related to the cultural landscape (*Samarbeidsgruppa i Sogn og Fjordane* 1987). The term cultural landscape was here being used in the narrow sense of agricultural landscape. The editor of the Norwegian Journal of Geography, *Norsk Geografisk Tidsskrift*, was present at the seminar, and, finding that the term was being used rather differently from its common usage in geography, encouraged me to write an article discussing this. To this end I examined the use of the term cultural landscape in Norwegian literature from its first use by Tschudi (1934-1935) and identified seven principal definitions or usages (Jones 1988):

1. as an areal category (contrasting with natural landscape and sometimes urban landscape);
2. as a chronological stage of development;
3. as the human components in any landscape;
4. as countryside;
5. as heritage;
6. as scenery with esthetical qualities;

7. as elements in the landscape with meaning for human activity.

The seven usages were further grouped into three main forms of understanding (Jones 1988; 1991):

- as landscape modified or influenced by human activity (the first three usages above);
- as valued features of the human landscape that are threatened by change or disappearance (the fourth to sixth usages above);
- as elements in the landscape with meaning for a human group in a given cultural or socio-economic context (the seventh usage above).

4.2 *Cultural landscape* under administration

With the establishment of a government-sponsored Research Programme on Cultural Landscape in 1991, the varying ways in which the term *cultural landscape* was being used proved sufficiently confusing to lead to a study of the different usages of the concept in public administration and management (Daugstad & Jones 1994; 1998; Jones & Daugstad 1997). An analysis of some 40 administrative and applied research documents from the period 1987–1993 identified the following meanings given to the concept, reflecting to a certain extent previous usages:

1. as an areal category (modified by human use, often limited to rural or agrarian landscape);
2. as physical traces of human activity through time;
3. as selected elements in the landscape (frequently threatened semi-natural vegetation, historic buildings or cultural heritage);
4. as managed landscapes (e.g. the cultural landscape of agriculture, natural and cultural heritage);
5. as beliefs and traditions associated with certain localities.

This study further identified four discourses on cultural landscape, which, while initially relatively independent of one another, were increasingly coming together in a common discourse.

The first discourse was within the agricultural sector, where an initial understanding of cultural landscape as meaning cultivated landscape (producing food and fibers) was changing towards a broader understanding of the cultural landscape of agriculture as also producing environmental goods. While the underlying values remained primarily economic – maintaining agricultural land in production – increasing attention began to be paid to ecological and cultural historical values – protecting rest biotopes and cultural heritage.

The second discourse was within the nature conservation sector, where cultural landscape tended to be interpreted as meaning primarily human-influenced or semi-natural ecosystems of significance for the maintenance of biodiversity. Here ecological values were primary, but there was also a certain focus on esthetic and cultural-historical values.

The third discourse was within the cultural heritage sector, where cultural landscape tended to be defined in the classical geographical meaning as all landscapes that have been modified or influenced by human activity, although with an emphasis on ancient monuments, historic buildings and other built structures. The problem of selection faced by cultural management led to discussions of conservation theory. Awareness that the significance of physical landscape cannot be divorced from the meanings attached to it led to the extension of the concept of cultural landscape to include mental landscape. For the cultural heritage sector, cultural historical values came first, followed to a lesser extent by esthetic values.

The fourth discourse identified was in local physical planning, where cultural landscape tended to be associated with trees and green areas, and related to esthetic concepts of landscape design. Thus esthetic values were stressed, followed by cultural-historical and ecological values.

Further work analyzed the debate on cultural landscape that continued in the 1990s as a result of the interaction of the management-oriented bureaucracy, applied research and critical research. Three strands in this debate were identified: a critical discussion of the dichotomy between nature and culture contained in the distinction between natural landscape and cultural landscape; the application of research for the purpose of assisting the management of cultural landscape by public agencies; and a critique of the use of the cultural landscape concept in planning and public management (Jones 1998a).

My own position on the use of the concept of *cultural landscape* is that it is useful in focusing on the role of humans in processes of change in our physical surroundings, and the cultural meaning of landscape (Jones & Daugstad 1997). Cultural landscape thus has both a physical dimension – traces of human activity in the landscape as humans have influenced and modified 'nature' through time; and a cognitive dimension – the cultural meanings that humans attach to their physical surroundings, both natural components and human components (Jones 1998b). In practice, nearly all landscapes show signs of the work both of natural forces and of human agency and society, and these two sets of influence are often difficult to disentangle from one another. At the same time all landscapes are cognitive, with varying meanings for different groups of people. The way in which the physical surroundings, whether of natural or human provenance, are structured and shaped in people's minds begins with the naming of landscape features and places, itself a very culture-specific practice. Hence cultural landscape is more a perspective than a 'thing'.

5. *CULTURAL LANDSCAPE*: HISTORIOGRAPHICAL NARRATIVES IN GEOGRAPHY

In the following, the development of cultural landscape as a perspective in research will be illustrated within the discipline of geography, where the term first appeared and where it has been most persistently used. This illustration is based on selected passages from a few selected historiographical narratives. It should be kept in mind that all historiographies (including this essay), however much they are presented and accepted as 'fact', reflect their author's particular perspective on and selection of information from disciplinary history.

5.1 From Ratzel to Sauer and Hartshorne

"Cultural landscape" has been a fundamental concept for geographers since its first use in Germany in the 1890s. Ratzel (1895-1896) defined *Kulturlandschaft* as an area modified by human activity, as opposed to the primeval natural landscape. Otto Schlüter (1903) emphasized the formative influence of human activity in shaping the landscape in reaction against explanations of natural determinism (Schick 1982). Norbert Krebs (1922) defined *Kulturlandschaft* as the *Ökumene* that is the settled region transformed by human activity, as opposed to the *Anökumene*, the non-settled region where human life was subordinate to nature.

Introducing the term to the English-speaking world, Sauer (1925) defined the cultural landscape as an area successively altered by humans through their cultural activities, and fashioned by particular cultural groups from the natural landscape that preceded human activity. In his famous essay, *The morphology of landscape*, Sauer wrote:

> "The area prior to the introduction of man's activity is represented by one body of morphologic facts. The forms that man has introduced are another set. We may call the former, with reference to man, the original, natural landscape. In its entirety it no longer exists in many parts of the world, but its reconstruction and understanding are the first part of formal morphology.
>
> ...
>
> The works of man express themselves in the cultural landscape. There may be a succession of these landscapes with a succession of cultures. They are expressed in each case from the natural landscape, man expressing his place in nature as a distinct agent of modification. ... The

cultural landscape then is subject to change either by the development of a culture or by a replacement of cultures. ... The division of forms into natural and cultural is the necessary basis for determining the areal importance and character of man's activity" (Sauer 1925: 37).

In a much-quoted passage (e.g. Billinge 1981; Cosgrove 2000), Sauer wrote further:

"The cultural landscape is fashioned out of a natural landscape by a culture group. Culture is the agent, the natural area is the medium, the cultural landscape is the result. Under the influence of a given culture, itself changing through time, the landscape undergoes development, passing though phases, and probably reaching the end of its cycle of development. With the introduction of a different, that is, alien culture, a rejuvenation of the cultural landscape sets in, or a new landscape is superimposed on remnants of an older one. The natural landscape is if course of fundamental importance, for it supplies the materials out which the cultural landscape is formed. The shaping force, however, lies in culture itself" (Sauer 1925: 46).

Richard Hartshorne (1939) criticized Sauer and the idea of cultural landscape, and more particularly its opposite natural landscape. Hartshorne found a division of the material features of any particular area into separate natural and cultural landscapes illogical, since the cultural elements in the landscape, even when considered separately, merely constituted a collection of parts of the total landscape. He referred to J.O.M. Broek (1938), who, recognizing this difficulty, suggested it was sufficient to distinguish between natural and cultural elements in the landscape. According to Hartshorne (1939: 346): "Most American geographers who use the term 'cultural landscape' mean simply the present landscape of any inhabited region". His view was that, strictly, the natural landscape would be found only in areas never disturbed by human hand. Hence in inhabited areas the concept was purely theoretical. Moreover, in the past, humans have influenced the vegetation more anciently and more universally than generally thought. Nor is nature static, so that the original, unknown natural landscape was not the same as the present theoretical natural landscape. Hartshorne (1939: 348) suggested that: "unaltered natural landscapes and those altered but uncontrolled by man" might be called "wild landscapes", in contrast to "tamed" or "cultivated" landscapes. He remained, nonetheless, critical of Sauer's emphasis of *landscape* as the focus of geography.

5.2 *Cultural landscape* in American human geography

In his historiographical overview, Rowntree (1996) traces the intellectual heritage of the cultural landscape concept in American human geography by first examining its definitional ambiguity:

> "To some, a cultural landscape is simply an environment modified by human action and the research emphasis, then, is to document empirically the consequences or process of that interaction, be it subtle or overt. The landscape thus is treated as an apparently objective by-product or artifact of human habits and habitation. This is fairly close to Sauer's early intentions ... and is still common today, particularly in Europe" (Rowntree 1996: 129).

Rowntree refers here to Simmons (1988), interested in the earliest cultural landscapes of England as the impact of prehistoric people on baseline natural landscapes. Rowntree refers also to a collection of articles documenting ecological changes resulting from historical habitation and agriculture in northern Europe, published to commemorate the 100th anniversary of the Botanical Museum in Bergen, now part of Bergen University (Birks et al. 1988). Rowntree points out:

> "While both Simmons and Birks use *cultural landscape* in their titles, notions of subjectivity, scenery, and interactivity between people and environment (beyond our immediate subsistence and habitation needs) are not explicit concerns in their research. The term refers instead only to an altered nature" (1996: 129).

More common than 'this narrow ecological use', continues Rowntree (ibid.): "is a broad, sometimes all-encompassing conceptualization that implies a larger degree of subjective interaction between humans and their surroundings". Christopher L. Salter (1978: 71), for example, defines cultural landscape as "that segment of earth space which lies between the viewer's eye and his or her horizon," referring to both the material forms in the landscape and human responses to them. Rowntree refers further to J.B. Jackson's (1986) engagement with *vernacular landscape*, the landscape not so much looked at as lived in and shaped by ordinary people, and to Cosgrove's (1985) contrasting criticism of landscape as a historically elitist 'way of seeing' allowing appropriation of the world. Rowntree finds that:

> "... there is no precise definition of *cultural landscape*. Instead, there is a certain definitional ambiguity that allows varied uses and research emphases. If this is bothersome for some, it provides creative license for others to explore the complicated interface between humans and our varied environments" (Rowntree 1996: 130).

Rowntree discusses the significance of Sauer's essay *The morphology of landscape* (1925) and Hartshorne's critique of the landscape concept. Despite Hartshorne's criticism, which paved the way for the rise of the idea of geography as a spatial science, Sauer and the Berkeley school as well as landscape geographers in other American universities continued to be influential. Studies of cultural landscape examined environmental transformation by humans and the ecological consequences of this. Geographers as well as cultural anthropologists came together at the Princeton symposium of 1955 on *Man's Role in Changing the Face of the Earth* (Thomas 1956). It can be mentioned here that the symposium was dedicated to the geographer George Perkins Marsh, whose book *Man and Nature; Or, Physical Geography as Modified by Human Action* had appeared in 1864 and inspired the environmental movement (Olwig 1980). A new symposium held in 1990 was entitled *The Earth Transformed by Human Action* (Turner et al. 1990).

In the 1960s, in reaction against the positivism of the spatial science school, "a humanistic geography arose that resurrected the cultural landscape concept as a major vehicle for analyzing the ties between culture and environment" (Rowntree 1996: 133-134). Rowntree identifies two ways in which landscape was studied by means of historical and interpretive methods in humanistic geography. The first examined the visual and material details of landscape, documenting tangible elements such as house types and field patterns, and explaining them in terms of diffusion of technologies and other social processes. The second examined cultural perceptions and visual preferences, aiming at understanding people's cognition of and response to their environment. In the second approach, influential works were by David Lowenthal (1961, 1968), in part together with Hugh Prince (Lowenthal & Prince 1964, 1965), on 'landscape tastes', and Yi-Fu Tuan (1974, 1976) on 'landscapes of the mind'. Rowntree (1996) goes on to identify nine directions that further landscape studies have taken: landscape as ecological artifact, landscape as evidence for origins and diffusion; landscape as material culture; urban landscapes; art, literature and landscape meaning; landscape as visual resource; landscape as ideology; landscape's role in the production and maintenance of social categories; and landscape as text, symbols and signs. Rowntree concludes:

> "*Cultural landscape* will remain a fundamental concept in human geography more because of its momentum than because of its conceptual clarity. *Landscape's* definitional or methodological shortcomings notwithstanding, the term has been used too long and in too many ways to be radically revised. Furthermore, a receptive lay audience awaits our writings about *landscape* because it strikes intuitive chords about the visual and material environment. A seductive ambiguity provides

> hospitable refuge for all; *landscape*, it seems, embraces whatever we ask from our curiosity about human-environmental interactions. This conceptual conviviality is at once an asset and a liability" (Rowntree 1996: 147).

Rowntree thus conflates without further discussion the concepts of cultural landscape and landscape.

A similar conflation is evident in Cosgrove's (2000) entry on *Cultural landscape* in the fourth edition of *The Dictionary of Human Geography*. The entry is considerably expanded from Mark D. Billinge's entry in the first edition of the dictionary from 1981, which is confined to the approach of Sauer and the Berkeley school. Criticizing Sauer's conceptualization, Cosgrove (2000: 139) writes: "All landscapes are thus equally natural landscapes and cultural landscapes, according to the questions we ask of them and the processes we choose to examine in relation to them". Cosgrove then outlines new approaches, including the work of writers such as himself (Cosgrove 1992), Duncan (1990) and others (Daniels & Cosgrove 1988) emphasizing social construction, representation, power and contestation in studies of landscape. Finally criticisms of an overfocus on the representational aspects of landscape are referred to, in particular Kenneth Olwig's (1996) rediscovery of 'the substantive meaning of landscape' in the concept's early Nordic origins, and the works of W.J.T. Mitchell (1994) and Don Mitchell (1996) demonstrating the ability of landscapes to naturalize social and political power. Cosgrove (2000: 140) concludes: "... a wealth of cultural landscape studies are appearing, and while the genealogy of the Sauerian concept remains fertile, the usage of the term cultural landscape within geography no longer implies filial attachment to a narrow Berkeleyan model". Surprisingly, Cosgrove's entry does not contain a cross-reference to Jim Duncan's (2000) entry in the same dictionary on *landscape*, despite the latter going over some of the same ground as Cosgrove concerning Sauer and the Berkeley school, and then presenting new directions in landscape interpretation in what is termed 'the new cultural geography' (Duncan's entry does, however, contain a cross-reference to *cultural landscape*).

5.3 *Cultural landscape* in European geography

In Cosgrove's chapter on *Cultural landscapes* in *A European Geography* (1998), the chapter title is the only time the term is used as such. The chapter examines the "social and aesthetic unity between land and human life" in five sections. The first, inspired by Olwig (1996), traces the meaning of landscape in the Germanic languages as: "expressing the experience and intention of a social group tied by bonds of custom and law to a determined territory" (Cosgrove 1998: 66). This is contrasted with the equivalent term in the Romance languages, derived from Latin *pagus*, which provides an urban and imperial view of the land and its agrarian communities. Cosgrove notes that the complexities of European agrarian history and land law are as significant as the variation in the physical geography for the appearance of European agrarian regions. This has led geographers: "to identify, map and describe the characteristic features of local landscapes, seeking to explain their emergence from long historical continuities of local settlement and interactions between land and life" (Cosgrove 1998: 67). As these landscapes became affected in different ways by modernization, the fear of their loss gave such descriptions an ideological imperative. Second, the idea of landscape as scenery giving an esthetic experience emerged as feudal structures of economic and social life in Europe yielded to modern ones through long-distance trade and the organization of agricultural production for urban markets, with cities controlling the surrounding countryside. Landscape became in the 15th and 16th centuries associated with landscape painting and landscape gardening intended for the esthetic pleasure of landowners. The esthetic view of scenery provided the antecedents for modern, urbanized populations to: "demand strict controls over the 'look' of the land" (Cosgrove 1998: 69). The third theme is the techniques of representation and expression of landscape in art and design through geometry, from perspective drawing to produce 'prospects' to land surveying and cadastral mapping, and their reinscription on the land through the geometries of heathland reclamation, enclosure of common lands, and draining of polders:

> "Rules and conventions have been developed from painting, theatre and gardening to create a 'discourse' of landscape which has been communicated and debated across Europe through original and reproduced works of artists and designers, and in theoretical writings on landscape. This discourse continues to affect not only environmental design, ... but also the ways that ordinary people enjoy landscape, cultivate and design their domestic garden spaces, and record places in photographs, videos and sketches" (Cosgrove 1998: 70-71).

A fourth theme is the role of landscape ideology in constructions of territory and identity as Romantic landscape visions and the ideals of urban design played a significant role in 19th and 20th century nationalism. The modern territorial state "has come to replace local landscapes as the political expression of collective socio-spatial attachment" (Cosgrove 1998: 71). Allegiance and identity are constructed within an 'imagined community' of nationhood defined and represented by language and landscape. Landscape became a weapon in the aggressive nationalism of Nazi Germany. Postwar reconstruction produced modernistic landscapes throughout Europe, implemented though state planning in which principles of urban design reflected the ideals of economic rationality. Resistance to planned landscapes since the 1970s has, however, led to a struggle between the demands of modernization and demands for traditional landscapes in the interests of heritage, tourism and environmentally gentle forms of land use, in which "the 'appearance' of the landscape is a powerful weapon, wielded generally on the side of tradition and identity" (Cosgrove 1998: 76). Cosgrove's final theme is "landscape as container and medium of collective memory" in the form of monuments and memorials, and as a repository of imagined pasts.

These examples of historiographical narratives concerning cultural landscape in geography indicate a development over a hundred-year long period from definitions that distinguish between natural and cultural landscapes as means of attempting to separate the roles of natural forces and human agency in landscape change, to a conflation between landscape and cultural landscape in which both are seen expressions of a complex interaction between human ideas, social structures and the physical features of the human environment, in which natural forces and human agency are inextricably entwined.

6. *CULTURAL LANDSCAPES*: A CONSERVATION NARRATIVE

6.1 *Cultural landscape* in World Heritage

At the same time as problems of conceptualization have led in the discipline of geography to a certain retreat from using the concept of cultural landscape, it has come to the fore in international conservation. In 1995 a volume entitled *Cultural Landscapes of Universal Value*, edited by Bernd von Droste, Harald Plachter & Mechtild Rössler, was published under the auspices of UNESCO. Its purpose was to provide guidance for the identification and nomination of such landscapes after the decision of the World Heritage Committee to include 'outstanding cultural landscapes' on the World Heritage List. The work opens with a broad definition of the concept of cultural landscape:

> "Cultural landscapes reflect the interactions between people and their natural environment over space and time. ... A cultural landscape is a complex phenomenon with a tangible and an intangible identity. The intangible component arises from ideas and interactions which have an impact on the perceptions and shaping of a landscape, such as sacred beliefs closely linked to the landscape and the way it has been perceived

over time. Cultural landscapes mirror the cultures which created them" (Plachter & Rössler 1995: 15).

The protection of designated cultural landscapes is intended to overcome the lack of success of the World Heritage Convention of 1972 to avoid the designation of natural heritage and cultural heritage separately from one another. According to Plachter & Rössler (1995: 16): "cultural heritage protection was dominated by the ideas of art historians and focused on single monuments, while nature protection reflected the goal of natural scientists to protect threatened species and "untouched" nature". Plachter & Rössler (ibid.) criticize the aim "to preserve or reconstruct single, unique objects at some clearly defined phase of their existence" as reflecting a museum-like attitude that excluded dynamic processes and context. Cultural objects were perceived independently of their cultural context and landscape environment; natural areas were similarly separated from their surrounding context, including human activities. The emphasis in nature conservation on nature reserves and species protection saw humans as "a nuisance"; human influence meant a loss in value. "Nature modified by humans seemed beside the point..., had little value, and was not recognized as a genuine problem for conservation" (ibid.).

The problem of protecting nature and culture together became evident when the nomination of the English Lake District as a World Heritage Site was deferred in 1986 and again in 1989. Ensuing discussions in UNESCO, and the establishment of the World Heritage Centre in Paris in 1992, paved the way for the inclusion at the Santa Fe meeting of the World Heritage Committee in December 1992 of cultural landscapes in the operational guidelines (revised in 1995) for the implementation of the World Heritage Convention (Rössler 1995). Cultural landscapes representing 'the combined works of nature and man' could hereafter be designated in accordance with Article 1 of the Convention. They were to illustrate the evolution of human society and settlement over time, under the influence of constraints and opportunities presented by the natural environment, and of successive social, economic and cultural forces. They were to be selected on the basis of their 'outstanding universal value' and their representativity as illustrations of 'the essential and distinct cultural elements' of clearly defined geo-cultural regions. The guidelines specified that: "the term 'cultural landscape' embraces a diversity of manifestations of the interaction between humankind and its natural environment". The following categories of cultural landscape were recognized:

1. Clearly defined landscapes designed and intentionally created by man, embracing garden and park landscapes constructed for esthetic reasons;
2. Organically evolved landscapes resulting from an initial social, economic, administrative and/or religious imperative that have developed

their present form by association with and in response to their natural environment, including:

a) Relict or fossil landscapes where evolutionary processes came to an end at some time in the past, their distinguishing processes still being visible in material form;

b) Continuing landscapes that retain an active social role in contemporary society closely associated with the traditional way of life, and in which the evolutionary process is still in progress;

3. Associative cultural landscapes where powerful religious, artistic or cultural associations relate to natural elements rather than material cultural evidence, which may be insignificant or even absent (von Droste et al. 1995: 431-432).

The first two cultural landscapes to be inscribed on the World Heritage List, Tongariro National Park in New Zealand (in 1993), and Uluṟu–Kata Tjuṯa National Park in Australia (in 1994), are associative cultural landscapes of particular significance for the indigenous peoples of these two countries (Rössler 1995).

Discussions of the term cultural landscape are found in several chapters of *Cultural Landscapes of Universal Value*. Wolfgang Haber (1995), discussing the concept, origin and meaning of landscape, notes that the land can be shaped both by natural forces, resulting in a natural landscape, and by humans, who create – not always intentionally – a cultural landscape. The term *culture*, he notes further, stems from the Latin word *colere*, which has various meanings, from the cultivation of fields to large cities without agricultural connotations, as well as careful maintenance, adornment and even veneration. He does not directly comment further the significance of the last of these meanings for an understanding of cultural landscape conservation as a form of worship, but he does see both nature protection and cultural landscape protection as a form of cultivation:

> "We have begun to 'cultivate' islands of genuine nature in our cultural landscape or, where true nature has either disappeared or been 'humanized', we cultivate islands or areas of what we feel to be a harmonious, pleasing amalgamation of natural and cultural traits" (Haber 1995: 39).

A landscape ecologist in the tradition of the German geographer and botanist Carl Troll, Haber (1995: 38) emphasizes: "the gradient from natural or quasi-natural to human-made or cultural landscape elements". He is also concerned that the preservation of valuable cultural landscapes is not restricted to agricultural or rural landscapes.

Adrian Phillips (1995: 380) notes: "Cultural landscapes are to be found in every part of the populated world" and "represent a rich and almost infinitely varied part of the human heritage". He is at pains to state that, although often much changed from their natural state, they are important for the conservation of nature and biodiversity, and many ecosystems are the result of and dependent on human intervention. Such landscapes also contain evidence of human history – archaeology, vernacular buildings and past land use – and as such contribute to human health, inspire writers and artists, and provide places for enjoyment and learning. Finally, they reflect the presence of indigenous peoples and local communities and are important as repositories of sustainable land use. However, Phillips points out the semantic difficulty of the term *cultural landscape*:

> "It is now recognized that nearly all landscapes bear the imprint of past or present human activity or influence. The more we learn about the evolution of landscapes, the more apparent it has become that human influences are very widespread indeed, and often of great antiquity – many landscapes, which had previously been considered to be 'natural', turn out to have been modified by humans over centuries, even millennia. ... Often, indeed, the assertion that a landscape is natural is a by-product of an imperialistic view of the world which found it hard to recognize the influence of cultures which came before those of the Europeans.
>
> If the influence of landscapes has indeed been so universally pervasive, then the adjective 'cultural' ought correctly to be applied so widely that it is of little value" (Phillips 1995: 381).

Plachter (1995: 393) also observes that in a "broad definition almost all landscapes of the world can be viewed as 'cultural landscapes'". Looking for a functional definition of cultural landscapes, he states: "Cultural landscapes or seascapes might be defined as those areas on earth, where man exposed a substantial influence on nature and thereby changed its image significantly", but finds this to be insufficient with regard to practical reglementations and measures. Concerned that cultural landscapes by definition result from the interaction of man and nature, he restricts cultural landscapes to areas of the Earth's surface where culture and nature have shaped one another, where human influence was consciously creative, and where ecological mechanisms are still at work in interaction with humans. Hence he excludes the urban landscapes of the modern metropolis; although they contain elements of nature and biodiversity, man has emancipated himself totally from the limits of nature through his buildings, techniques and the import of resources from the outside and thus no longer needs to take advantage of natural ecological mechanisms but determines the structure and internal function of the remaining parts of nature. Similarly he excludes modern, 'industrialized' agricultural and forestry landscapes with man made ecosystems (Plachter 1995). He presents a functional-ecological gradient of landscapes:

– *natural and semi-natural landscapes*, which, even where changed by humans, retain the natural qualities of self-regulation and self-development;
– *traditional cultural landscapes*, where humans are dependent on the limits of the natural system while shaping the landscape in line with specific technical, visual or religious concepts;
– *modern agricultural landscapes*, which depend on huge imports of resources and energy; and
– *urban landscapes*, which have an even higher import of resources and energy and the local population is no longer functionally dependent on urban nature (Plachter 1995).

The implication is that only the second category should be considered as cultural landscape.

This limitation differs considerably from the broader geographical concept of cultural landscape presented by John Aitchison (1995) in a brief descriptive overview of cultural landscapes in Europe tracing in broad terms human impact on the continent's physical features. He begins with the classical definition of Philip L. Wagner & Marvin W. Mikesell:

> "The cultural landscape ... is a concrete and characteristic product of the complicated interplay between a given human community, embodying certain cultural preferences and potentials, and a particular set of natural circumstances. It is heritage pf many eras of natural evolution and of many generations of human effort" (Wagner & Mikesell 1962: 11).

For Aitchison (1995: 272): "cultural landscapes are in a state of 'being' and 'becoming'"; they are also 'ways of seeing', demanding both "'idealist and materialist explanation'", referring to Cosgrove (1984). Aitchison maintains that it is too much to expect consensus on definitions, meanings and methodologies; cultural landscapes will always remain elusive. Nonetheless, the World Heritage Committee is faced with the task, in identifying cultural landscapes of universal value, of reaching agreement on the 'ways of seeing' (Aitchison 1995).

6.2 The Council of Europe and *cultural landscape*

The designation of World Heritage Sites is in its essence based on expert assessments; the philosophical issue of whether universal values are possible is not tackled, while representativity is treated as a technical matter (see Pocock 1997). A somewhat different procedure is laid out in the Recommendation of the Committee of Ministers to Member States on the

Integrated Conservation of Cultural Landscapes, adopted by the Committee of Ministers of the Council of Europe in September 1995. *Cultural landscape areas* are defined as follows:

> "Cultural landscape areas: specific topographically delimited parts of the landscape, formed by various combinations of human and natural agencies, which illustrate the evolution of human society, its settlement and character in time and space and which have acquired socially and culturally recognized values at various territorial levels, because of the presence of physical remains reflecting past land use and activities, skills or distinctive traditions, or depiction in literary and artistic works, or the fact that historic events took place there" (Council of Europe 1995: 2).

A comprehensive process for identifying cultural landscape areas for necessary legal protection is outlined:

> "These identification procedures should be conducted:
>
> a) under the responsibility of the competent authorities at the appropriate territorial level;
> b) by independent experts appointed in the light of the aspects to be dealt with (for example in the sectors of architecture, landscape design, archaeology, geography, town planning, history, ethnology, anthropology, geology, agronomy, economics, sociology, ecology, natural sciences and law);
> c) with the participation of the local community:
>   – elected members and members of the authorities involved;
>   – representatives of the main professions concerned with social and economic activity in the relevant area, for example farmers, foresters, craftspeople, industrialists and tourist officers;
>   – representatives of residents, associations competent in the protection of the cultural and natural heritage and other associations" (Council of Europe 1995: 4)".

Although undoubtedly demanding in implementation, this approach nevertheless provides a blueprint that attempts to combine top-down and bottom-up approaches. Importantly, it recognizes that cultural landscape is not something that is given. It has both material and immaterial dimensions, and the identification of areas for protection can only be done after extensive consultation and, implicitly, negotiation, involving government authorities at different levels, experts with their varying specialisms and competence, and, not least, representatives of the local people whose surroundings the landscape constitutes.

7. CONCLUSION: ALTERNATIVE NARRATIVES OF *CULTURAL LANDSCAPES*

In conclusion, I will suggest that the terms natural landscape and cultural landscape can be seen as embracing a number of perspectives on the relationship between humans and their physical surroundings. Although showing varying degrees if interdependence, these perspectives also display certain differences from one another, related to different conceptualizations, different methods of investigation, and different narratives that are told about the landscape.

The usefulness of the term natural landscape has been criticized on the grounds that it is impossible to know fully what the 'original' natural landscape was like, as nature is continually changing, and humans were present in the landscape for so long in prehistory that it is impossible to know all the ways in which they have physically affected the landscape. However, natural scientists – for example, geologists, geomorphologists, hydrologists, zoologists and botanists – do study natural forces operating in affecting the physical features of the landscape. On the basis of a large array of scientific methods aiming at objective observation and measurement, empirical data is collected on which scientific interpretations are made. This perspective tells narratives of the influence of natural forces on the landscape or particular landscape features. Human beings tend to be regarded as interrupting influences, having an impact that is not infrequently seen as negative in relation to natural processes. The term natural landscape remains, nonetheless, useful to distinguish a research focus on natural processes that contribute to the formation of our physical surroundings, even if these are to varying extents affected by human activity, in contrast to research that focuses on the human contribution to the shaping of landscape. This approach is not overly concerned with the philosophical problem of ideas of nature as a social construction that varies in time and from society to society.

The classical use of the concept of cultural landscape is to describe the role of humans in changing the physical surface of the earth. The realization that the human contribution has been more extensive through time than often thought previously has led to some natural scientists, such as botanists and landscape ecologists, to promote the term cultural landscape as a means of drawing attention to the significance of the human impact, as it is still generally termed. Narrative, based on classical scientific methodology, indicates that the impacts of humans and their domestic animals cannot necessarily always be deemed negative but can also historically have had a positive influence, for example, through producing particular, valued compositions of biodiversity in semi-natural or other culturally influenced ecosystems.

Within historical geography and landscape history, there may similarly be reason to use the term cultural landscape in historical narratives telling of the many types of transformation of material landscapes that have resulted from human activity through time, thus distinguishing this from the processes described by geologists, geomorphologists and others. Generally, however, once the context of the study is established, it will be sufficient to talk of landscape when describing changes, whether due to hunting and gathering, livestock rearing, cultivation, manufacturing, urbanization, recreation, tourism or other human activities. Such narratives, which can be based on evidence found both in the field and in archives, are accompanied by a wide range of interpretations and explanations, which might variously emphasize human intentionality, functional systems or social structures. The arguments sustaining such narratives might vary from themes of improvement and progress to themes of social oppression and conflict.

Humanistic approaches in geography have brought to the fore the study of landscape meanings, involving ways in which landscapes are perceived, understood and mentally structured by different groups in society. The meanings that people give landscapes are bound up with culture, and the very act of naming gives shape to physical features of the landscape in people's cognition and communication. The term cultural landscape has sometimes been applied to what are also called cognitive landscapes or mental landscapes in order to emphasize that even natural features of the landscape are culturally interpreted. Often, however, the term landscape is deemed sufficient in the context of cultural historical studies. Methods of study are diverse, including text analysis, iconography, interviews and participant observation. Representations of landscape, in the form particularly of landscape paintings, but also photographs, film, descriptions and terminology, have proved fruitful sources of analysis of ideas and conceptualizations of landscape. Narratives provide accounts of, for example, the historical development of and social relations implicated in elitist or popular views of landscape as scenery. Historiographical narratives present the development of academic ideas of landscape. Interviews can provide a means of recording informants' own narratives of attitudes towards and practices regarding landscapes. Listening – to personal narratives or to the retelling of the collective narratives of local communities or indigenous peoples – can be an alternative to viewing the landscape. Participant observation can be a means of gaining information about people's activities and practices in relation to landscape, be they farmers or recreationists. Such methods provide the basis for alternative narratives to those constructed by empirical science.

It has been seen that cultural landscape has gained a renaissance as a new category of protected landscape. Here cultural landscape denotes landscapes that, after assessment, are protected because they illustrate the interplay of natural and human factors in their formation, or because they reflect strong religious, artistic or cultural associations with natural features. The protection of such landscapes implicates methods of identification, assessment, selection, legal designation and monitoring. The purpose of the new category of protected landscape is specifically to integrate nature and culture, which previously have largely been treated as separate categories in conservation. Official conservation authorities, supported by academic competence in natural science and cultural history, have had a dominating role in describing such landscapes, although participatory approaches are now being given some attention. Historical narratives are important in documenting the development of these landscapes over time, at the same asserting the value judgments and attachments that make them venerated landscapes, along with the perceived threats to them that are deemed to make protection necessary.

A final perspective on landscape that will be mentioned here is one that emphasizes the cultural role of law and custom as the institutions whereby members of a community regulate their relations to each other in using the resources of their physical environment. The term cultural landscape has in practice been rarely used in this context, the emphasis being instead on the etymological roots of the word landscape in the Germanic languages. Using methods of literary analysis, the narratives told here are of the origins and changing usages of the term landscape, the role of continuity and change in the legal and customary institutions regulating the use of and hence shaping over time the landscape as a cultural manifestation, and, in the modern context, providing critical analysis of official bureaucracies that maintain their power by promoting a distanced, scenic view of landscape against the locally centered perceptions of those living in local landscapes.

The narratives told by each of these perspectives are bound up with differing conceptualizations of landscape and cultural landscape. The appearance, adoption and varying use at different times of the term cultural landscape reflect particular research agendas and power interests. The same can be said when use of the term in certain contexts declines, or attempts are made to disavow its use as unnecessary. Rather than suggesting that particular usages of a term are necessary or unnecessary, and thereby further implying that particular narratives derived from those usages are correct or incorrect, the different interpretations and accounts can be compared and inform one another as a means of producing a deeper understanding.

REFERENCES

Aitchison, J. (1995). Cultural landscapes in Europe: a geographical perspective. In B. von Droste, H. Plachter, & M. Rössler (Eds.), *Cultural Landscapes of Universal Value – Components of a Global Strategy* (pp. 272-288). Jena: Gustav Fischer in cooperation with UNESCO.

Aitchison, J. (1996). The World Heritage Convention: cultural sites and landscapes. In F.H.A. Aalen (Ed.) *Landscape Study and Management* (pp. 183-193). Dublin: Trinity College Dublin Department of Geography, The Office of Public Works & Boole Press.

Asheim, V. (1975). *Kulturlandskapets historie i Østlandets jord- og skogbruksbygder.* Ås-NLH: Landbruksbokhandelen.

Asheim, V. (1978). *Kulturlandskapets historie. Jord- og skogbruksområdene slik det var og slik det er i flatbygdene på Østlandet.* Oslo: Universitetsforlaget.

Austad, I. (1983). Det åpne kulturlandskap In Norsk Naturforvaltningskandidatlag & Norsk Naturforvalterforening (Eds.) *Skjøtsel av bevaringsverdig natur og landskap i Norge* (pp. 153-166). Oslo: Landbruksforlaget.

Barnes, T.J. (2000). Situated knowledge. In R.J. Johnston, D. Gregory, G. Pratt, & M. Watts (Eds.) *The Dictionary of Human Geography* (4th ed., pp. 742-743). Oxford: Blackwell.

Barnes, T.J. & Duncan, J.S. (1992). Introduction. Writing worlds. In T.J. Barnes & J.S. Duncan (Eds.) *Writing Worlds. Discourse, Text and Metaphor in the Representation of Landscape* (pp. 1-17). London: Routledge.

Billinge, M.D. (1981). Cultural landscape. In R.J. Johnston, D. Gregory, P. Haggett, D. Smith & D.R. Stoddart (Eds.) *The Dictionary of Human Geography* (pp. 67-68). Oxford: Blackwell.

Birks, H.H., Birks, H.J.B., Kaland, P.E., & Moe, D. (Eds.) (1988). *The Cultural Landscape – Past, Present and Future.* Cambridge: Cambridge University Press.

Broek, J.O.M. (1938). The concept landscape in human geography. *Comptes rendus du Congrès International du Géographie Amsterdam, 2:3a*, 103-109.

Bruun, M. (1967). Landskap – natur og menneskeverk. *Bygg, 5*, 110-115.

Campbell, Å. (1936). *Kulturlandskapet.* Studentförening Verdandis Småskrifter, 387. Stockholm: Bonniers.

Christensen, A.L. (2002). *Det norske landskapet. Om landskap og landskapsforståelse i kulturhistorisk perspektiv.* Oslo: Pax.

Cosgrove, D. (1984). *Social Formation and Symbolic Landscape.* London: Croom Helm.

Cosgrove, D. (1985). Prospect, perspective and the evolution of the landscape idea. *Transactions of the Institute of British Geographers, New Series 10*, 45-62.

Cosgrove, D. (1992). *The Palladian Landscape. Geographical Change and its Cultural Representation in Sixteenth-Century Italy.* Leicester: Leicester University Press.

Cosgrove, D. (1998). Cultural landscapes. In T. Unwin (Ed.) *A European Geography*(pp. 65-72, 76-77, 80-81). London: Longman.

Cosgrove, D. (2000). Cultural landscape. In R.J. Johnston, D. Gregory, G. Pratt, & M. Watts (Eds.) *The Dictionary of Human Geography* (4th ed., pp. 138-141). Oxford: Blackwell.

Council of Europe. (1995). Recommendation No. R (95) 9 on the Integrated Conservation of Cultural Landscapes Areas as Part of Landscape Policies (Adopted by the Committee of Ministers on 11 September 1995 at the 543rd Meeting of the Ministers' Deputies). http://www.cm.coe.int/ta/rec/1995/95r9.htm

Daniels, S., & Cosgrove, D. (Eds.). (1988). *The Iconography of landscape. Essays on the symbolic representation, design and use of past environments.* Cambridge: Cambridge University Press.

Darby, H.C. (1951). The changing English landscape. *Geographical Journal, 117*, 377-394.

Darvill, T. (1996). European heritage and the recognition of cultural landscapes areas. In F.H.A. Aalen (Ed.) *Landscape Study and Management* (pp. 173-182). Dublin: Trinity College Dublin Department of Geography, The Office of Public Works & Boole Press.

Daugstad, K. & Jones, M. (1994). *Kulturlandskap i forvaltning. En begrepsutredning.* Trondheim: Geografisk institutt, Universitetet i Trondheim.

Daugstad, K., & Jones, M. (1998). Kulturlandskap i forvaltningen – begrepsbruk og verdisyn. In E. Framstad, I.B. Lid, A. Moen, R.A. Ims, & M. Jones (Eds.) *Jordbrukets kulturlandskap. Forvaltning av miljøverdier* (pp. 158-165). Oslo: Universitetsforlaget.

Duncan, J.S. (1990). *The City as text: The politics of landscape interpretation in the Kandyan Kingdom.* Cambridge: Cambridge University Press.

Duncan, J. (2000). Landscape. In R.J. Johnston, D. Gregory, G. Pratt, & M. Watts (Eds.), *The Dictionary of Human Geography* (4th ed., pp. 429-431). Oxford: Blackwell.

Eidsvik, H. (1993). Cultural landscapes – the need for a political perspective. In J.M. Fladmark (Ed.) *Heritage – Conservation, Interpretation and Enterprise. Papers Presented at the Robert Gordon University Heritage Convention 1993* (pp. 43-49). Oxford: Donhead.

Gabrielsen, E. (1969). *Landskapsarkitektur*, I. Vollebekk: Institutt for hagekunst.

Gregory, D. (2000). Discourse. In R.J. Johnston, D. Gregory, G. Pratt, & M. Watts (Eds.), *The Dictionary of Human Geography* (4th ed., pp. 180-181). Oxford: Blackwell.

Haber, W. (1995). Concept, origin and meaning of "landscape". In B. von Droste, H. Plachter, & M. Rössler (Eds.), *Cultural Landscapes of Universal Value – Components of a Global Strategy* (pp. 38-41). Jena: Gustav Fischer in cooperation with UNESCO.

Hartshorne, R. (1939). The nature of geography. A critical survey of current thought in the light of the past. *Annals of the Association of American Geography*, *29*, 173-644.

Jackson, J.B. (1986). The vernacular landscape (pp. 65-81). In E.C. Penning-Rowsell, & D. Lowenthal (Eds.), *Landscape Meanings and Values.* London: Allen and Unwin.

Jacques, D. (1992). The welcome complexity of cherished landscapes. *Paysages & amènagement*, 21, Special issue – Blois Conference "Landscapes in a new Europe: Unity in diversity?" together with *Landscape Research*, 83-91.

Jones, M. (1972). *Some Responses in Human Geography to Land Uplift in the Vasa Area, Finland (with Special Reference to Maxmo Commune).* Ph.D. thesis, University of London.

Jones, M. (1977). *Finland, Daughter of the Sea.* Folkestone: Dawson, & Hamden, Connecticut: Archon Books.

Jones, M. (1985). Datakilder, datainnsamling og verdisyn (pp. 57-84). In B. Dale, M. Jones, & W. Martinussen (Eds.), *Metode på tvers. Samfunnsvitenskapelige forskningsstrategier som kombinerer metoder og analysenivåer.* Trondheim: Tapir.

Jones, M. (1987). *Landhöjning, jordägoförhållanden och kulturlandskap i Maxmo.* Bidrag till kännedom av Finlands natur och folk, 135. Helsingfors: Societas Scientiarum Fennica.

Jones, M. (1988). Progress in Norwegian cultural landscape studies. *Norsk Geografisk Tidsskrift*, *42*, 153-169.

Jones, M. (1991). The elusive reality of landscape. Concepts and approaches in landscape research. *Norsk Geografisk Tidsskrift*, *45*, 229-244.

Jones, M. (1993). Om tverrfaglig samarbeid – kommunikasjon, verdisyn og begrepsdrøfting. *Jordbrukets kulturlandskap. Forskerkonferansen 1992, 26.–27. oktober – Sundvollen Hotell.* (pp. 84-86). Ås: Forskningsprogram om kulturlandskapet, Norges Forskningsråd.

Jones, M. (1998a). Kulturlandskapsdebatten i Norge 1987–1996 (pp. 47-64). In A. Norderhaug (Ed.), *Nordisk landskapsseminar Sogndal 1996.* Høgskulen i Sogn og Fjordane, rapportnr. 7/98. Sogndal: Høgskulen i Sogn og Fjordane, Avdeling for naturfag.

Jones, M. (1998b). Kan landskapsopplevelser forvaltes? In E. Framstad, I.B. Lid, A. Moen, R.A. Ims, & M. Jones (Eds.) *Jordbrukets kulturlandskap. Forvaltning av miljøverdier* (pp. 182-190). Oslo: Universitetsforlaget.

Jones, M. (2001). "Hva er geografi uten grenser?" Om geografi og rettsgeografi. *Det Kgl. Norske Videnskabers Forhandlinger 2001* (pp. 225-281). Trondheim: DKNVS.

Jones, M., & Daugstad, K. (1997). Usages of the "cultural landscape" concept in Norwegian and Nordic landscape administration. *Landscape Research, 22*, 267-281.

Jones, M. (2001). Norsk geografi – sett utenfra og innenfra. *Norsk Geografisk Tidsskrift–Norwegian Journal of Geography, 55*, 176-180.

Keisteri, T, (1990). The study of changes in cultural landscapes. *Fennia, 168* (1), 31-115.

Kjeldstadli, K. (1999). *Fortida er ikke hva den en gang var. En innføring i historiefaget*, 2. utgave. Oslo: Universitetsforlaget.

Krebs, N. (1923). Natur und Kulturlandschaft. *Zeitschrift der Gesellschaft für Erdkunde zu Berlin, 3-4*, 81-94.

Langdalen, E. (1965). Natur og menneskeverk i fjellet. In A. Sømme, E. Langdalen, O. Einevoll, A. Aase, A. Thormodsæter, S. Øvstedal, et al (Eds.) *Fjellbygd og feriefjell* (pp. 25-53). Oslo: J.W. Cappelen.

Livingstone, D.N. (1992). *The Geographical Tradition. Episodes in the History of a Contested Enterprise.* Oxford: Blackwell.

Lowenthal, D. (1961). Geography, experience, and imagination: towards a geographical epistomology. *Annals of the Association of American Geographers, 51*, 214-260.

Lowenthal, D. (1986). The American scene. *Geographical Review, 58*, 61-88.

Lowenthal, D., & Prince, H. (1964). The English landscape. *Geographical Review, 54*, 309-346.

Lowenthal, D., & Prince, H. (1965). English landscape tastes. *Geographical Review, 55*, 186-222.

Marsh, G.P. (1864). *Man and Nature; Or, Physical Geography as Modified by Human Action.* New York: Charles Scribner & Co.

Mitchell, D. (1984). *The Lie of the Land: Migrant Workers and the California Landscape.* Minneapolis: Minnesota University Press.

Mitchell, W.J. T. (Ed.). (1994). *Landscape and Power.* Chicago: University of Chicago Press.

Olwig, K. (1980). Historical geography and the society/nature "problematic": the perspective of J.F. Schouw, G.P. Marsh and E. Reclus. *Journal of Historical Geography, 6*, 29-45.

Olwig, K. (1996). Rediscovering the substantive meaning of landscape. *Annals of the Association of American Geographers, 86*, 630-653.

Phillips, A. (1995). Cultural landscapes: an IUCN perspective. In B. von Droste, H. Plachter, & M. Rössler (Eds.), *Cultural Landscapes of Universal Value – Components of a Global Strategy* (pp. 380-392). Jena: Gustav Fischer in cooperation with UNESCO.

Plachter, H. (1995). Functional criteria for the assessment of cultural landscapes. In: B. von Droste, H. Plachter, & M. Rössler (Eds.), *Cultural Landscapes of Universal Value – Components of a Global Strategy* (pp. 393-404). Jena: Gustav Fischer in cooperation with UNESCO.

Plachter, H., & Rössler, M. (1995). Cultural landscapes: reconnecting culture and nature. In B. von Droste, H. Plachter, & M. Rössler (Eds.), *Cultural Landscapes of Universal Value – Components of a Global Strategy* (pp. 15-18). Jena: Gustav Fischer in cooperation with UNESCO.

Pocock, D. (1997). Some reflections on World Heritage. *Area, 29*, 260-268.

Ratzel, F. (1895-96). Die deutsche Landschaft. *Halbmonatshefte der Deutche Rundschau, 4*, 407-428.

Rössler, M. (1995). UNESCO and cultural landscape protection. In B. von Droste, H. Plachter, & M. Rössler (Eds.), *Cultural Landscapes of Universal Value – Components of a Global Strategy* (pp. 42-49). Jena: Gustav Fischer in cooperation with UNESCO.

Rowntree, L.B. (1996). The cultural landscape concept in American human geography. In C. Earle, K. Mathewson & M.S. Kenzer (Eds.) *Concepts in Human Geography* (pp. 127-159). Lanham: Rowman & Littlefield.

Salter, C.L. (1981). *The Cultural Landscape*. Belmont: Duxbury Press.

Samarbeidsgruppa i Sogn og Fjordane (Ed.) (1987). *Konferanserapport. Kulturlandskap. Bruk og vern. Sogndal 15.–17. september*. Sogndal.

Sauer, C.O. (1925). The morphology of landscape. *University of California Publications in Geography*, 2:2, 19-53.

Sayer, A. (1992). *Methods in Social Science. A Realist Approach*, Second Edition. London: Routledge. (Original work published 1984).

Schick, M. (1982). Otto Schlüter 1872–1959. In T.W. Freeman (Ed.), *Geographers. Biobibliographical Studies*, *6* (pp. 117-122). London: Mansell.

Schlüter, O. (1903) *Die Siedelungen im nordöstlichen Thüringen. Ein Beispiel für die Behandlung Siedelungsgeographischer Fragen.* Berlin.

Simmons, I.G. (1988). The earliest cultural landscapes of England. *Environmental Review*, *12*, 105-116.

Smeds, H. (1935). *Malaxbygden. Bebyggelse och hushållning i södra delen av Österbottens svenskbygd. En studie i människans och näringslivets geografi*. Helsingfors: Ernst Ingelius.

Surnevik, P. (1970). Vern om kulturlandskap. In P. Surnevik, & J.B. Godal (Eds.), *Vern om kulturlandskap og kulturjord* (pp. 5-37). Olso: Landbruksforlaget.

Thomas, W.L. (Ed.). (1956). *Man's Role in Changing the Face of the Earth*. Chicago: University of Chicago Press.

Tschudi, A. B. (1934-35). Avfolkningen i Vest-Agder og nedleggingen av heiegårdene, særlig i Sør-Audnedal og Spangereid. *Norsk Geografisk Tidsskrift*, *5*, 207-248.

Tuan, Y.-F. (1974). *Topophilia: A Study of Environmental Perception, Attitudes, and Values*. Englewood Cliffs, New Jersey: Prentice-Hall.

Tuan, Y.-F. (1976). Humanistic geography. *Annals of the Association of American Geographers*, *66*, 266-276.

Tuan, Y.-F. (1991). Language and the making of place: a narrative-descriptive approach. *Annals of the Association of American Geographers*, *81*, 684-696.

Turner II, B.L., Clark, W.C., Kates, R.W., Richards, J.F., Mathews, J.T., Meyer, W.B. (Ed.). (1990). *The Earth as Transformed by Human Action: Global and Regional Changes in the Biosphere over the Last 300 Years*. Cambridge: Cambridge University Press.

Vandsemb, B.H. (1995). The place of narrative in the study of third world migration: the case of spontaneous rural migration in Sri Lanka. *The Professional Geographer*, *47*, 411-425.

von Droste, B., Plachter, H., & Rössler, M. (Eds.). (1995). *Cultural Landscapes of Universal Value – Components of a Global Strategy*. Jena: Gustav Fischer in cooperation with UNESCO.

Wagner, P.L., & Mikesell, M.W. (Eds.). (1962). *Readings in Cultural Geography*. Chicago: University of Chicago Press.

Widgren, M. (1997). Landskap eller objekt: Kring kulturminnesvårdens problem att hantera landskapets historia. In J. Brendalsmo, M. Jones, K. Olwig, & M. Widgren (Eds.), *Landskapet som historie* (pp. 5-16). NIKU Temahefte 4. Oslo: Norsk institutt for kulturminneforskning.

Chapter 4

A COMPARATIVE STUDY ON TREES AND HEDGEROWS IN JAPAN AND ENGLAND

Katsue Fukamachi[1], Hirokazu Oku[1] & Oliver Rackham[2]
[1] *Kansai Research Center, Forestry and Forest Products Research Institute, Japan;*
[2] *Corpus Christi College, United Kingdom*

1. INTRODUCTION

Japan and England have distinctive rural landscape patterns, resulting from the interaction of natural and cultural influences. Despite long-standing differences in land use and cultivation, there are many parallels in the infrastructure of their landscapes. In the 20th century, landscapes in the two countries underwent similar changes such as urbanization, abandonment of traditional land-management practices, depopulation of remote areas, and destruction of woodland and forest. The disappearance of old grasslands, trees and hedgerows on farmlands, and the destruction of trees by disease epidemics are familiar problems.

Both in England and Japan, the cultural landscape has drastically changed, being affected by the mechanization of land management and by fashions for amalgamating fields and for not using organic resources. In both countries the agricultural population has become older in demography and fewer in numbers. In England, moreover, livestock and arable farming have become separated, so that it is difficult for each to continue to use the by-products of the other.

The main issues in current landscape ecology are the conservation of traditional rural landscapes and their indigenous elements as a cultural heritage, rehabilitation of traditional practices such as coppice management, promotion of biodiversity, and development of recreational use of rural areas (Zonneveld & Forman 1989; Buckley 1992; Green 1996; Fukamachi & Sakuma 1998; Okada 1999). It is one of the most important landscape

H. Palang and G. Fry (eds.), Landscape Interfaces, 53–69.

ecology concepts that human beings and the landscape we make are influential part of ecosystems led us to realize that we could not prescribe what future landscapes should be without consulting human values (Nassauer 1997).

Japanese traditional rural landscapes, called *satoyama,* are the managed areas, the compact mosaics of cultivation, open woodland, coppice-woods and forests which surround villages and hamlets (Fukamachi et al. 2001). They were sustained by methods that were exclusively organic, involving the woods as well as the fields. *Satoyama* landscapes, closely related to the peculiar cultural and ecological characteristics of their regions, are rapidly deteriorating. They are famed for their beauty and cultural and spiritual values; moreover about half of the country's many endangered species are found in *satoyama.* Their conservation is an important political issue in Japan. The Ministry of the Environment, in its publication *The National Biodiversity Strategy of Japan*, underlines the importance of *satoyama* for the conservation of biodiversity (Nature Conservation Bureau 2002).

English rural landscapes have for centuries been appreciated for their beauty, and their cultural significance is now well understood. There are many local variations with distinctive ecological and cultural associations. They have been the objective of conservation efforts for a hundred years by voluntary bodies such as the National Trust, and by a notable number of farmers (Evans 1992). The Countryside Agency, established in 1999 from the merger of the Rural Development Commission and the Countryside Commission, is a statutory body with the duty of conserving and enhancing the natural beauty of the countryside. Its publications, for example, explain hedgerows as a defining feature of the English countryside, creating a characteristic structure and pattern of landscape (The Countryside Agency 1999, 2000).

It is important to gather a wide variety of scientific and cultural data on recent rural landscape patterns that have a long history of influence by human land use. Our study presents a comparison between Japanese and English rural landscapes, focused on their cultural infrastructures, especially non-woodland trees and hedgerows. We seek to explain similarities and differences of distribution, function and management of trees and hedgerows in cultural landscapes, and to understand the characteristics and changing trends of each landscape. We compare rural landscape patterns, and draw conclusions on how they should be conserved, based on the cultural and social backgrounds that are behind them. Our results are based on field surveys and the study of documents, maps, and aerial photographs (Japan: Ebisawa 1982, 1996, 2000; Shimizu & Nakamura 1990; Fukamachi 1992; Mitani 1994; Imamori 1995; Ochiai 1999; Nagasawa 2001; England:

Hoskins 1955; Pollard et al. 1974; Woodell 1985; Rackham 1986, 1998; Vickery 1995) in lowland Japan and England.

2. NATURAL AND SOCIAL CHARACTERISTICS OF JAPAN AND ENGLAND

The Japanese archipelago extends over 3,000 km from north to south, on the same latitude as the Mediterranean. A chain of volcanic and alpine mountains, 2,000-3,000 meters in height, forms the backbone of the islands. The climate varies from subtropical to cool temperate. As elsewhere in east Asia, the monsoon brings summer rain, but on the west side rain and snow, amounting to 1,700 mm or more, can occur throughout the year.

Japan has a sharp distinction between mountains and plains. It has not been much glaciated, but uplift of the mountains and deposition of volcanic ash resulted in huge extents of alleviation during the Pleistocene. The mountains, comprising two-thirds of the country, are nearly all forest, although parts of them have been farmland in the past. Agriculture, cities, towns and industry compete for the one-third of the country which is plains.

England is on the same latitude as northern Sakhalin. It is less mountainous than Japan, with a very varied but not volcanic geology. The climate is cool temperate, unusually warm for such a high latitude. Rainfall ranges from 500 to over 2,000 mm, evenly distributed throughout the year. Geography and vegetation have been greatly affected by Pleistocene glaciations. Most of the country is farmland, and nearly all has been farmland in the more or less distant past. Forests comprise about one-twelfth of the country, part of which is ancient woodland which has been in existence for many centuries.

Although the genera of plants and animals are similar in the two countries, the species are all different, and Japan has larger numbers of species. England has about 1,500 'indigenous' vascular plants, only a handful of which are endemic, whereas Japan has about 5,500 species in one-and-a-half times the area, one-third of which are endemic (World Resources 2000-2001). An important distinction is that Japan has giant bamboo (*Phyllostachys* species), a grass as tall as a big tree, which has no parallel in Europe.

An important cultural difference of the two nations concerns diet. The English, like other Europeans, get their calories from dry-land crops; their protein comes from meat and cheese, so they keep livestock. Fish, although important in the past, have declined in the last fifty years. The Japanese, like many other Asians, traditionally get their calories from rice, which is a wetland crop, and their protein mainly from fish. They keep few livestock, although stock-rearing and dry-land cultivation were more significant in the past. Japanese farmers, therefore, have a tradition of turning dry land into wetlands on which to grow rice; English, like most European, farmers have a tradition of turning wetlands into dry land on which to grow wheat, barley, and potatoes. Another difference is that Japan, by tradition, has been reluctant to import rice, whereas England has for centuries not grown all its own wheat.

In England, governments have for centuries exercised an influence on agriculture, and indirectly on the infrastructure of the landscape, through subsidies, taxes, and tariffs. Modernization of agriculture has been going on for 300 years, and some of the earlier phases already affected the infrastructure. After World War II, subsidies of production grew to become a large part of the income of most farmers; they were taken over by the European Union when Britain joined. For a time the destruction of traditional infrastructure was directly promoted, for example by grants for grubbing out hedges.

Agriculture continues to enjoy a subsidized prosperity in the cereal-growing parts of England, where from 1950 to 1980 destruction of the infrastructure of landscape was particularly severe (Pollard et al. 1974; Rackham 1986). Hedges and farmland trees became very unfashionable in that period of excessive mechanization and rising labor costs. Depending on the whims of individual farmers, many hedges were grubbed out; others were neglected and grew up into tall trees. It was a matter of common observation that hedges, if not neglected, passed through a period of excessive managed (trimmed mechanically and hastily every year, instead of carefully every few years) and became low and fragmented. In the north and west, where more of the infrastructure still exists, agriculture is now in a state of decline (subsidies notwithstanding), though not yet to the point of land abandonment.

In England, government subsidies, designation of areas is important for conservation, and even direct legislation. It is now (in theory – but there seem to be few prosecutions) a criminal offence to destroy certain types of hedgerow without authorization. Significant also is the pride of some individual farmers in the appearance of their properties, the greed of others, public interest in wildlife (especially farmland birds), education of the public (especially landowners) in how landscapes function. Agriculture itself is declining in importance of agriculture; tourism is growing, and in some areas forms a large part of farmers' income.

Between 1950 and 1980, whatever the motives for these changes, their effect worked, though very unevenly, in favor of destruction of the landscape (Rackham 1985). Almost throughout midland, eastern, and south-east England, and in parts of the north and west, comparison of successive editors of Ordnance Survey maps, or of aerial photographs taken in World War II with more recent ones, shows a severe loss of hedges and trees between those years. After 1980, the balance of factors has shifted, though very unevenly, in favor of conservation and even restoration. It is now unusual for a hedge to be destroyed, and new hedges are commonly planted.

In Japan, too, governments have for centuries exercised an influence on agriculture, and indirectly on the infrastructure of the landscape, through subsidies and taxes. A hundred years ago the government began to discourage farmland trees, which were considered to decrease agricultural productivity, and to legislate for the removal of trees on river levees. However, even though the farmers were required to cut down the trees, they replanted new saplings, which were needed for everyday farming practice. At the same time agriculture itself retreated from the mountains: depopulation of the remoter villages began a century ago and has continued steadily.

The rapid modernization of agriculture changed land use and management as well as the lifestyle of farmers. What was called 'field improvement' – for the sole objective of increasing productivity – converted small, irregular-shaped rice-fields into big quadrangular ones, without winding lanes, trees, or hedges. Landscapes devoid of trees and hedges have become the proto-landscapes for the children of contemporary Japan.

To this day, the Japanese Ministry of Agriculture, Forestry, and Fishery implements a policy of 'improvement of land for production efficiency' (Ministry of Agriculture, Forestry and Fisheries 2000). However, this policy conflicts with conservation of biodiversity and cultural landscape, since it involves interference in rural areas where properties are small and land use is mixed (Moriyama 1997).

However, since the early 1990s some Japanese have recognized the ecological importance of conserving traditional landscapes. A number of groups and communities have put great efforts into conserving the landscapes of their towns and villages.

Other changes are outside the influence of politicians. A conspicuous change in Japan is the subtraction of pine (principally *Pinus densiflora*), killed gradually over the last 70 years by an introduced nematode. Pine was one of the main components of Japanese historic landscape. Its loss is comparable to the effects of two fungi in Europe, *Ceratocystis* on elms and *Endothia* on chestnuts (the latter not in Britain). However, European elms may, and chestnuts do, recover from their diseases, whereas the loss of Japanese pine is likely to be permanent (Rackham 1986; Kamada & Nakagoshi 1996; Fukamachi et al. 2001). Pines are killed completely, and even if the nematode problem were to be solved there is little hope of reintroduction because of the loss of the specific land management that encouraged pine.

3. TREES AND HEDGEROWS IN JAPAN AND ENGLAND

3.1 Trees and hedgerows in English landscapes

Hedges are a defining feature of the English countryside, creating the characteristic pattern and structure of the landscape. There are many local variations with distinctive ecological and cultural associations, often dating from widely different periods. Many hedges have trees in them at intervals. Hedgerow trees and trees scattered in fields, meadows, and pastures form the category of non-woodland trees.

Figure 4-1. Hedges of hawthorn and hazel between fields, and between a field and a pond, Gloucester, England

Hedges act as property boundaries; they prevent livestock from escaping or getting into sown crops; and they provide wood and other products, both from the hedge itself and from the hedgerow trees. Hedges along roads define a grassy verge which was a minor source of hay, or could be grazed by tethered animals. Along streams, rows of alders and willows help to keep cattle out of the stream and thus to protect the water quality and reduce erosion.

Figure 4-2. Scene with numerous non-woodland trees in hedges and fields, Gloucester, England

Figure 4-3. Landscape with straight hedges, probably of late date, and hedgerow trees, Oxford, England

Many hedges (in parts of England the majority of hedges) are at least 400 years old; some are over 1000 years old. While some hedges are as old as the ancient woods that they connect, others come from post-1600 enclosures and later 'rationalizations' of infrastructure. Hedges dating from after 1800 tend to have only one or two shrub or tree species like hawthorn. The oldest hedges are a mixture of species, including maple and spindle (*Euonymus*). Hazel, although valued for its edible nuts as well as its wood, occurs mainly in hedges that have arisen naturally, especially those of medieval date. Figure 1 presents hedgerows of hawthorn or hazel between the fields, and between fields and a pond.

Figure 4-4. Row of poplars planted on a field boundary, probably intended as a windbreak, Gloucester, England

The non-woodland trees of England include ash, elm, oak, maple, hawthorn, and hazel. They behave in different ways. Elm regenerates by suckering and forms clonal patches; the others coppice – they sprout from the stump – when cut down. Each of these can form the hedge itself; ash, elm, oak, and maple can also occur as big trees in a hedge made of the same of other species. They all enter into folklore, especially ash and hawthorn. Special individual trees, venerated for their historic or sacred character, are uncommon in the general countryside (e.g. Queen Elizabeth's Oak at Huntingfield, Suffolk). They are more common in settlements, parks, and especially churchyards (for example there are several hundred ancient yews (*Taxus baccata*) beside churches). However holly (*Ilex aquifolium*) is an abundant tree in hedges in Norfolk because all hollies being bad luck to those who cut them down.

Besides trees such as oak, ash, and elm in hedges, there are, or were, trees such as oak and black poplar scattered in fields (Figure 2). Black poplar normally reproduces by falling down and rooting from the prostrate trunk.

Figure 4-5. Mixed hedge between fields, and row of willows and alders along a stream, Oxford, England

Farmland trees (Figure 3) were often regenerated by natural seedings or suckers. They served various functions: shade trees, shelter from wind, timber, wood for fuel and other purposes, boundary markers, and less often leaves to feed livestock. Those used for wood or leaves were cropped by pollarding, being periodically cut at 2-5 meters above ground, so that farm animals could not eat the young regrowth. Oak was the preferred tree for general-purpose timber; ash, elm, and poplar were less expensive timbers but also had their own special uses. Non-woodland trees could provide a considerable proportion of the wood and timber, even on estates which also possessed woodland.

These informal farmland trees should be distinguished from formal, deliberately planted features such as ornamental avenues of limes or windbreak lines of poplars (Figure 4).

Figure 5 shows examples of mixed hedge between fields, and row of willows and alders along a stream. They include many trees or herb species which regenerate by natural seedlings, and which are expected to have additional functions and benefits such as the protection against erosion and the improvement of stream water quality.

Figure 6 is an English lane between hedges. It is a *holloway,* sunk below the surrounding fields by centuries of erosion by traffic. The vegetated banks are periodically trimmed. The grassy verges of roads were a minor source of hay, and could be grazed by tethered animals. Chipped or trimmed shrubs and plants were used for feeding livestock or as organic fertilizer for the farmlands.

Figure 4-6. An English lane between hedges, Kent, England. It is a holloway, sunk below the surrounding fields by centuries of erosion by traffic. The vegetated banks are periodically trimmed

3.2 Trees and hedgerows in Japanese landscapes

In Japan hedges are not associated with enclosures or livestock farming, and rarely form a complete circuit. Land is valuable, and rice is a very light-demanding crop which does not tolerate shade. Hedges occur as isolated lengths, mainly along rivers or roads. The network of field boundaries is formed instead by canals, banks, levees and ditches between rice-fields. These are property boundaries and footpaths, and are mown for hay which is fed to cattle.

There are many local traditions in non-woodland trees and hedgerows, each with its distinctive species composition and its functional, ecological, and cultural associations. Common non-woodland trees include persimmon (*Diospyros kaki*), black pine (*Pinus thunbergii*), cherry (*Prunus yedoensis*), *hannoki* (*Alnus japonica*) and the elm-like *keyaki* (*Zelkova serrata*); there are also the bamboos *Phyllostachys heterocycla* and *Ph. bambusoides,* grasses as tall as a tree. All these coppice or sucker when cut down, except pine which regenerates only by seed.

Figure 4-7. Persimmon tree and statues of Bodhisattva, protectors of children and travelers, Tochigi, Japan

All six species can form the hedge itself, but except bamboo seldom do so. Persimmon, pine, *keyaki,* and cherry sometimes occur as a single big or old tree in a hedge, and are preserved as landmarks or sacred trees. They are linked to human life in various material and spiritual ways. Figure 7 shows a persimmon tree and statues of Bodhisattva, protectors of children and travelers.

Fruit trees like persimmon, chestnut (*Castanea crenata*) and apricot (*Prunus mume*) are also planted in the middle or at the edge of a field for shade and fruit or nuts. Persimmons supply food, lacquer for coating paper, and other materials. Pollarded trees such as *Quercus acutissima* or *Alnus japonica,* growing on the raised banks between rice-fields (Figure 8), were used to hang up tools or to hang rice to dry; edible mushrooms or insects such as beetles were gathered from their roots or stems. Larches (*Larix kaempferi*) and pines form windbreaks along fields (Figure 9); they are also timber trees. Where the tea-bush (*Camellia sinensis*) grows, in regions such as Kansai, it forms hedges between and within properties. Farmers grow their own tea by plucking the hedges.

Figure 4-8. Small pollard trees of kunugi (*Quercus acutissima*) on banks between rice-fields, Shiga, Japan

Figure 4-9. Windbreak of larch, Hokkaido, Japan

Figure 4-10. Hedges of giant bamboo along canals that irrigate rice fields, Tochigi, Japan. They protect the fields against side-cutting by the canals

Figure 4-11. Sacred grove, Shiga, Japan. Sacred trees are closely related to cultivation

Among the common deciduous oaks, *kunugi* (*Quercus acutissima*), is an introduction from China. In woods it is less common than the native oaks *konara* (*Q. serrata*) and *mizunara* (*Q. crispula* (*mongolica*)), and is usually pollarded, whereas the native oaks are coppiced. It is grown as a pollard among rice-fields, providing firewood and being a place where farmers gather for breaks during work.

Giant bamboos form hedges between fields and along canals (Figure 10). Their stems have many uses for structures (including houses) and implements; they can be harvested without destroying the hedge. Giant bamboos have much increased and have invaded forests, forming a zone between the cultivated plains and the native forest on the mountains.

Throughout Japan, trees (whether useful or ornamental) draw attention to wayside statues and monuments. Special individual trees and groves have spiritual or legendary significance; they shelter divinities and indicate a sacred place, and people conserve, propagate, and replace them. Pairs of trees are a symbol of harmony. Sacredness does not attach to trees at random but most often to individuals of certain species such as *sugi* (the great conifer, *Cryptomeria japonica*), ginkgo (*Ginkgo biloba*), *keyaki* (*Zelkova serrata*), and some broadleaved evergreens. Big and old trees in sacred or important places such as the entrances of villages, the vicinity of statues, and around temples and shrines tend to acquire sacred significance, and are preserved by local residents, religious institutions, and secular authorities. When a sacred tree dies, it does not cease to be sacred, but a new sapling is often planted alongside.

Figure 4-12. A group of trees centered on a sacred or legendary willow, which is set off by the flowering cherries, Tochigi, Japan

Sacred groves called *yamanokami* (Figure 11) consist often of evergreen trees such as *tabunoki* (*Machilus thunbergii*) and *yabutubaki* (*Camellia japonica*). These are the homes of the gods that protect the rice-fields and the woods. Every January farmers pray for good harvests, safe work, and their families' prosperity. Sacred groves tend to be maintained even where cultivation has been abandoned. Riverside willows are usually a source of wood, but can also be sacred trees, especially when they grow in the middle of rice-fields (Figure 12).

Developers usually take care not to fell sacred trees, since such an action is said to bring bad luck in Japanese tradition (Himeda 1980). Sometimes they move a sacred tree to another site, or replace it with a sapling.

4. CONCLUSIONS

In both countries hedges and free-standing trees play a role as landmarks, property boundaries, fences, shade trees, windbreaks, containment of streams, a habitat or passage for wildlife, and a spiritual symbol, as well as a source of wood and other products. Their traditions differ from those of woodland trees. Hedgerows and free-standing trees are conspicuous in agricultural landscapes around the world (Forman & Godron 1986). The hedgerow networks have the complex spatial and temporal dynamics, and are needed to take into account spatial attribute to evaluate more accurately over all network quality (Schmucki et al. 2002). Besides linear units of the landscape such as hedgerows, small patches of free-standing trees or small tree groves will provide both cultural and ecological significance.

In England, big trees derived from ancient wood-pasture, and scenic trees planted at times when they were in fashion, are widespread and keep alive a particular cultural landscape. Hedges have developed over a long period through human action and human default. They form (or are the relics of) a network of enclosures. The timber trees in them are a parallel to the standard trees scattered through coppice-woods.

In Japan hedges are less ubiquitous than in England, and have little relationship with livestock, but have a close relationship with everyday life and religion; they consist of few species, such as bamboo. Farmland trees are various, and sometimes pollarded or coppiced. Groves are planted or preserved in uncultivable spots or on boundaries or places of social significance.

In England sacred or legendary trees, though more abundant than in the rest of Europe, play a lesser part in the ordinary countryside than in Japan. In Japan, places with spiritual or legendary associations occur all over the country. They are often marked by special trees, either alone or emphasizing some other object. Those that indicate a spiritual relationship with people are preserved.

In both countries trees used to have various functions in the economy, society, or spiritual life of the human community. These uses and links have become weaker; the practices that sustained the trees have partly faded away; many traditional management techniques have not been handed down to new generations; and this has already affected the survival of the trees and hedges themselves (Rackham 1998). Changes in agricultural system and common practices of trees and hedgerows will strongly affect species richness at the landscape level (Burel & Baudry 1989; Burel 1992).

In Japan trees and hedges with spiritual relationships tend to be conserved at a local level; those valued for wildlife or tourism are the objective of national conservation. Many special trees or groves are protected as cultural assets or national monuments under the National Treasure Preservation Act. In England, Tree Preservation Orders, locally administered, give legal protection to a large number of designated trees – rarely to hedges or woodland. Hedges seldom come under the designation of Sites of Special Scientific Interest. Only very recently has national legislation given some legal protection to specific hedges.

Landscapes that can be designated as cultural heritage were created by the interaction of people and nature. Their conservation problems are diffused on a large scale, rather than being specific to especially important areas. Conservation needs to be based on investigating and identifying the biological, historical, and anthropological aspects together. Only thus can the remaining farmers, other local residents, wildlife and archaeological trusts, and national and international conservation organizations have a solid basis on which to promote the management of landscapes.

REFERENCES

Buckley, G.P. (Ed.). (1992). *Ecology and management of coppice woodlands*. London: Chapman & Hall.

Burel, F. (1992). Effect of landscape structure and dynamics on species diversity in hedgerow networks. *Landscape Ecology, 6* (3), 161-174.

Burel, F., & Baudry, J. (1989). Hedgerow network patterns and processes in France. In I.S. Zonneveld & R.T.T. Forman (Eds.) *Changing landscapes: An ecological perspective* (pp. 99-120). New York: Springer-Verlag.

Ebisawa, H. (1982). On lane-side trees in paddy fields (1). *Forest Culture Research, 3*, 133-144. In Japanese.

Ebisawa, H. (1996). On lane-side trees in paddy fields (2). Lane-side trees of the double cropping area in Nagahama City, Shiga Prefecture. *Forest Culture Research, 17*, 211-223. In Japanese.

Ebisawa, H. (2000). On lane-side trees in paddy fields (3). The type of distribution and tree forms of the lane-side trees in Shibaharaminami, Yokaichi City, Shiga Prefecture. *Forest Culture Research, 20*, 123-134. In Japanese.

Evans, D. (1992). *A history of nature conservation in Britain*. London: Routledge.

Forman, R.T.T., & Godron, M. (1986). *Landscape ecology*. New York: John Wiley & Sons.

Fukamachi, K. (1992). *A study on the significance of rural landscape trees in Ashino area.* Master of Agriculture Thesis, University of Tokyo. In Japanese.

Fukamachi, K. & Sakuma, D. (1998). Advances in satoyama studies: Looking for the planning of the interface between people and nature. *J. J. Landscape Research, 61* (4), 276-280. In Japanese.
Fukamachi, K., Oku, H. & Nakashizuka, T. (2001). The change of a satoyama landscape and its causality in Kamiseya, Kyoto Prefecture, Japan between 1970 and 1995. *Landscape Ecology, 16*, 703-717.
Green, B.H. (1996). *Countryside Conservation – Landscape ecology, planning and management*. Routledge.
Himeda, T. (1980). *Topography of trees. Trees and Japanese*. Tokyo: Miraisya. In Japanese.
Hoskins, W.G. (1955). *The making of English landscape*. London: Penguin Books.
Imamori, M. (1995). *SATOYAMA In harmony with neighboring nature*. Tokyo: Shinho-sya. In Japanese.
Kamada, M. & Nakagoshi, N. (1996). Landscape structure and the disturbance regime at three rural regions in Hiroshima Prefecture, Japan. *Landscape Ecology, 11*, 15-25.
Ministry of Agriculture, Forestry and Fisheries (2000). *Annual Report on Food, Agriculture and Rural Areas in Japan FY 1999*.
Mitani, Y. (1994). *A background to English literature. The countryside and natural landscape*. Tokyo: Senpu-sya. In Japanese.
Moriyama, H. (1997). Biodiversity in rural area. *Journal of rural planning association, 16* (3), 199-204. In Japanese.
Nagasawa, T. (2001). Plant-lore. Hoseidaigaku-syuppannkyoku, Tokyo. (In Japanese)
Nassauer, J.I. (Ed.). (1997). *Placing nature. Culture and landscape ecology*. Washington: Island Press.
Nature Conservation Bureau. (2002). *Living with Nature*. Ministry of the Environment, Government of Japan.
Ochiai, M. & Takahashi, T. (1999). A study on the preservation of levee trees in rural landscapes. *J. Rural Planning 55-60*. In Japanese with English abstract.
Okada, I. (1999). Restoration and management of coppices in Japan. *Ecological Restoration, 17* (1-2), 31-38.
Pollard, E. Hooper, M.D. & Moore, N.W. (1974). *Hedges*. London: Collins.
Rackham, O. (1985). Ancient woodland and hedges in England. In S.R.J. Woodell (Ed.), *The English landscape: past, present, and future* (pp. 68-105). Oxford.
Rackham, O. (1986). *The history of countryside*. Phoenix.
Rackham, O. (1990). *Trees & woodland in the British landscape* (2nd ed.). London: Dent.
Schmucki, R., Blois, S. & Bouchard, A. (2002). Spatial and temporal dynamics of hedgerows in tree agricultural landscapes of southern Quebec, Canada. *Environmental Management, 30* (5), 651-664.
Shimizu, Y., & Nakamura, Y. (1990). A study on historic landscape in SETAGAYA-ku, Tokyo. *J. J. Landscape Research, 53* (5), 205-209. In Japanese with English abstract.
The Countryside Agency (1999). *National trails; An introduction to National Trail routes in England and Wales*. Countryside Agency Postal Sales.
The Countryside Agency. (2000). *Hedgerows of England*. Countryside Agency Publications.
Vickery, R. (1995). *A dictionary of plant-lore*. Oxford University Press.
Woodell, S.R.J. (Ed.). (1985). *The English landscape: past, present, and future*. Oxford.
Zonneveld, I.S. & Forman, R.T.T. (Eds.). (1989). *Changing landscapes: An ecological perspective*. New York: Springer-Verlag.

Chapter 5

TRANSFORMATIONS OF CULTURAL LANDSCAPE

The Case of the Polish-Ukrainian Borderland

Ewa Skowronek, Renata Krukowska & Andrzej Swieca
Department of Regional Geography, Earth Science Institute, University of Maria Curie-Sklodowska, Poland

1. INTRODUCTION

The area of eastern Poland, referred to as 'the eastern wall', is a rarity on a European scale. Unique landscapes have been preserved here, shaped in the past by the representatives of the nations, religions and cultures, which coexisted in this area. In this part of Poland – currently almost homogenous in terms of ethnical structure – elements that document stormy periods of historic development have been permanently inscribed in the landscape. In the present study, we perceive landscape as a historically shaped fragment of geographic space, which is the outcome of environmental and cultural interactions. The picture forms a specific structure, which manifests itself in the individual regional character of physiognomy. It is a synthesis of the natural environment and man's activities. A natural environment comprises the whole of the natural conditions that provide a basis for man's economic activity.

The discussion to follow, concerning the development and condition of the cultural landscape in the Polish-Ukrainian borderland, deals exclusively with the Polish part of the borderland. We present the results of research carried out at the end of the 1990s in the area of Roztocze. The research marks a stage in the study of landscape changes (in a historic perspective) in mid-eastern Poland.

Roztocze is a geographic region clearly pronounced in terms of appearance, covering areas in Poland and the Ukraine. Its spatial extent is

H. Palang and G. Fry (eds.), Landscape Interfaces, 71–89.

demarcated by two towns: Krasnik (situated about 50 km to the SW of Lublin) on the Polish side and Lviv on the Ukrainian side. According to historical and archaeological regional divisions, the area of Roztocze (within its Polish boundaries) belongs to a historical-ethnographic region known as Lubelszczyzna (the Lublin Region), which forms a part of the historical province of Malopolska (Little Poland).

The choice of Roztocze as the subject of study was not incidental. The findings of research into the character and condition of some elements of the natural environment point to a relatively low intensity of human pressure (Swieca 1998). Areas of unique natural and cultural values have survived in this region characterized by poor industrialization, low urbanization rate and the prevalence of agriculture in land management. Field and forest patterns have hardly changed over the last two centuries and traditional land management has been maintained. Moreover, there are objects that testify the cohabitation and intermingling of the Polish, Ukrainian and Jewish populations. Contemporary landscapes bear traces of man's interference in the surrounding environment. It was the main objective of the present study to evaluate the landscape transformations produced by man's economic activity, with a special emphasis on the contribution of the nationalities inhabiting the area in the past to the shaping of the present-day picture of the region.

2. NATURAL CONDITIONS

Roztocze occupies the southern part of the Vistula and Bug interfluve. The region is marked as a clear range of elevations, composed of plateaus and hummocks rising up to 400 m a.s.l. The range, which follows the NW-SE orientation, is situated in the southern part of the Lublin Region, in the Polish-Ukrainian borderland (Figure 1). Within the borders of Poland, it covers an area of 2,391 km^2 and the population is approximately 148,000. The rich geological past of the region, and particularly the variety of Cretaceous and Tertiary rocks and Quaternary deposits, have largely contributed to the regional diversification of the natural conditions (Buraczynski 2002) and the division of the area into four subregions: the Goraj (758 km^2), the Szczebrzeszyn (218 km^2), the Tomaszow (985 km^2) and the Rawa (430 km^2) Roztocze.

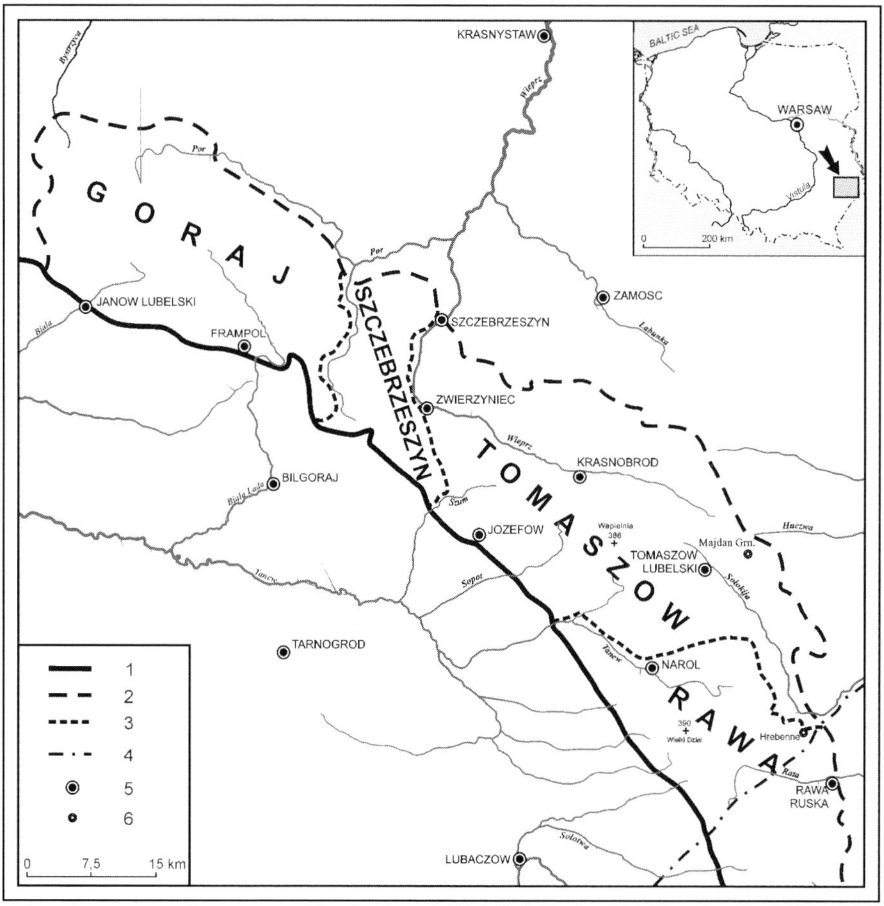

Figure 5-1. Location of the study area. Boundaries of the Roztocze (Buraczynski 1997): 1 – first range, 2 – second range, 3 – third range; 4 – state border, 5 – towns, 6 – other localities

In the western part of Roztocze (the Goraj and Szczebrzeszyn Roztocze), upper Cretaceous rocks (marls, opokas) occur in the basal complex, overlain by loess. Typical surface features include deeply incised dry valleys, which dissect the hummocks into a number of elevations and plateau ridges rising up to 300-340 m a.s.l. The thick loess mantle in this part of Roztocze determines the occurrence of loess relief. Numerous deep-cut (up to 20 m) gullies form a branched system several kilometers in length. Mean density of the gully network amounts to approximately 5 km/km^2, while the maximum density – to 11 km/km^2 (Buraczynski 1977). Arable land dominates in the structure of land management. This is due to the occurrence of fertile fawn combined with brown soils formed on loess and loessic deposits (Turski et al. 1993). Forest complexes (hornbeam-beech forests) have been preserved on the scarps and steep slopes of the gullies. It is the most densely populated part of the region.

The central part of Roztocze, the Tomaszow Roztocze, is built of upper Cretaceous rocks (mainly opokas and gaizes), and covered in places with patches of Tertiary deposits. Vast interfluve areas with individual relic mountains rising up in places are characteristic of land relief here. Soils formed from Cretaceous rocks and sands, were not favorable to the development of agriculture and settlement, thus cultivated land occupies smaller area. The Tomaszow Roztocze is characterized by poor settlement network and high afforestation. Among forest communities, fir and pine coniferous forests predominate (Szafer 1959).

The southern part of Roztocze, situated in the immediate vicinity of the Ukraine, is known as the Rawa Roztocze. Tertiary deposits overlying Cretaceous rocks prevail in this part of the region. Tertiary rock remnants (Wielki Dzial 390 m a.s.l., Dlugi Goraj 391.5 m a.s.l.) are the highest elevations in the Polish part of Roztocze. Economic activities in the Rawa Roztocze have been influenced by political considerations due to its situation in the border zone; it is characterized by considerable afforestation.

3. HISTORICAL CHANGES IN SOCIO-ECONOMIC CONDITIONS

The diversified natural environment conditions in Roztocze were favorable to the process of settlement and economic activity. The earliest traces of human presence in Roztocze date from the late Vistulian period (12000-8000 BC). In the Mesolithic period (8000-4000 BC), hunters belonging to the Komornica culture and the Janislawice culture entered the area (Libera 1992). During the Neolithic age (about 3100 BC) the land was used by shepherds and farmers from the Funnel Beaker, the Globular Amphora and the Corded Ware cultures (Nosek 1957). Numerous findings and a number of sites of the Trzciniec culture date back to the early Bronze Age (1450-1200 BC) (Machnik 1961). In the middle period of that age (1200-1000 BC), the people of the Lusatian culture appeared in Roztocze. Their presence has been documented by findings in the area of Bondyrz and Guciow. Towards the end of the Bronze Age, settlement in the area began to gradually recede. The biggest depopulation took place at the end of the 5th century, during migrations of nations. The settlement network was only restored in the next century, at the beginning of the Middle Ages. At that time, a dozen settlements were set up in Roztocze (Gurba 1965; Kutylowska 1990) (Figure 2).

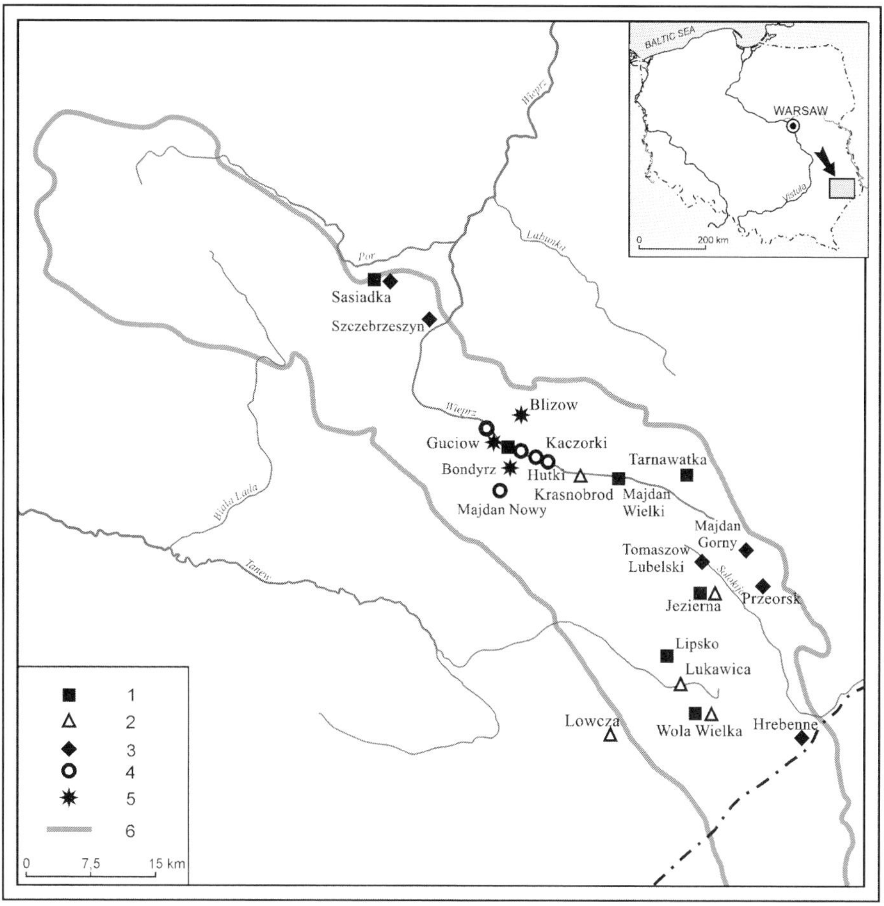

Figure 5-2. Settlement traces in the Roztocze region before 6th century. 1 – late Paleolithic, 2 – Mesolithic, 3 – Neolith, 4 – Bronze Age, 5 – Iron Age; 6 – boundary of Roztocze (Buraczynski 1997)

Between the 10th and the 13th centuries, a large part of the area was occupied by the complex of Czerwiensk strongholds, for which Poland competed with Ruthenia. Their significance was enhanced by the trade-route Kiev-Krakow-Prague (Wasowicz 1961). Due to frequent invasions by, among others, the Scythes, the Ruthenians, the Tartars and the Lithuanians, the existing settlement network was destroyed in the 13th century. The early Medieval strongholds in Guciow (10th century) and in Sasiadka, preserved in the Roztocze landscape, are a testimony to the olden times. In the 11th century, Sąsiadka played the role of a fort on the Polish-Russian border established at the time (Wartolowska 1958). The rich past of the region is also illustrated by the hills where the Szczebrzeszyn and Batorz strongholds were situated, as well as by numerous necropolises – grave-mounds (7th-11th century) in the neighborhood of Guciow (Zoll-Adamikowa 1974).

The reconstruction of the settlement network was begun in the 14th century. Then, a huge complex of the Gorajskis' estate, comprising several towns and villages, was built in the western part of Roztocze. In 1388, in the east of the region, Ziemowit IV received the Belz land and initiated Wallachian settlement there. Settlement activity contributed to the changing of the region's picture. Large forest complexes, especially those in the Goraj and Rawa Roztocze, were significantly reduced in area. Numerous new settlements were set up in the clearings and the size of the population of the region increased. Agriculture and shepherding were the main spheres of activity of the inhabitants (Skowronek 1996, 1999a).

Ethnical diversification of the population of Roztocze manifested itself as early as the 14th century. The eastern part of the region was mainly inhabited by the Ruthenians, while the western part by the Polish. It should be emphasised that the policy of the Polish state at the time was trending towards respect for various religious denominations, which is apparent in the royal privilege bestowed on Lviv in 1356, which warranted the Ruthenians, the Armenians, the Jews, the Saracens, and the Tartars retention of their rights and religions (Baczkowski 1999). Tolerance and freedom were conducive to settlement.

3.1 Settlement systems

The regional diversification of Roztocze is also reflected in the spatial arrangement of towns and villages. In the western part of the region (the Goraj and the Szczebrzeszyn Roztocze), the development of both towns and villages was largely based on German law. The so-called chain-villages and street-villages, situated in river valleys, still predominate in the area. In the east – in the Rawa Roztocze – settlement followed a different pattern. It comprised scattered large villages surrounded by many hamlets. The conditions of the natural environment also determined the different forms of land management in Roztocze. In its western part, fertile soils formed on loess and land cultivation was the occupation of the inhabitants. In the east – in the Rawa Roztocze – soils were poorer and shepherding prevailed.

Settlement processes greatly intensified in the 16th century (Figure 3). The number of towns and villages in the Polish part of Roztocze increased from 23 to over 100 at that time. It was accompanied by the development of local industries, supported by the newly established Zamoyski Estate.

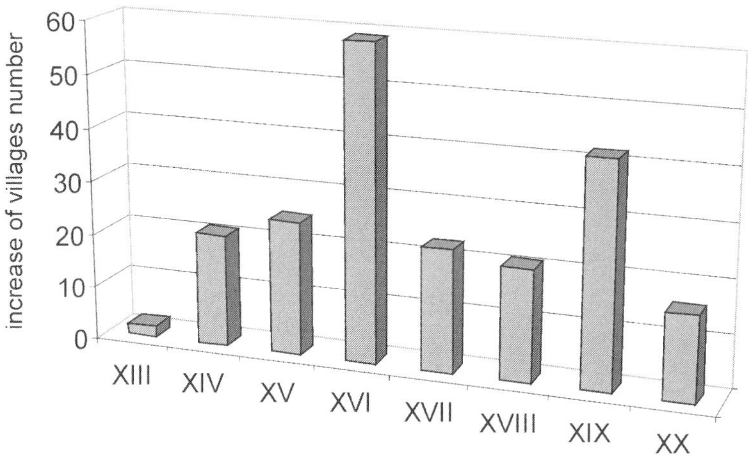

Figure 5-3. Number of settlements in the Roztocze region between 13th and 19th centuries

The Estate was founded in 1589 by Jan Zamoyski. Initially, it comprised two towns and 39 villages. Several years later, the numbers increased to 6 and 149, respectively, and the Estate occupied the area of 3,830 km^2. In 1847, 9 towns, 291 villages and 116 granges were to be found in the area of the Estate (Chlebowski 1919; Bartys 1963) (Figure 4). A large part of the area was occupied by forests. They were protected by professional foresters as early as the 18th century (Pawlowska-Wielgus 1971). In addition to the settlement development and forest husbandry, ore mining, metallurgy, cloth production, agricultural and food industry, timber industry, and others developed in the Estate. Towards the end of the 19th century, about 150 small and medium production plants, which employed about 1,300 workers, operated within the Zamoyski Estate. The Zamoyski Estate began to decline in the inter-war period. The final liquidation of the Estate followed the proclamation of land reform by the communist government in 1944.

It has to be emphasized that it was the measures taken by the Zamoyski Estate that brought about the beginning of settlement in the forested uninhabited central part of Roztocze (the Tomaszow Roztocze) in the 16th century. Owing to poor sandy soils in the area, craft, on a par with land cultivation, was an occupation of the inhabitants of the newly-established villages who also worked in forests and small industrial plants (steelworks, paper mills) owned by the Zamoyski Family. However, the economic growth was hindered by lack of convenient highways and low productivity (Iskrzycki 1956). Despite that, the settlement and economic activity has significantly influenced and shaped the present-day picture of the region.

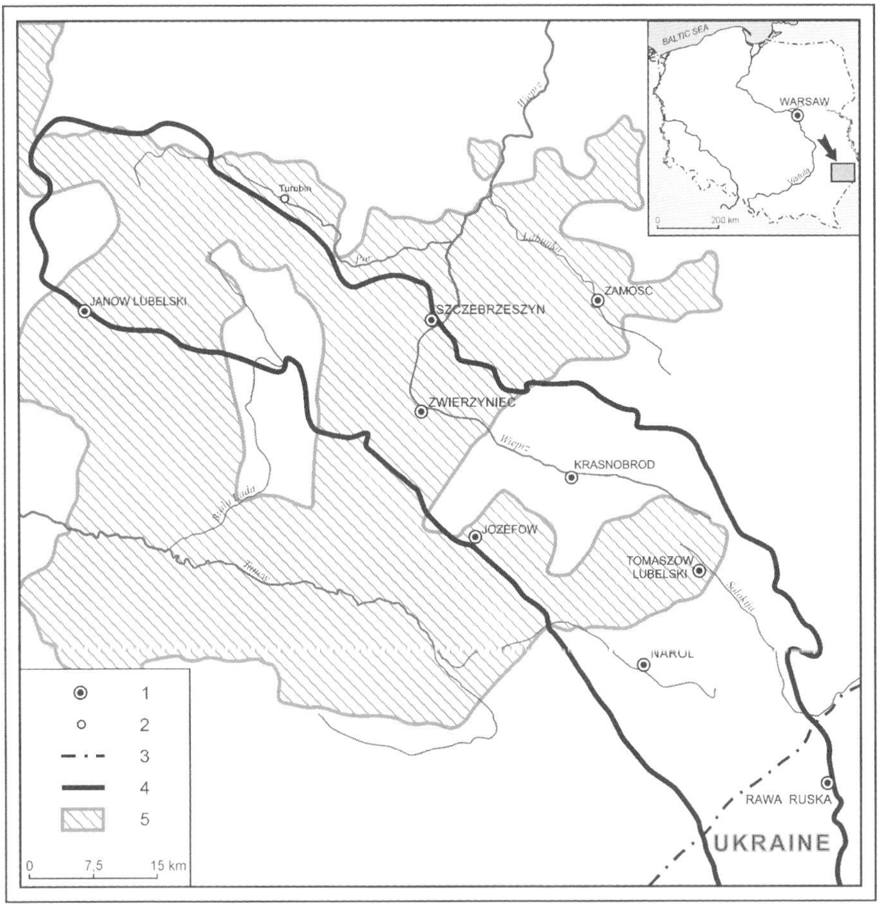

Figure 5-4. Location of the Zamoyski Estate in the Roztocze region. 1 – towns, 2 – other localities, 3 – state border, 4 – boundary of the Roztocze (Buraczynski 1997), 5 – territory of the Zamoyski Estate

3.2 Ethnic composition

Along with the accelerated growth of settlement, agriculture and industry, Jews and Armenians arrived in the area of eastern Lublin Region in the 16th century attracted by numerous privileges aimed at enhancing trade. The Armenians only settled in the capital city of the Estate – Zamosc – where they engaged in trade. The Jews inhabited almost all towns and villages, opting for smaller and larger towns (Jagusiewicz 1990). Their basic occupations included trade, craft, maintenance of inns and usury. When settled, they built synagogues, bathhouses, butchers' stalls, set up schools and Jewish cemeteries (*kirkuts*).

It has to be pointed out that during the whole period of prosperity of independent Poland (between the 14th and the 18th centuries) the Jews were a separate social class of a distinct language, custom and religion, as well as rights and social status. Organized into autonomous communities (*kahalas*), they had well-developed self-government structures and a central representation of all the Jews in the Republic, called *Waad Arba Aracot* (Parliament of Four Lands). The parliament was in function 1581-1764. They also set up one of the oldest systems of compulsory general education and religious formation of the youth in Europe (Trzcinski not dated; Jagusiewicz 1990).

Between the 16th century and World War II, three ethnic groups shaped the cultural landscape of the area: the Polish, the Ruthenians and the Jews. During the period of loss of independence (1772-1918), the level of economic development of the region remained low. Many factors contributed to the situation, among others, the low standing of trade and craft and lack of convenient transport routes and markets for industrial output and agricultural produce. Roztocze was then distinguished by large population density (Skowronek 1999a), which exceeded 80 persons/km^2. In the 19th century the number of settlements approximated 200. Agriculture was the basic source of income of the inhabitants of both little towns and villages. Villages were overpopulated, farms were small and arable land partitioned. The arrangement of houses, which faced the road with their narrower sidewall due to the narrowness of land plots, has been preserved till nowadays.

After Poland regained independence in 1918, no significant changes in the ethnical structure or living standards of the Roztocze population took place. In the light of statistical data, it is evident that the Polish population dominated in the Goraj, the Szczebrzeszyn and the Tomaszow Roztocze 1918-1939. Data from 1931 concerning the Lublin Region, which encompassed the western and central part of Roztocze, document a marked preponderance of the Polish people (Chalupczak & Browarek 1998). The Poles accounted for 85.1% of the total population of the voivodship (up to 90.3% in villages). The second largest group was made up of the Jewish people – 10.5% (4.5% in villages), the third – of the Ukrainians – 3% (3% of the village population). The prevalence of the Polish in the Tomaszow Roztocze is confirmed by data for the Tomaszow administrative district from that period: the Polish people accounted for 84% of its population, the Ukrainian population formed 8.7%, and the Jewish population – 7.6%. The religious structure reflected the ethnical one (Town Index for the Polish Republic 1924): 62.4% of inhabitants were Catholics, 24.1% – members of the Russian Orthodox Church, 0.4% – Greek Catholics, 0.01% – protestants and 13.1% – Jews (Figure 5).

A considerably different ethnical picture formed in the eastern part of the area, the Rawa Roztocze. The 1921 census indicates a prevalence of the Ukrainian people (over 60% of the total population). The Polish accounted for about 29% of the district's population, and the percentage of the Jewish population amounted to 8%. In the religious structure, Greek-Catholics dominated (Town Index for the Polish Republic, 1924) (Figure 5).

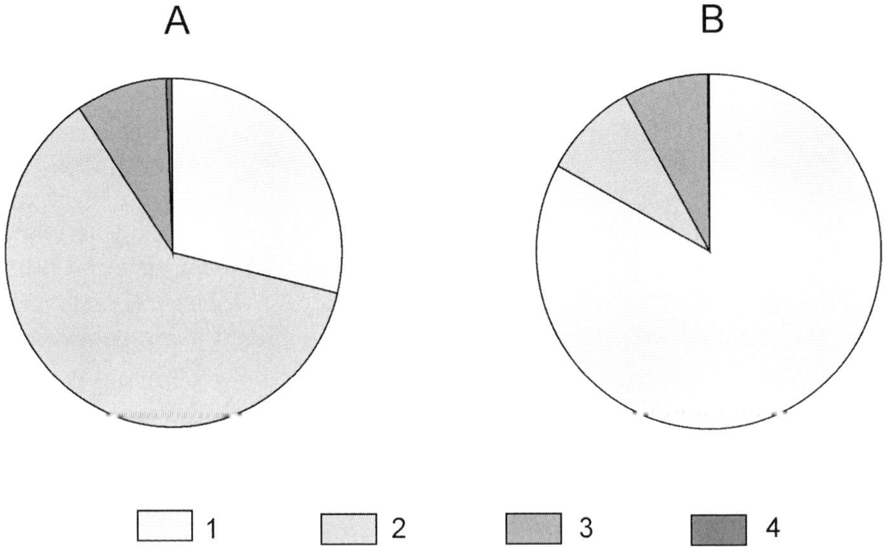

Figure 5-5. Ethnic diversity of people in the Roztocze region in 1921. A – Rawa Ruska district, B – Tomaszow Lubelski district, 1 – Poles, 2 – Jews, 3 – Ukrainians, 4 – others

The structure of the Roztocze population changed dramatically in consequence of the events of World War II. Jewish settlement in Poland virtually ceased to exist. The fascism of Hitler's III Reich brought about the extermination of almost all of the 3.5 million Polish Jews. After the war, 150,000 Jews remained in the area of Poland. However, the new socialist state did not favor the people. As a result of repressive measures a large number of Jews abandoned the country. Their present number is estimated at 10,000. They are scattered and mostly live in towns (Chalupczak & Browarek 1998).

The situation of the Ukrainian people was also transformed. After the war, due to border changes, the main centers of the Ukrainian people were incorporated into the Ukrainian Socialist Soviet Republic on the strength of a treaty on population exchange from September 9, 1944. Those staying were resettled by Polish communist authorities to the Western and Northern Lands (the area of the western and northern part of Poland) during the so-called Vistula action. Because of those events, several dozen villages and hamlets in Roztocze became totally depopulated. The number of inhabitants dropped dramatically in the area. The displacement of the Ukrainian population entailed considerable material losses. As a result of political changes, large villages and hamlets typical of the pre-war period ceased to exist in the region. They were replaced by small settlements established in the vicinity of PGRs (state-owned land holdings). Service and industrial outlets were destroyed, along with the monuments of spiritual and material culture. Large areas, previously densely populated, were re-afforested. The decline in the population number was permanent. A comparison of the present-day number of inhabitants in selected Roztocze villages in the former administrative district of Rawa Ruska with the corresponding data from 1924 reveals a large depopulation of the area: the total number of people is almost twice smaller (a decrease from 64,500 to 27,000); mean rate of population density decreased from 93 persons/km^2 to 39 persons/km^2 (Town Index for the Polish Republic 1924; Lublin Voivodship 1999).

The changes did not affect other parts of Roztocze – the Goraj, the Szczebrzeszyn and the Tomaszow Roztocze – to such an extent. The settlement structure which existed before the war has remained virtually unchanged. The functions of the region have also been preserved. However, the number of the population decreased.

In conclusion, it may be stated that the contemporary Roztocze is an area characterized by low population density, irregular settlement network, low industrialization rate and poor development of urbanization. It is an extensively forested area, of a typical rural character (the countryside accounts for as much as 90% of the surface area and is inhabited by 73% of the population). Arable land and forests are the main types of land management, small individual farms dominate and land is strongly partitioned. The bad condition of the region is worsened by an unfavorable economic situation of the country and poor accessibility of the area. The municipal and rural districts are affected by stagnation or population outflow. No new investments are made here. The local people engage in individual agriculture, forestry and services which are the occupation of over 2/3 of the population.

4. CHANGES IN THE CULTURAL LANDSCAPES OF ROZTOCZE

The contemporary picture of Roztocze is to a great extent the outcome of man-induced environment transformations. An analysis of the sequence of events makes it possible to distinguish periods of varying intensity of human pressure.

On the basis of research findings obtained so far, it can be concluded that the transformations of the natural environment in Roztocze did not begin until the 14th century. They were produced by settlement actions aimed at attracting settlers to uninhabited areas and economic activation of the newly established estates. It is typical of Roztocze that the actions were carried out differently in the different parts of the region. In the Goraj Roztocze and the Szczebrzeszyn Roztocze (the western part) town foundations were modeled upon west-European countries – the German law, while in the Rawa Roztocze (the eastern part) the settler groups were mainly composed of Wallachian shepherds arriving from the south-east. They set up settlements which, in time, merged and expanded forming vast villages with numerous hamlets.

Intense clearing of land accompanied the settlement in Roztocze, which was linked to construction work, demand for arable land and the utilisation of forested areas for the needs of local industry. The functions of the settlements changed in the process. Small industry and craft developed parallel with agriculture. As early as the 14th century, iron and glass were produced from local ore and sands.

As a result of settlement actions and economic activity, significant changes in the proportion of natural landscape elements (e.g. forests and meadows) to anthropogenic ones (fields, pastures, built-in areas, roads) took place in Roztocze between the 14th and 16th centuries. It is estimated that the size of the area occupied by anthropogenic elements in the Roztocze landscape increased from 19% in the 14th century to approximately 33% in the 16th century (Skowronek 1999a, 1999b). Also in the successive centuries, the growth tendency of the proportion of anthropogenic elements was marked; at the beginning of the 20th century it was higher than 50%. A culmination of anthropogenic pressure on Roztocze landscape took place during the twenty years of Poland's independence after World War I. It was then that the greatest ever number of people inhabited the region. An increase in population density brought about considerable partitioning of land and contributed to a decrease in wooded area. The processes are reflected in the existing settlement network, as well as in the remaining industrial facilities (ruins of paper-mills, quarries, mills, breweries, sugar-plants) and in the field pattern characteristic of the region (field-patches).

The growth tendency in the share of anthropogenic elements declined during 1939-1945. After World War II, human pressure diminished. It was caused by a decrease in the size of the population, particularly in the Rawa Roztocze, and a change in the activity of the population, which, under the socialist economy, was 'double-occupational'. It was common for country-dwellers to combine work in towns and cities with the maintenance of small farms.

Another turning point in the influence of man on the environment occurred in the early 1990s. A change of the politico-economic system led to a decline of many industrial plants and institutions. Unemployment resulted, and, as a consequence, migration of the local people. The youth were leaving for big towns seeking work and villages became depopulated. The above changes in the Roztocze landscape are reflected in the systematic shrinking of the surface of arable land (Figure 6). It gives way to spontaneously appearing or planned forests. The area of wasteland has also increased (Gardziel et al. 2001). Due to these phenomena, the proportion of cultural elements in the landscape at the end of the 20th century decreased to approximately 40%, which corresponds to the level recorded at the end of the 19th century, in spite of a twofold increase in the number of inhabitants during that period (Skowronek 1999a).

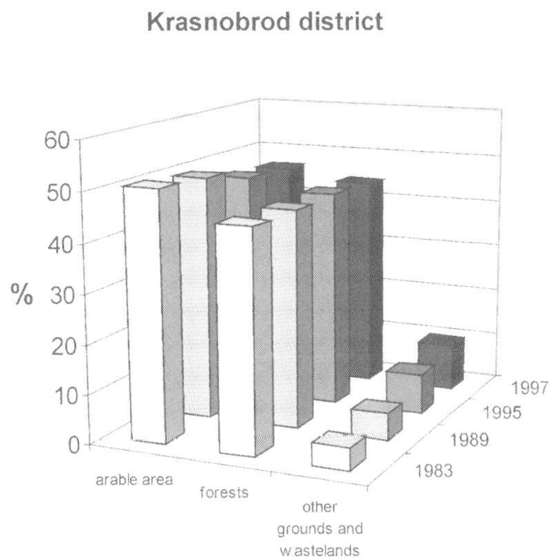

Figure 5-6. Changes in land use in the Roztocze region 1984-1998 – example of the Krasnobrod district

There is yet another aspect to the cultural landscape of Roztocze. It was here, in the borderland area, that the eastern and western Slavonic cultures, many styles and beliefs intermingled. Distinct architecture and a unique form of settlement network, different from the one in the west of the country, took shape. Although in the beginning of the 21st century, the social, national and religious landscape of the Roztocze region bears no resemblance to the picture recorded in population censuses from 1921 and 1931, it has to be emphasized that numerous precious monuments have been preserved. They testify to the rich history, varied culture, religion and cohabitation of the former inhabitants of the area. They include temples, synagogues, belfries, cemeteries, roadside shrines, martyrdom sites, which persist in defiance of the destiny which the events of history have brought to their creators.

Roztocze is now mainly inhabited by the Polish people of Roman-Catholic faith. Their churches (over 90) are scattered throughout the region, mostly in towns and big villages. They coexist with the temples of Greek-Catholics, Russian Orthodox churches and synagogues, which testify to the rich history of the region (Figure 7). In the borderland zone, in the Tomaszow and the Rawa Roztocze formerly inhabited by the Ukrainians, Uniate churches, roadside crosses and cemeteries can be found. The objects, in different state of preservation, occur in over 30 towns of the borderland zone. The most magnificent wooden Greek-Catholic churches are to be found in Hrebenne and Radruz. The temples are often accompanied by Uniate cemeteries like those preserved in Kniazie and Kornie. The only Orthodox Church remaining in Roztocze is situated in Tomaszow Lubelski.

The few monuments of the Jewish culture which have survived the war destruction include synagogues and cemeteries (*kirkuts*). They occur throughout Roztocze. Buildings of former synagogues remain in Szczebrzeszyn and Jozefow, among other sites. They perform various functions, used as libraries, museums and cultural centers. None of them is a sacral building. Jewish cemeteries can only be found in three towns of the region.

Roztocze is also the location of places of remembrance commemorating tragic historical events and the martyrdom of the Jewish people during World War II. An interesting historical object dating back to the period of WW II can be found in the Rawa Roztocze. It is a belt of defenses with numerous concrete bunkers called the Molotov Line. They were constructed 1940-1941 by the Russian army along the German-Russian demarcation line from 1939, which cut across the eastern areas of Poland. The defenses were to protect the USSR borders against a German invasion. They failed to perform the task, however, since only a few had been finished and armed by the time of the attack in 1941.

The villages of Roztocze are an interesting and precious cultural asset, as they retain elements of folk culture characteristic of the region, such as: timber architecture, farmstead layout, wells, shrines, roadside crosses. The most valuable of them have been protected in different museums, including the Lublin Museum of the Village, the Ethnographic Museum in Bilgoraj and the skansen museums in Guciow and Krasnobrod.

Figure 5-7. Location of the cultural heritage objects in the Roztocze region. 1 – Orthodox church, 2 – Greek Catholic church, 3 – Greek Catholic cemetery, 4 – synagogue, 5 – Jewish cemetery, 6 – place of Jewish martyrology; 7 – state border, 8 – boundary of the Roztocze (Buraczynski 1997)

The towns of Roztocze also deserve attention. They are rarely inhabited by more than 5,000 people. Their buildings, spatial arrangement and functions (most often as markets for agricultural produce and handicraft) brilliantly reflect the lifestyle of their former inhabitants. Szczebrzeszyn is an interesting locality, which boasts Jewish houses from the pre-war period centered around the town hall, two 17th-century brick Roman Catholic churches, a Uniate church founded at the end of the 12th century and reconstructed in the 16th and the 18th centuries and a 16th-century synagogue and a *kirkut* – a cemetery of the Jewish community.

5. CONCLUSIONS

The present-day landscapes of Roztocze reflect the relations between economic activity and natural conditions. The regional diversification of the natural conditions in Roztocze (different soil quality, degree of afforestation, accessibility of the area) and the geopolitical situation of the region have determined the extent of settlement (since the 14th century) and the character of the spatial arrangement of localities. In the western part of Roztocze (the Goraj Roztocze and the Szczebrzeszyn Roztocze), the development of settlement was based on utilization of fertile soils, which is manifest in the preponderance of agricultural landscapes (dominance of arable land). This part of Roztocze was mainly inhabited by the Polish people who set up chain-villages and street-villages in river valleys. In the Rawa Roztocze, areas characterized by frequent occurrence of sandy covers and soils of poorer quality were utilized by the Wallachian people. Shepherding and cattle raising were the basis of the economic activity of this settlement group. A settlement pattern distinct from the one in the Western Roztocze took shape here, consisting of expanding and merging shepherds' settlements. The settlement development, and thereby the condition of the landscape in the central part of the region (the Tomaszow Roztocze) was largely influenced by the economic activities of the Zamoyski Estate. This area was the last one to undergo the settlement process and up till now is the least populated part of Roztocze. The owners' care of the forests situated within the Zamoyski Estate allowed the natural character of the region's landscape to be preserved.

Estate forming in Roztocze largely contributed to the changes in land use. The area occupied by man-induced elements (fields, buildings, roads) gradually increased. The area of forests and meadows decreased. The growth tendency of the share of anthropogenic elements in Roztocze landscape was marked until the outbreak of WW II. The largest landscape transformations took place in the inter-war period of twenty years and were the result of historical events and socio economic processes. At the time, Roztocze was inhabited by the largest population which was diversified in ethnical and denominational terms. The Polish, the Ukrainians and the Jews made up a mosaic of nationalities. The distribution of the Jewish people was almost uniform in the whole of Roztocze; the Polish people prevailed in its western part, and the Ukrainians – in the eastern part. The occupations of the population shaped the ethnical structure of Roztocze's towns and villages. The Jews engaged in trade, craft and usury and were the most numerous in towns. The Polish and the Ukrainians – engaging in traditional land management – formed rural societies.

World War II and its effects (depopulation, material destruction) impeded the intensity of demographic and economic processes in the region. Since then, the rate and size of man-induced changes in the landscape has been weaker. Also, the ethnical structure of Roztocze has changed irreversibly. The largest changes affected the eastern part of the region (the Rawa Roztocze) from where the Ukrainian people were removed as a result of political actions. Holocaust brought about the extermination of almost total Jewish population. Those who survived left the area of Poland in successive years.

The political and socio-economic events which took place in the post-war period produced a relative stabilization of anthropogenic transformations of Roztocze. The proportion of the elements did not increase. Recent years shave shown a decrease of the proportion. Therefore, an increase in the area occupied by natural landscape elements, mainly forests, is noticeable in Roztocze. It defies common beliefs about the rate and character of the present-day transformation of the area's geographic environment. It needs to be stressed that the centuries-long evolution and interference of man have not led to the degradation of the region's unique landscape amenities. This makes Roztocze a rarity not only in Poland, but in the whole of Europe. Protection, analysis and documentation of this invaluable element of our heritage is therefore an urgent matter. A number of measures have already been taken in Poland to secure appropriate functioning and utilization of Roztocze. It was relatively early taken under protection. In 1974, the Roztocze National Park was founded, and in successive years, four landscape parks were set up: South Roztocze, Solski Forest, Szczebrzeszyn and Krasnobrod, accompanied by areas of protected landscape. Reserves and nature monuments are also numerous in the region.

Contemporary trends in the economic development of Roztocze seem to favor actions aimed at utilization of its landscape diversity. The tourist attractiveness of the region, mainly the wealth and variety of its natural environment, high recreational, sightseeing, salubrious and aesthetic value are emphasized in promotional publications. The recognition and promotion of the value of the region may enhance the development of a new function of Roztocze – tourism, which, along with the traditional land management, will provide a chance for the development of the region and contribute to the utilization of its recreational, cultural and educational assets.

REFERENCES

Bartys, J. (1963). *Lowland sheep-farming in the 2nd half of the 18th century and in the 1st half of the 19th century as exemplified by the Zamoyski Family Estate* (pp.13-30). Wroclaw: IHKM PAN, Ossolineum.

Baczkowski, K. (1999). History of late medieval Poland (1370-1506). *Great History of Poland, 3*, 12-15.

Buraczynski, J. (1977). Intensity of gully erosion and soil erosion in Goraj Roztocze. *Zeszyty Probl. Post. Nauk Rolniczych, 193*, (pp. 91-99). Warsaw.

Buraczynski, J. (1997). *Roztocze. Structure - relief- landscape*. Lublin: ZGR UMCS.

Chalupczak, H., Browarek, T. (1998). *National minorities in Poland 1918-1995* (pp. 55-92, 163-201). Lublin: UMCS Press.

Chlebowski, B. (1919). Zamosc. *The Zamoyski Estate and the Zamosc district*. Zamosc: Zygmunt Pomaranski &Co.

Eastern Roztocze. A tourist-topographical map 1:50,000. Raszyn: Wydawnictwo Oswiatowe ORTUS, Lubaczow: Wydawnictwo Naukowe, Turystyczne i Edukacyjne.

Fuks, M., Hoffman, Z., Horn, M., & Tomaszewski. J. (1982). *Polish Jews*. Warsaw: History and Culture.

Gurba, J. (1965). From the problems of early medieval settlement in the Lublin Upland. *Annales UMCS* (pp. 44-58). Lublin.

Gurba, J. (1985). An outline of the history of the Roztocze National Park. In T. Wilgat (Ed.), *Roztocze National Park* (pp. 7-9). Lublin: KAW.

Gorak, J. (1984). *Former Orthodox churches in the Zamosc Voivodship*. Zamosc: Biuro Badan i Dokumentacji Zabytkow w Zamosciu.

Gradziel, T., Janicki, G., & Skowronek, E. (2001). Transformations of the natural environment of the Guciow village in Central Roztocze under the influence of man's activity. In K. German (Ed.) *Transformations of the natural environment in Poland and its functioning. Problems of Landscape Ecology, 7*, (pp. 294-300). Krakow: PAEK, GUJ.

Jagusiewicz, A. (1990). *Ethnic values of Poland for Jewish tourism. Problems of Tourism, 3/4,* (pp. 92-109). Lodz: Institute of Tourism.

Kutylowska, I. (1990). *The development of Lublin in the 6th-14th centuries against the background of the urbanisation of the middle Vistula and Bug interfluve.* Dissertation, (manuscript), UMCS, Lublin.

Libera, J. (1992). Late Paleolithic and Mesolithic periods in south-eastern Poland, *Archaeological Letters, no. 2*. Lublin.

Machnik, J. (1961). Archaeological research in Lublin Roztocze in 1959. *Archaeological Reports, Vol. 12*, Wroclaw-Warsaw-Krakow: Ossolineum.

Nosek, S. (1957). Materials for the study of ancient and early medieval history of the Vistula and Bug interfluve. *Annales UMCS*. Lublin.

Pawlowska-Wielgus, H. (1971). *An inventory of cartographic materials of AOZ from Zwierzyniec. Management of AP*. Lublin.

Statistical Year-book for the Zamosc Voivodship. 1984. Warsaw: GUS.

Statistical Year-book for the Zamosc Voivodship. 1990. Warsaw: GUS.

Statistical Year-book for the Zamosc Voivodship. 1996. Warsaw: GUS.

Town Index for the Polish Republic, 4. 1924. Lublin Voivodship, Warsaw: GUS.

Town index for the Polish Republic, 13. 1924. Lviv Voivodship, Warsaw: GUS.

Skowronek, E. (1996). Settlement development in the Lublin Upland and Roztocze, *Geographic Magazine, LXVII, 2*, 209–223.

Skowronek, E. (1999a). *Anthropogenic transformations of the Roztocze landscape as exemplified by the Biala Lada and upper Wieprz drainage basins*. Doctoral dissertation, manuscript, Biblioteka Glowna UMCS. Lublin.

Skowronek, E. (1999b). Changes in the landscape of the upper Wieprz drainage basin caused by man's activity over the last millennium. *Annales UMCS, B, Vol. 54*, 14, 279-295.

Szafer, W. (1959). Land: Roztocze. Vol. 2. *Vegetation cover in Poland* (pp. 144-165). Warsaw: PWN.

Swieca, A. (1998). *The influence of anthropogenic factors on the river runoff of solutions and suspensions in the Vistula and Bug interfluve*. Lublin: UMCS Press.

Trzcinski, A. (not dated). *Following the traces of the Jewish culture in the Lublin Region*. Lublin: WOIT.

Turski, R., Uziak, S., & Zawadzki, S. (1993). *Soils. Natural Environment of the Lublin Region* (pp. 1-106). Lublin: Lublin Scientific Society Press.

Wartolowska, Z. (1958). Sutiejsk Czerwiensk stronghold in the Polish-Ruthenian borderland. *Swiatowid*, (Vol. 12). *Selected data on the Lublin Voivodship*. Statistical data. Lublin: GUS.

Wasowicz, T. (1961). Notes on the early medieval settlement in the Lublin Region. *Archaeology of Poland*, *6*, 203-250.

Zoll-Adamikowa, H. (1974). Findings of preliminary research on the early-medieval settlement complex in Guciow in Zamosc district. *Archaeological Reports*, *26*, 115-171.

Chapter 6

THE ROLE OF CULTURAL VALUES IN MODERN LANDSCAPES

The Flemish Example

Marc Antrop
Department of Geography, Ghent University, Belgium

1. INTRODUCTION

This chapter discusses the way cultural landscape values were revalorized to make a survey and assessment of the whole landscape in the Flemish region as a base for a new policy that aims a more integrated landscape management in the future.

Landscape protection and management is a slow and difficult process in the densely populated Flemish region of Belgium. Since the first law on the protection of monuments, sites and landscapes was issued in 1931, only 2.7% of the territory was legally protected in 2001 (Van Hoorick 2000; De Borgher 2001). The procedure was slow and cumbersome. Most of the areas protected as landscape were selected for their natural qualities. One of the reasons is that the public opinion became aware about the ecological deterioration of the environment. As the lobby of the nature conservation was becoming more powerful, natural landscape values were easily accepted and helped the protection of such sites. In many cases, the cultural, historical and aesthetic qualities were added as non-vital complementary ones. Also, the descriptions and motivations of the cultural, historical and aesthetic qualities were rather vague, lacked consistency and theoretical background. Meanwhile, devastating changes happened at an accelerating speed, in particular since the 1960s. During the 1960s the law on physical and urban planning created district master plans that fixed the possible land use for the coming ten years. However, a long societal and political debate followed and slowed down the finalization of these plans. Very often autonomous changes

H. Palang and G. Fry (eds.), Landscape Interfaces, 91–108.

occurred faster than could be enforced by the law. Nevertheless, these master plans guaranteed some protection for vast areas of rural landscape as the plans indicated zones referred to as 'valuable landscape', which should receive special attention when development was planned in these areas. These zones were based upon a first National Survey of the landscapes (Delaunois 1960), which was mainly descriptive. No strict rules were given and a clear description was lacking about what was valuable in these areas and which criteria were used to delineate them. The procedure for landscape protection was too slow and the regulations for the management of valuable landscapes as formulated in the law of physical planning were too soft to oppose the hard economic demands. Only since 1995 an important change in policy occurred and placed the integrated landscape management as an objective. Simultaneously, the cultural, historical and aesthetic qualities of the traditional cultural landscapes were revalorized and used in a new integrated inventory of the landscape resulting in the Landscape atlas of Flanders. This stimulated the implementation of a new legislation that aims an integrated management of all landscapes according to the suggestions of the European Landscape Convention.

2. DEFINING CULTURAL VALUES

2.1 Landscapes

In common language, landscape bears multiple meanings (Claval 2002). It is a noun referring to a perceivable part of the earth's surface that has a dominant rural or natural character. It applies to the countryside where the coherence between the natural settings and human organization of the land can be recognized easily. In English landscape is also used as a verb referring to embellish the land by landscape architecture, thus performing landscaping as a special aesthetical intentioned organization of the land. The focus upon the countryside caused also the emergence of complementary words such as townscape, cityscape and seascape. The definition given in the European Landscape Convention somewhat widens the concept as: "landscape means an area, as perceived by people, whose character is the result of the action and interaction of natural and/or human factors" and "covers natural, rural, urban and peri-urban areas. It includes land, inland water and marine areas. It concerns landscapes that might be considered outstanding as well as everyday or degraded landscapes" (Council of Europe 2000).

Landscape and *land* are no synonyms. Land refers to soil, ground (Zonneveld 1995), thus to property and territory. Landscape is considered as a common heritage and a collective identity (Lowenthal 1997). Landscape covers many properties, which makes that no one really owns it, but also makes it confusing regarding who should take care of the landscape (Antrop 2000a). A landscape is bound by views and not by administrative borders, unless it received a special status such as a legal protection. Nevertheless, the etymology of landscape does not only refer the scenery of the countryside, but also to a homeland and a collectively organized territory (Keisteri 1990; Groth 1997). This is expressed by the suffix *-scape*, or *-schap* in Dutch, which refers to creating, shaping as well as to character of the area. A similar relationship exists between the French *paysage* (landscape) and *pays* (land referring to a regional identity), which is different from *terre* (land as soil, ground). The interaction between the perceived scenery with its cognitive and aesthetical qualities and the organization of the land and the shaping of the landscape is strong and forms an essential property of the cultural landscape (Cosgrove 1984; Keisteri 1990; Groth 1997; Pedroli 2000; Claval 2002; Olwig 2002). The landscape is the manifestation of a complex of interacting features, some of which consist of material elements. Many of these witness from past structures and processes such as land organization practices (Muir 2000; Yorke 2001). Thus landscape is an important source of information of forgotten knowledge (Austad 2000). Understanding these relics is essential for their preservation and integration in future landscape design (Fry & Sarlöv-Herlin 1995). The growing integration between research in landscape ecology, spatial analysis and archaeology illustrates this also (Fry et al. 2001; Vermeulen & Antrop 2001; Grove & Rackham 2001).

2.2 Traditional and modern landscapes

The actual concern about the degradation of our cultural landscapes refers mainly to the 'past' rural landscapes and not to the new emerging modern cultural landscapes. The concept of traditional (cultural) landscapes is therefore meaningful as is the clear distinction between traditional and modern landscapes. Maybe the names are not completely adequate, as some also refer to postmodern landscapes and consider the term 'traditional' too heavy loaded with nostalgia and romanticism. The essential difference is the rather sudden break of the pace of landscape changes and transformation that occurred during the 19th century. The end of the Ancien Régime in France, which led to the French revolution that affected the whole west-European civilization, happened synchronically with the Industrial Revolution. Both initiated profound changes in society and politics and gradually a new way of living. Most of these changes were not peaceful at all, so one can easily speak of a Revolutions' Age that might be continuing still. Change in mobility patterns and growing urbanization and globalization are important driving forces in the landscape changes that were initiated. The traditional rural character of the countryside before these changes is one of slow transition and great stability and an inherent sustainability. Grandparents and grandchildren lived in the same environment and knew and spoke about the same landscape. The landscape was a stable reference in their lives. Since the Revolutions' Age, the pace of changes in our environment has increased. Today, each of us will live in completely different environments and landscapes during his lifetime. Consequently, attitudes towards the landscape changes as well. The stable reference is not the landscape anymore, but becomes oneself, the individual. Thus the differences between traditional and modern landscapes are both a different attitude of people towards their landscape, and different sources of information and methods to study the landscape.

There is also a more practical reason to situate a break between the traditional and modern landscapes in the Revolutions' Age starting in the 18th century. Most of the nation states in Europe were established then and the first topographical maps covering the whole country appeared. These are extremely valuable references of times as they cover large areas systematically; they are easily comparable and show the situation of the landscapes before most of the devastating changes started. The maps contain valuable references to a remote past that were wiped out since the Revolutions' Age. Fortunately, since the 19th century, regular and systematic recordings of the land occurred so that many of the fast changes can be followed as well. Many historical geographical studies focused upon that period (Verhoeve & Vervloet 1992).

The modern landscapes that have been formed since are still continuously changing at an increasing pace. For Europe, the trends of change are clear and haven been described (Vos & Klijn 2000). In general a clear polarization can be observed. Geographical space is reorganized in more intensively and more extensively used areas, thus forming new spatial patterns.

2.3 Culture

The etymology of culture refers to our civilization that emerged in the Neolithic – reshaping and organizing the land to allow agriculture and the creation of the urban settlement form. Jordan-Bychkov and Bychkova Jordan (2002) characterize the European culture by three important components: religion, language and race or ethnicity. Combined these give a rich cultural diversity that is also expressed in the diversity of the cultural landscape. These components belong to the mentifacts as used by the biologist Julian Huxley as one of the three components of culture, the two others being sociofacts and artifacts (Haggett 1975, 2001). Consequently, cultural landscapes must be seen as expressions of the different aspects of one or many successive cultures affecting a place or region (Groth 1997). Also, landscapes do not only consist of a spatial arrangement of artifacts, but carry information, meaning and witness of more mental aspects. They are characterized by a clear regional identity (Fry 2000; Pinto-Correia 2000). The landscape identity is defined in a large extent by repetitive similar elements having a typical character, such as architecture, type of field borders as hedges and stonewalls, etc. Many of the elements are important sources of knowledge of the past (Austad 2000; Muir 2000; Yorke 2001) and build the landscape memory (Lowenthal 1975; Fry 1998). Also, place names and symbolic values given to particular landscape elements add to the landscape memory and character (Cosgrove 1984). The new landscapes became also an important subject in the societal and philosophical debate (Kolen & Lemaire 1999; Cosgrove 1984, 1993; Lorzing 2001; Claval 2002). Many examples show the relation between the cultural landscape changes and environmental change (e.g. Head 2000).

2.4 Criteria for evaluation

In general, three main groups of qualities are generally used for evaluating landscapes: natural, cultural and aesthetical ones. The natural ones often refer to natural sciences, where biology and ecology are dominant aspects. The cultural group includes history, social and economical values as well as religious, symbolic and linguistic aspects. The aesthetical qualities are often restricted to visual or scenic qualities and sensorial ones such as tranquility.

Table 6-1. Legal values for landscape protection in Flanders

Values	Examples
Natural and scientific value	Ecologically valuable area with characteristic flora and fauna, undisturbed land form, soil catena
Historical value	Historical field pattern, settlement type, archaeological site, traditional land use, typical architectural form
Social-cultural value	Site or scenery frequently represented in paintings
Aesthetical value	Panoramic openness, non-urbanized scenery, landmarks

According to the actual Flemish law (decree on the Landscapes of 1996), four groups of values can be used for selecting landscapes to protect (Table 1). These are the natural-scientific value, historical value, socio-cultural value and aesthetical value. Very different criteria are used to evaluate each of them. Following quotations are extracts and literal translations of statements in the legal dossiers for protection of landscapes and sites that refer to these qualities (Antrop & Van Damme 1995).

> The geological-pedological value of the area is determined by its unique history of landscape formation. … forms a exceptional rare physical entity. … unique fauna, … one of the last refuges in the valley of the IJzer for birds passing and staying during winter. … an exceptional representation of the medieval landscape. Because the permanent use as grassland the relics of the sites have been preserved perfectly. The landscape structure … was determinant for the future evolution. … also the original 13th-15th century cadastral organization has been preserved to a large extent, which is rather unique. … the picturesque setting of the village (polder, Lampernisse, protected as landscape 1993).

> The Barbierbeek is a meandering or twisted brook which path developed without important human disturbances so its geomorphologic value can be considered as high. … In Flanders such 'natural' evolving brooks are extremely rare (Kruibeke, Barbierbeek, fase I, protected as landscape 1994).

> The occurrence of many characteristics bear witness the way humans managed the wood centuries ago. … the small-scale wood, the presence of different wood associations in a small area … offer a high aesthetic value (Muggenbos, Nieuwkerken, protected as landscape 1993).

> The historical value is given by the occurrence of one of the rare preserved dune houses. … The dune 'Mieke Hill' and her house are inseparable. … the most characteristic and most exceptional of such dune houses is that the house 'Mieke Hill' still has a sea view. … The aesthetical value is given by the dominance of the dune and the house in the scenery, forming a landmark (Dune Mieke Hill, Koksijde, protected as landscape 1993).

These examples of the description of qualities and motivation for protection illustrate well the vagueness and the broad scope of values used. Also the lacking of references to landscape ecological aspects is missing even for these rather recent protection dossiers. References to natural values are mainly supported by list of rare or endangered species and habitats. Direct indications to archaeological findings are rare as well. Nevertheless, the way the values are described using internal cross-references and repetitive style, illustrates the holistic approach to the landscape as well. Although different values are described according to the legal system, it is their combination that makes the exceptional quality for their protection. Each of these values is described and assessed using a set of criteria which are referred to by similar names but receives different meaning according to the context. A content analysis of the terms is used in these dossiers to evaluate cultural, historical and aesthetical qualities, revealed a set of concise criteria. Table 2 summarizes some of these criteria.

Table 6-2. Criteria for the assessment of cultural, historical and aesthetic landscape values

Criteria	*Description*	*Conditions*
Cultural-historical		
Nature – kind	Refers to the type of the landscape (element) according to a classification system.	Reference date or period; Reference situation or condition.
Coherence	The strength of the relations between the composing elements forming the whole landscape or ensemble.	Ecological; Functional; Genetic; Chronologic; Context.
Information	The degree of (potential) information about the past.	Representativity; Uniqueness: Extraordinary; Rarity.
Quality – condition	The preservation or conservation of an element or landscape structure.	Intrinsic; Extrinsic – context; Soundness; Completeness.
Aesthetical		
Legibility	The degree by which parts and the whole structure can be recognized and understood.	Identity; Orientation; Coherence; Story-telling potential.
Quality	The degree the scenic composition is harmonic according to perceptive an psychological rules; the lack of disturbing elements.	Intrinsic; Extrinsic – context; Soundness; Completeness.
Naturalness	The degree human impact is (seen) missing; opposed to intensive land uses.	Degree of built-up; fragmentation by infrastructure; Environmental condition.
Authenticity	The degree the original condition or situation is preserved.	
Stewardship	Degree of maintenance and care of the landscape site or element.	Order; Cleanness.
Diversity	The degree of variation in contrasting elements or parts (areas) of different nature; extreme low or high degree of diversity are experiences. less optimal.	Contrast; Mystery; Variation.
Heterogeneity	Combination of diversity and	Complexity;

Criteria	*Description*	*Conditions*
	the spatial arrangement of the elements in relation to fragmentation.	Order – chaos.
Atmosphere	Refers to the sensation and feeling of experiencing a place using all senses combined. Related and described by moods and affections.	Separate descriptions according to senses: noise, color, movement, etc.
Utility		
Accessibility	The possibility to enter an area or move trough a terrain.	Borders; passing rights; Soil and terrain condition.
Freedom of land use	The number and nature of legal constraints in using or transforming the land.	Potential uses; Possible uses.
Money value	Actual ground price.	Land use; Geographical situation.

3. TRADITIONAL LANDSCAPES OF FLANDERS

During the 1970s and 1980s successive constitutional reforms made Belgium to become a federal state. The regional governments got the authority on spatial planning, environment, nature and landscape issues. As a consequence, new maps were made showing only the regions instead of the whole country. In this context the concept of traditional landscapes was introduced in 1985 (Antrop, in Knops 1985). These were defined as the combined result of natural capabilities (relief, soil, geology etc.) and human land occupation through history before the modern changes starting in the Revolutions' Age (end of the 18th century). The choice of the reference period to distinguish between traditional and modern was made on two grounds. First, a detailed historical map – the Ferraris map – was available from the end of the 18th century, dating from the Austrian period just before the French revolution and the important changes that started in Belgium in the late 19th century.

The map of the traditional landscapes (Figure 1) showed 77-zoned landscape units consistently grouped according to natural regions and history. The great diversity in the rather small area of Flanders was obvious and the classification proved to be a helpful framework for the new spatial planning policy (Antrop 1997). However, there was a need to have a finer delineation and more detailed descriptions of the ideal characteristics, and conditions of these pre-industrial landscapes. A first adaptation was published in 1989 (Antrop 1989) and a more extensive one including criteria and case studies for mapping landscape types and relics was reported in 1995 (Antrop & Van Damme 1995). This formed the basis for the landscape atlas of Flanders, which allowed also refining the classification and delineation on a map at a working scale of 1:50,000.

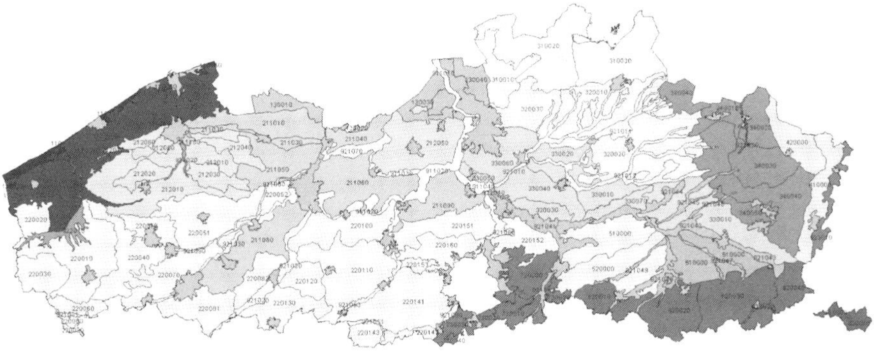

Figure 6-1. The traditional landscapes of Flanders

4. THE LANDSCAPE ATLAS OF FLANDERS

4.1 History

The inventory of the relics of the traditional landscapes in Flanders region started in 1995 on a provincial basis. Although no GIS facilities were implemented in the Flemish administration at that time, the concept of the atlas was conceived entirely as an open GIS-database. The initial plan was to complete the inventory for the entire Flemish region in an extremely short period of four years. Several very practical reasons caused this. First, the parallel development of the new spatial planning policy in Flanders was asking urgently for basic information about landscapes to be included in the integrated process. In particular, the discrepancy between the available information about nature and ecology and the cultural and historical assessment became unacceptable. Second, the increasingly faster changes in the landscape and environment demanded a rapid response. Third, the short term of the project was politically attractive and feasible during the time of one legislative system. However, the scientific and methodological consequences of these constraints were important (Antrop 2001). Only existing knowledge could be used and available data needed to be integrated. In order to fill in the many gaps in scientific knowledge that were expected, the whole system had to be flexible and open for further extension.

Furthermore, the method to assess the actual condition of the landscape had to be holistic and had to use available datasets of very different nature. Fortunately, Belgium possesses a rich collection of detailed topographical maps of high quality dating from the end of the 18th century. Also, several sets of aerial photographs were extremely important. At the start of the project in 1995 most of these were still analogue data and maps. During the completion of the atlas, all data were gradually digitalized and integrated into a GIS.

4.2 The method

The Ferraris map dating from about 1770 was used as the reference for indication the transition of the traditional rural landscapes towards all new landscapes that emerged since. The analogue orthophoto map of 1990 was used as the reference to assess the actual condition of the landscape. The classification and description of the traditional landscapes served as a descriptive reference base of the ideal landscape types and character. The selection of the relics was based on the analysis of time series of maps and the visual interpretation of patterns on aerial photographs.

All relics had to fit into a classification scheme that could be implemented into an open GIS-database. New information was very likely to emerge and had to be integrated easily into the system. Also, many cultural layers and time periods were combined in most of the relics. The main categories were based upon the spatial dimension of the relics selected (Table 3). Relic zones (R) referred to larger areas and were represented by polygons with fuzzy borders. Particular unique complexes were classified as ensembles called anchor places (A) and delineated as polygons. Linear elements such as road segments, canals, rivers, ditches and military defense works were classified as line relics (L). Discrete objects such as buildings, solitary trees and landmarks were classified as point objects, coded as P when a description or name was available, otherwise coded as X. Finally, a preliminary attempt to map also particular viewpoint and sights was made (Z). Each of the relics received a unique identification code starting with the code letter of its type, followed by the statistical code of the province it is situated in and a serial number. This allowed to store the descriptions of each relic in a database and to link them to the map (Figure 2).

During the process, the GIS facilities increased in such a way that most thematic datasets became available in digital form and also the capacity of using electronic information in the administration became possible. The final realization of the atlas was achieved entirely in the GIS environment coordinated by GIS-Flanders (GIS Vlaanderen 2001). Technically, the atlas is based on the Arcview GIS 3.2 shape format linked to a Microsoft Access database. However, the files can be consulted using Arc Explorer or a stand-alone viewer in the Windows environment using Windows Internet Explorer (Antrop & De Vriese 2002). The atlas was widely distributed on CD-ROM accompanied by a book (Hofkens & Roossens 2001). Internet versions are planned in the near future.

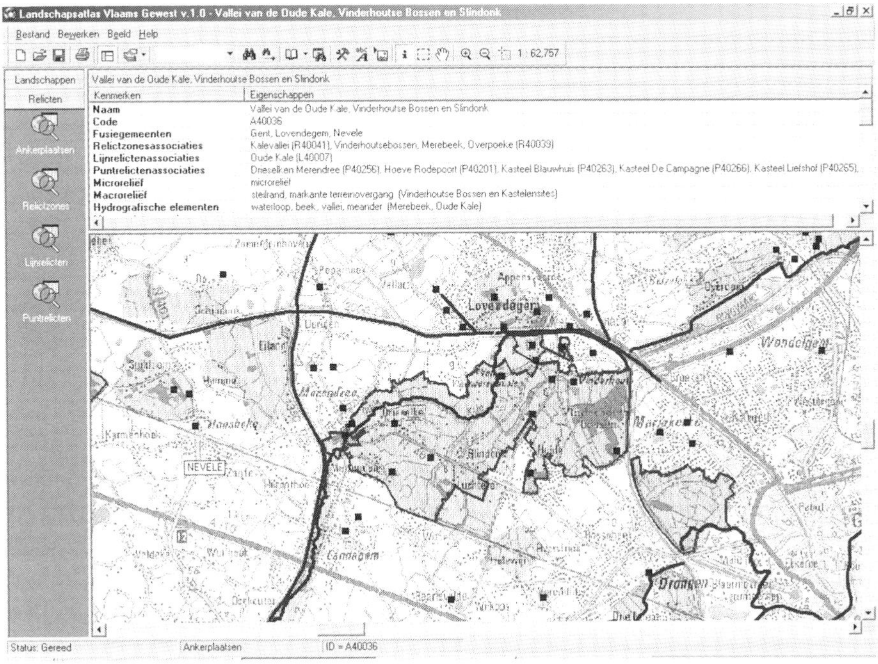

Figure 6-2. Example of a query of the GIS database of the Landscape Atlas of Flanders (GIS-Vlaanderen 2001)

4.3 Criteria and selection rules

The criteria used for the selection of relics and their evaluation were based upon the analysis of the ones used in protection dossiers as shown in Table 2 (Antrop & Van Damme 1995). The most important ones were coherence, legibility and preservation condition (soundness) of landscape structures such as settlement patterns and types, field systems, road networks, landscape type and land use. Additional criteria for the evaluation were scenic quality and information potential. For each of the types of relic a procedure and rules for the selection were described.

An area was selected as a relic zone if one or more of following properties were present:

- existence of recognizable structures and objects dating from the end of the 18th century up to the WW II;
- occurrence of geomorphologic features that structure the landscape or have a physical or natural monumental value;
- occurrence of concentrations of typical objects or linear elements;
- occurrence of known or expected concentration of important archaeological sites;
- occurrence of particular viewpoints and sight offering a good scenic view over the relic zone;
- landscape with high scenic value and not disturbed visually by modern constructions or infrastructures.

Relic zones are indicated on the map at scale 1:50,000 with fuzzy borders, indicating that a more detailed delineation should be done on a larger appropriate scale and after field surveying.

An ensemble was selected as an anchor place if one of the following criteria was fulfilled:

a) The ensemble is situated outside a relic zone:
   - a complex of clustered heritage elements;
   - elements having a distinct genetic, morphological or functional coherence;
   - elements that are representative for a particular period, style or type;
   - a sufficient preservation condition so the ensemble can be used as an ideal example.

b) The ensemble is situated within a relic zone:
   - a complex of clustered punctual or linear elements;
   - elements having a distinct genetic, morphological or functional coherence;
   - elements that are representative for a particular period, style or type;
   - possesses typical characteristics that are related to the relic zone;
   - is part of a larger geographical structure;
   - forms an aesthetical and not disturbed whole.

Selected anchor places were checked on the field and were delineated sharply on the map using landscape features that can be recognized easily in the field during 2000.

Linear features were selected as relics if:

- the element possesses a cultural or historical meaning as a connection or border that still can be recognized in the landscape;
- or the element has a physical and visual structural quality in the landscape;
- or the element is an ecologically valuable corridor;
- or the element is a natural ecotone or forms a visual gradient or transition between different landscapes;

- and the element can be clearly recognized in the landscape and forms a visual and structural element.

  The whole element is selected and the disturbed segments are marked.

  The rules for the selection of point relics are one of the following:
- objects that have an intrinsic natural, scientific, cultural, historical or aesthetical value;
- and has a good condition of preservation;
- and is situated in a undisturbed spatial context.

  If they have a legal protection status, this is indicated.

Table 6-3. Results from the atlas of the relics of the traditional landscapes of Flanders

Relic type	*Code*	*Number*	*Area or length*	*% of Flanders area*	*Average size*
Relic zones	R	515	530000 ha	39.0	1029 ha
Anchor places	A	381	221051 ha	16.3	580 ha
Linear relics	L	544	4851 km		Min: 5 m Average: 5.3 km Max : 77 km
Point relics	P, X	4607			

4.4 Results and discussion

The results of the inventory were surprising to many (Table 3 and Figures 3 to 5). Indeed, the highly urbanized and severely fragmented landscape of Flanders easily gives the impression that no valuable heritage and nice cultural landscapes are left. The Landscape Atlas revealed on the contrary that still many areas exist where the character of the traditional landscape is preserved. Clearly the selected relics refer mainly to the rural countryside and are complementary to the already existing biological and ecological assessment (Tables 4 and 5). The comparison with the actual land use also indicates that not only extraordinary and special landscape types have been selected, but the whole of the ordinary landscape is covered as well. For policy this is important as it indicates areas, which are lacking proper governance, as these are not covered by particular legislation such as nature conservation. Therefore, new legal regulation is proposed to consider the selected areas in a more appropriate way in the process of structural and physical planning in order to preserve the still existing landscape values.

Table 6-4. Relation of landscape relic zones with other classifications in Flanders (after Tack and Van den Brempt 2001)

Land use	%	*Master plan*	%	*Biological Valuation map*	%
Wetlands & water	1.7	Agrarian	60.1	Less valuable	65.9
Arable	34.4	Nature, forest	26.5	Valuable	20.5
Grassland	27.0	Other	12.8	Very valuable	8.2
Heath land, dunes	2.5			Not valuable	0.1
Woodland	19.2			Not mapped	4.5
Built land	1.1				
Other	5.1				

Table 6-5. Relation of landscape anchor places zones with other classifications in Flanders (after Tack and Van den Brempt 2001)

> 10% of land use is	%	*Master plan*	%	*Biological Valuation map*	%
Wetlands & water	17.7	Agrarian	53.2	Less valuable	34.4
Arable	65.7	Nature, forest	39.4	Valuable	14.5
Grassland	86.8	Other	7.4	Very valuable	11.1
Heath land, dunes	8.1			Not mapped	4.6
Woodland	73.3				

However, the number of relics, their average size and length as well as the maps shows that they form many fragmented and disconnected patches. Most of the relic zones are situated at the periphery of the former municipal territories, which indicates that they represent more the landscapes formed during the late 18th century on the former outfields. Clearly, most disturbances and degradation of the traditional landscapes are caused by urban sprawl around even smaller centers and following the dense road network. The maps (Figure 3) also show that larger relic areas are preserved in the periphery of Flanders and that the fragmentation is most severe in the center between the cities of Antwerp, Brussels and Ghent. This core area also contains a concentration of point relics, mainly monuments (Figure 5). This spatial pattern suggests different strategies and management to be developed according to the region and interference with other economical and societal demands. Models such as the one proposed by Wood & Handley (2001) could be useful.

Figure 6-3. The relic zones (light grey) of the traditional landscapes of Flanders and the anchor places (black) according to the Landscape Atlas of Flanders; reference situation 1990 for the relic zones and 2000 for the anchor places

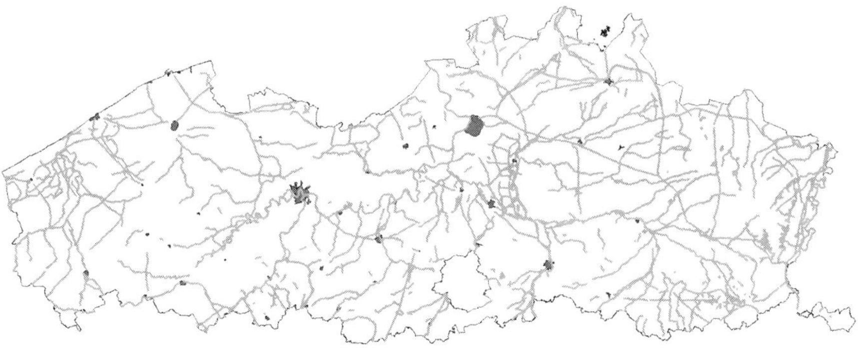

Figure 6-4. Linear landscape relics according to the Landscape Atlas of Flanders; reference situation 1990. The town areas are shown in the late 18th century period where the city walls still bordered the agglomeration. Brussels Capital region is not mapped

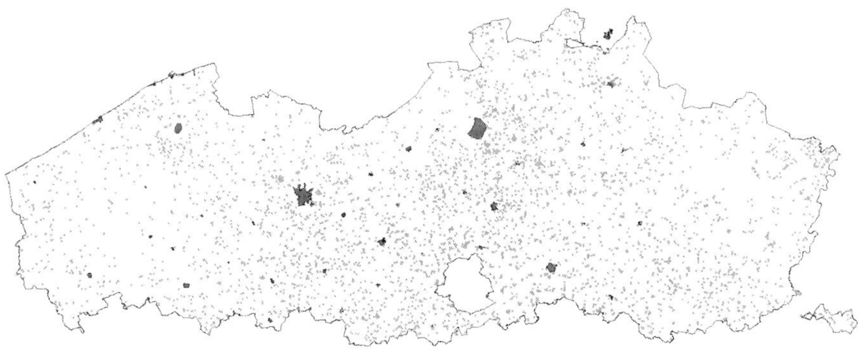

Figure 6-5. Distribution of the punctual landscape relics (sites, monuments) according to the Landscape Atlas of Flanders; reference situation 1990

5. CONCLUSIONS

Modern landscapes are highly dynamical and many changes affect or destroy the traditional cultural values. In most cases the identity of the traditional landscape becomes lost due to processes of fragmentation caused by urbanization and different kinds of transportation infrastructures. Upscale of farming and homogenization of the land use due to globalization of agriculture adds to this loss of identity. As a result, the traditional rural landscapes became fragmented into rather small patches of relics and scattered elements. The loss of the coherence between the composing parts is important. An observer walking or driving around might experience the actual situation as very bad and even get a feeling that 'everything is lost'. However, the case of Flanders shows that many valuable parts still remain behind all the built-up and infrastructures. Making a detailed inventory from a vertical perspective can be realized rather rapidly and indicates the remaining important and valuable zones where special care and attention should be given to. By describing cultural, historical and aesthetical values of landscapes we use mainly soft criteria, which can hardly compete to economical ones. However, a systematic and elaborated use of criteria that are linked to an existential dimension, to perceptive qualities and knowledge helps to enhance the value of the remaining relics and forms a basis for a better and more differentiated landscape management.

REFERENCES

Antrop M. (1989). Het landschap meervoudig bekeken. Antwerpen: Stichting Leefmilieu/DNB-Uitg. Pelckmans.

Antrop, M., & Van Damme, S. (1995). *Landschapszorg in Vlaanderen: onderzoek naar criteria en wenselijkheden voor een ruimtelijk beleid met betrekking tot cultuurhistorische en esthetische waarden van de landschappen in Vlaanderen.* Gent: Ghent University. In Dutch. Download from http://geoweb.rug.ac.be/services/

Antrop, M. (1997). The concept of traditional landscapes as a base for landscape evaluation and planning. The example of Flanders Region. *Landscape and Urban Planning, 38*, 105-117.

Antrop, M. (2000c). Geography and landscape science. Belgian Journal of Geography, *Belgeo special issue*, 29th International Geographical Congress, nrs. 1-4, 9-35

Antrop, M. (2000b). Changing patterns in the urbanized countryside of Western Europe. *Landscape Ecology, 15* (3), 257-270.

Antrop, M. (2000a). Background concepts for integrated landscape analysis. *Agriculture, Ecosystems and Environment, 77*, 17-28.

Antrop, M. (2001). De landschapsatlas, Methode. In E. Hofkens, & I. Roossens (Eds.), *Nieuwe impulsen voor de landschapszorg. De landschapsatlas, baken voor een verruimd beleid* (pp. 21-43). Brussel: Ministerie van de Vlaamse Gemeenschap, Afdeling Monumenten en Landschappen.

Antrop, M., & De Vriese, S. (2002). *Landschapsatlas toont aan hoe rijk het Vlaams landschappelijk erfgoed (nog) is.* VI Matrix, Jg. 10, 1, (pp. 20-22).

Austad, I. (2000). The future of traditional agriculture landscapes: retaining desirable qualities. In J. Klijn, & W. Vos (Eds.). (2000). *From Landscape Ecology to Landscape Science* (pp. 43-56). Wageningen: WLO, Kluwer.

Christians, C., Daels, L., & Verhoeve, A. (1992). Les paysages ruraux. In J. Denis (Ed.), *Géographie de la Belgique.* Brussel: Gemeentekrediet.

Christians, C., Daels, L. (1988). Belgium: an introduction to its regional diversity and cultural richness. *Société Géographique de Liège, 24.*

Claval, P. (2002). *The languages of rural landscapes.* Presenations at The 20th Session of The Permanent European Conference on the Study of the Rural Landscape, Tartu-Otepää, Estonia.

Cosgrove, D. (1993). *Landscapes and Myth, Gods and Humans. In B. Bender (Ed.), Landscape Politics and Perspectives. Explorations in Anthropology Series.* Oxford: Berg Publishers.

Cosgrove, D. (1984). *Social Formation and Symbolic Landscape.* London: Croom Helm.

Council of Europe (2000). *The European Landscape Convention.* Florence 20 October 2002.

De Borgher, M. (2002). Toelichting bij het decreet van 21 december 2001. Tot wijziging van het decreet van 16 april houdende bescherming van landschappen. *Monumenten, & Landschappen, 21, 1*, Binnenkrant, 2-4.

Delaunois, H. (1960). *Inventaris van de landschappen: provincie Antwerpen.* Brussel: Bestuur van de Ruimtelijke Ordening

Fry, G.L., & Sarlöv Herlin, I. (1995). Landscape design: how do we incorporate ecological, cultural and aesthetic values in landscape assessment and design principles? In G.H. Griffiths (Ed.), *Landscape ecology: theory and application* (pp. 51-60). Reading: Proceedings IALE-UK Conference.

Fry, G.L. (2000). The landscape character of Norway – landscape values today and tomorrow. In B. Pedroli (Ed.*) Landscape – our Home. Lebensraum Landschaft. Essays on the Culture of the European Landscape as a Task* (pp 93-100). Stuttgart: Indigo, Zeist–Freies Geistesleben.

Fry, G., Jerpåsen, G., Skar, B., & Stabbetorp, O. (2001). Combining landscape ecology with archaeology to manage cultural heritage interests in the changing countryside. In Ü. Mander, A. Printsmann, & H. Palang (Eds.), *Development of European Landscapes. Vol. 2.* Conference proceedings IALE European Conference 2001 (pp.198-202). Tartu: Publicationes Instituti Geographici Universitatis Tartuensis.

GIS-Vlaanderen, (2001). *Landschapsatlas. Brussel: Ondersteunend Centrum GIS-Vlaanderen*, CD-ROM.

Groth, P., & Bressi, T.W. (Eds.). (1997). *Understanding Ordinary Landscapes.* New Haven: Yale University Press.

Grove, A.T., & Rackham, O. (2001). *The Nature of Mediterranean Europe–An Ecological History.* London: Yale University Press.

Gysels, H, Baccaert, J., Beenhakker, A., & Caspers, T. (1993). *De landschappen van Vlaanderen en Zuidelijk Nederland. Een landschapsecologische studie.* Leuven Apeldoorn: Garant.

Haggett, P. (2001). *Geography: a Global Synthesis.* Prentice-Hall.

Haggett, P. (1975). *Geography a modern synthesis.* London: Harper & Row.

Head, L. (2000). *Cultural Landscapes and Environmental Change. Key Issues in Environmental Change.* London: Arnold.

Hofkens, E., & Roossens, I. (Eds.) (2001). *Nieuwe impulsen voor de landschapszorg. De landschapsatlas, baken voor een verruimd beleid*. Brussel: Ministerie van de Vlaamse Gemeenschap, Afdeling Monumenten en Landschappen.

Jordan-Bychkov, T.G. & Bychkova Jordan, B. (2002). *The European Culture Area. A systematic Geography*. New York: Rowman, & Littlefield.

Keisteri, T. (1990). The study of changes in cultural landscapes. *Fennia, 168 (1)*, 31-115.

Klijn, J., Vos, W. (Eds.) (2000). *From Landscape Ecology to Landscape Science*. Wageningen: WLO, Kluwer.

Knops, G. (Ed) (1985). *De Open Ruimte in Vlaanderen*. Brussel: Koning Boudewijnstichting.

Lowenthal, D. (1997). *European landscape transformations: the rural residue*. In P. Groth, & T.W. Bressi (Eds.) *Understanding Ordinary Landscapes* (pp.180-188). New Haven, CT: Yale University Press.

Lowenthal, D. (1975). *Past time present place: Landscape and memory*. The Geographical Review, 65 (1), 1-36.

Lorzing H. (2001). *The Nature of Landscape. A Personal Quest*. Rotterdam: 010Publishers.

Muir, R. (2000). *The NEW Reading the Landscape. Fieldwork in Landscape History*. Exeter: University.

Olwig, K. (2002). *Landscape, Nature and the Body Politic: from Britain's Renaissance to America's New World*. Madison: University of Wisconsin Press.

Pinto-Correia, T. (2000). Landscape identity, a key for integration. In B. Pedroli (Ed.), *Landscape – our Home. Lebensraum Landschaft. Essays on the Culture of the European Landscape as a Task* (pp. 145-150). Stuttgart: Indigo, Zeist – Freies Geistesleben.

Tack, G., & Van den Bremt, P. (2001). De Landschapsatlas: Resultaten. In E. Hofkens, & I. Roossens (Eds.), *Nieuwe impulsen voor de landschapszorg. De landschapsatlas, baken voor een verruimd beleid* (pp. 44-62). Brussel: Ministerie van de Vlaamse Gemeenschap, Afdeling Monumenten en Landschappen.

Van Hoorick, G. (2000). *Juridische aspecten van het natuurbehoud en de landschapszorg. Intersentia Rechtswetenschappen*. Antwerpen – Groningen.

Verhoeve, A., & Vervloet, J. (Eds.) (1992). The Transformation of the European Rural Landscape: Methodological issues and agrarian change 1770-1914. *Tijdschrift van de Belg.Ver. voor Aardrijkskundige Studies, LXI, 1*.

Verhulst, A., (1990). *Précis d'histoire rurale de la Belgique*. Bruxelles: Editions de l'ULB.

Vermeulen, F., & Antrop, M. (Eds.) (2001). *Ancient Lines in the Landscape. A Geo-Archaeological Study of Protohistoric and Roman Roads and Field Systems in Northwestern Gaul. Leuven*. Peeters.

Vos, W., & Klijn, J. (2000). Trends in European landscape development: prospects for a sustainable future. In J. Klijn, & W. Vos (Eds.), *From Landscape Ecology to Landscape Science* (pp. 13-30). Wageningen: WLO, Kluwer.

Wood, R. & Handley, J., 2001. Landscape Dynamics and the Management of Change. *Landscape research, 26 (1)*, 45-54.

Yorke, T. (2001). Tracing the history of villages. Newbury: Countryside Books.

Zonneveld I.S. (1995). *Land Ecology*. Amsterdam: SPB Academic Publishing bv.

Chapter 7

SHAPING THE FUTURE OF A CULTURAL LANDSCAPE: THE DOURO VALLEY WINE REGION

Teresa Andresen[1] & Maria José Curado[2]
[1] *Faculty of Sciences, University of Porto, Portugal*
[2] *Departamento de Ambiente e Ordenamento, University of Aveiro, Portugal*

1. INTRODUCTION

Cultural landscapes have inscribed in themselves the evolution of the relationship between man and nature through generations. Seriously threatened, they are like a live archive where one learns about the variety of ways how to manage the natural resources whether water and soil conservation or the enhancement of biological diversity and aesthetic quality.

The future of such landscapes is very important for humankind's life quality and so their preservation is becoming understood as a priority. Europe has undertaken two very important steps in this respect with the Pan-European Biological and Landscape Diversity Strategy (1996) and the European Landscape Convention (2000) while, in 1992, UNESCO had included the category of cultural landscapes within the World Heritage Convention. Cultural landscapes are considered combined works of nature and of man and are of an outstanding value.Our objective is to present and discuss the outcome of a recent process of nomination of a cultural landscape for the World Heritage List and the development of a management program to preserve the landscape. The site is the Alto Douro Wine Region, located in the interior of northern Portugal, and part of the Douro Demarcated Region (Figures 1 and 2). It was nominated under the category of evolved continuing cultural landscape.

H. Palang and G. Fry (eds.), Landscape Interfaces, 109–124.

Portugal's position between the Mediterranean and the Atlantic is responsible for a very significant and quite unique variety of cultural landscapes through a rather small stretch of land as it has an area of 90,000 km^2. The Alto Douro with an approximate area of 25,000 ha is a remarkable example of this transitional condition and of very significant cultural dimension.

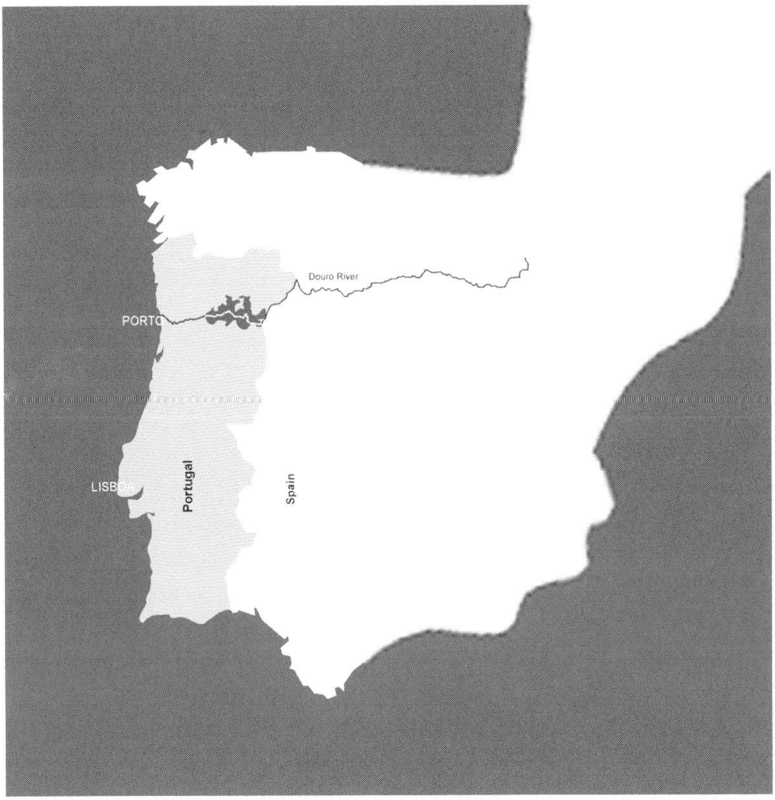

Figure 7-1. Location map

The concept of a cultural landscape is recent and has been the subject of in-depth discussion from various points of view (Rossler 2000). Nevertheless people are gradually accepting the fact that cultural landscapes must be seen as reference models or, in the words of Audrerie: "More than anything it is the idea that World Heritage Sites are not only a list of "exceptional" reserves that must be preserved in the midst of a world given over to arbitrariness but rather a model on which we must be inspired to conserve the harmony and balance of the planet that which make it pleasant or simply viable" (Audrerie et al. 1998: 101).

Figure 7-2. A view of the Douro Valley

So when we speak of living cultural landscapes, such as the Alto Douro, we are thinking of landscapes where the outcome of man's interaction with the natural elements has produced a collective work of art, which is the result of the work of various generations. Its preservation calls for a landscape policy strategy as well as for a landscape management plan, as presented forward by the European Landscape Convention. In this convention, in the chapter containing definitions, we can find: "Landscape policy means an expression by the competent public authorities of general principles, strategies and guidelines that permit the taking of specific measures aimed at the protection, management and planning of landscapes", and "Landscape management means action, from a perspective of sustainable development, to ensure the regular upkeep of a landscape, so as to guide and harmonize changes which are brought about social, economic and environmental processes".

Thus this paper, after presenting the character of the Alto Douro landscape and the process of inscription in the World Heritage List, discusses the management program that was proposed and is now about to be implemented.

2. THE CHARACTER OF THE LANDSCAPE

One of the most relevant aspects of the concept of cultural landscapes is the definition of the character or, what makes it unique. The Douro Demarcated Region, with an area of 250,000 ha, is the oldest demarcated and regulated wine region in the world due to the vision of the Marquis of Pombal (1757-1761), the Prime Minister at the time of King José I. Wine growing in this region is very ancient and today we still find roman vestiges but it reached a high point during the time of the presence of the Cistercian monastic orders in the valley in medieval days. The boom in the 18th century derived from trade agreements between England and Portugal (Bianchi Aguiar et al. 2000).

The intra-regional differences that one sees in the Douro Demarcated Region are a result both of the natural conditions and of the vicissitudes of history and activities of man. Nevertheless, these differences that distinguish between the Lower Corgo, the Upper Corgo and the Upper Douro sub-regions consolidate, rather than attenuate, the marked cultural identity of the region, shaped as it has been by a willingness to espouse interchanging human values and by the renewed centralism of a winemaking culture profoundly rooted in history. A major landmark in this landscape is the *quintas* – estates cantered around a continuum of buildings from whence the vineyards spread up the hillsides (Figure 3). The groups of farm buildings and wineries that crowd around the main house easily identify the *quintas*. Although notably present throughout the region, they are particularly evident in the Upper Corgo and in the Upper Douro, where in terms of land ownership the larger estates prevail. In the Lower Corgo, *quintas* often intermingle with *casais,* smaller size properties. Here, the architecture differs as granite is more frequently used for building, an indication of their proximity to this source of rock and of the influence of the long-ago golden days of trade with Brazil.

In itself, the winemaking culture represents a profound cultural mix, created not only by the peoples of the Douro but also by itinerant laborers from poor settlements on the other side of the mountains to the North and to the South and by Portuguese and British merchants and their agents.

The unique character of this cultural landscape is based upon a relationship guided by the scarcity and adversity of the natural elements: water and soil are scarce, steepness is a dominant feature and schist is the predominant geological formation. The wise relationship inscribed in the region reflects the deep knowledge of the adaptation of the Mediterranean crops to adversity and scarcity of the natural elements. Vineyards, olive trees and almond trees climb up the hills supported by kilometers and kilometers of stonewalls. The river Douro, coming form Spain, is the backbone of the region and for centuries it was the major route of the valley. Late in the 19th century, thc railroad was of a tremendous impact for the openness of the valley and form the 1950s till the 1980s the river was dammed. The area inscribed in the World Heritage List corresponds to 10% of the region and runs along the river and its main tributaries. It is considered that it gathers the most significant and well kept elements representative of the character of the Douro Demarcated Region while carrying the presence of the different generations through the centuries. Here the vernacular heritage related with the history of the valley and wine and olive oil production is most significant, as well, even if of a modest character as it conveys and gives meaning to the relationship with the natural elements.

Figure 7-3. A *quinta* in the Douro Valley

The Douro Demarcated Region landscape, with its Mediterranean-like environment and poor and rugged soil offering major constraints to human settlement and development is, above all, an outstanding example of man's unique relationship with the natural environment. Throughout the centuries, man has created and perfected specific techniques of land use that have made it possible for him to cultivate vines on the steepest slopes and built row upon row of *socalcos* – terraces buttressed by walls of schistous stone. *Socalcos* today stand side by side with new methods to plant vineyards based on the technological advancements of machinery, the so-called *patamares* and vertically planted vines.

The oldest *socalcos* have low walls and narrow and irregularly-shaped surfaces – *geios* – that sometimes only accommodate a single row of vines and which wind, willy-nilly around the curves of the land and rocky outcrops. These are the pre-phylloxera *socalcos*, built before the devastation of the Douro vineyards during the last third of the 19th century. Many of these were abandoned then and today form *mortórios* – or morgues of dead vines – that have become overrun with wild shrubs. Many pre-phylloxera *socalcos* have, however, been recuperated and replanted and still keep their old schist retaining walls. Some also bear traces of even more primitive forms of planting such as that of creating *pilheiros* – holes in the walls where vines were planted, thus leaving the surface free for other crops. Other *socalcos* have tall, straight solid walls that support wide surfaces planted with four, five or more rows of vines. These are the post-phylloxera *socalcos*, most of which were built between the end of the 19th century and the mid-1930s, when most of the vineyards in the region underwent reconstruction (Figure 4).

Figure 7-4. On the left pre-phylloxera *socalcos* and on the right post-phylloxera *socalcos*

The construction of the *patamares* began in the 1970s, on land occupied by *mortórios* and pre-phylloxera *socalcos* and, occasionally, on land that had not previously been planted with vines. This operation involved great scarifying of land and the destruction of old walls. This heralded the appearance of *patamares* – large plots of slightly sloping, earth-banked land usually planted with two rows of vines, laid out to facilitate the mechanization of the vineyard.

Thus, the architecture of the winemaking landscape of the Douro, with its mix of different methods for organizing the land, is indeed complex. Side by side with vineyards on *patamares* and vertically planted vines dating from the end of the 20th century, one finds several hundreds of kilometers of ancient *socalcos* whose shape varies according to the slope of the land and to the terracing techniques employed at different periods. Trials of other systems are continuing with a view to finding alternatives to the *patamares* and to minimize the impact of the new methods on the landscape.

By adapting the natural environment to the demands of an ancestral productive system, man fashioned a winemaking landscape whose unmistakable appearance, dramatic and dynamic configuration and universal uniqueness are recognized worldwide. Only the creation of a unique, lucrative natural product of exceptional quality such as Port Wine could ever justify the superhuman effort illustrated by this collective work of art.

Here, economy-culture-landscape form an indisputable unit, particularly as regards the terraced vineyards that appear to have anticipated more recent concepts of sustainable land use, in addition to the fact that they illustrate an optimal environmental solution from the viewpoint of conservation and use of rare resources – water and soil. The people of the Douro have embraced the natural environment they fashioned over the centuries by adapting their houses and villages to the constraints and demands of the environment and the productive process, not just of the grape but also of other Mediterranean cultivates such as the olive and the almond, as well as fruit and vegetables on more fertile alluvial soil and amongst the woodlands on high land.

3. THE INSCRIPTION OF THE ALTO DOURO IN THE WORLD HERITAGE LIST

The World Heritage Convention derives from the United Nations Conference on the Human Environment, which was held in Stockholm in 1972. It is a reference text but it is mainly an instrument designed to create actions to protect the heritage of the world, acting in conjunction with other international conventions. The Convention arose when the world became aware of the inestimable and irreplaceable value of the assets not only belonging to each nation but to all mankind and that their loss through degradation or disappearance would impoverish the world heritage as a whole (UNESCO 1972).

The category of cultural landscape was approved twenty years later, during the 16th meeting in Santa Fe, USA, in 1992 although the process had begun during the 8th meeting of the Commission in 1984 (Andresen 1999). A working group made up of representatives of the World Conservation Union, the International Council of Monuments and Sites and the International Federation of Landscape Architects, was then set up. Registering an asset in the World Heritage List implies that it must meet a number of criteria to show that it is of outstanding universal value. Its authenticity must be proved and legal mechanisms to ensure its conservation must be shown to exist.

The criteria to be obeyed in designating cultural landscapes were elaborated in 1992. Paragraph 36 of the Operational Guidelines for the Implementation of the World Heritage Convention (1999) states that: "Cultural landscapes represent the 'combined works of nature and of man' designated in Article 1 of the Convention. They are illustrative of the evolution of human society and settlement over time, under the influence of the physical constraints and/or opportunities presented by their natural environment and of successive social, economic and cultural forces, both external and internal. They should be selected on the basis both of their outstanding universal value and of their representativity in terms of a clearly defined geo-cultural region and also for their capacity to illustrate the essential and distinct cultural elements of such regions".

Cultural landscapes are classified into three main categories: landscapes designed and created intentionally by man (gardens and parkland landscapes); organically evolved landscape (relict or fossil and continuing) and associative cultural landscapes. The 'evolved continuing cultural landscape' is understood as the result of social, economic, administrative or religious imperative, which has achieved its present form by association and in response to the natural environment. The evolutive process is reflected both in the form and in the component features of the landscape. 'Evolved cultural landscapes' are subdivided into two categories: a) a fossil landscape – one which has undergone an evolutive process but has stopped, either abruptly or over a period, and b) a continuing landscape – which retains an active social role in contemporary society and closely associated to traditional way of life, where the evolutive process is ongoing. It simultaneously presents manifest proof of its evolution through the ages.

The criteria for classification of a cultural landscape are the same as those applied to the inscription of other assets (UNESCO 1999). In other words, it must:

> " i) represent a masterpiece of human creative genius, or ii) exhibit an important interchange of human values, over a span of time or within a cultural area of the world, on developments in architecture or technology, monumental arts, town planning or landscape design or, iii) bear a unique

or at least exceptional testimony to a cultural tradition or to a civilization which is living or which has disappeared; or iv) be an outstanding example of a type of building or architectural, or technological ensemble or landscape which illustrates (a) significant stage(s) in human history; or v) be an outstanding example of a traditional human settlement or land-use which is representative of a culture (or cultures), especially when it has become vulnerable under the impact of irreversible change; or vi) be directly or tangibly associated with events or living traditions, with ideas, or with beliefs, with artistic and literary works of outstanding universal significance ...; and meet the test of authenticity in design, material, workmanship or setting and in the case of cultural landscapes their distinctive character and components ... and have adequate legal and/or contractual and/or traditional protection and management mechanisms to ensure the conservation of the nominated cultural properties or cultural landscapes".

The last two criteria – authenticity of the asset, and demonstrating legal protection and management mechanisms – are decisive. Note that in terms of cultural landscape authenticity is determined based upon the character of the landscape or its distinguishing components. The character of the landscape is clearly an integrative parameter, directly meeting the very meaning of landscape as a physical manifestation of a meaningful relationship between human beings and nature throughout generations.

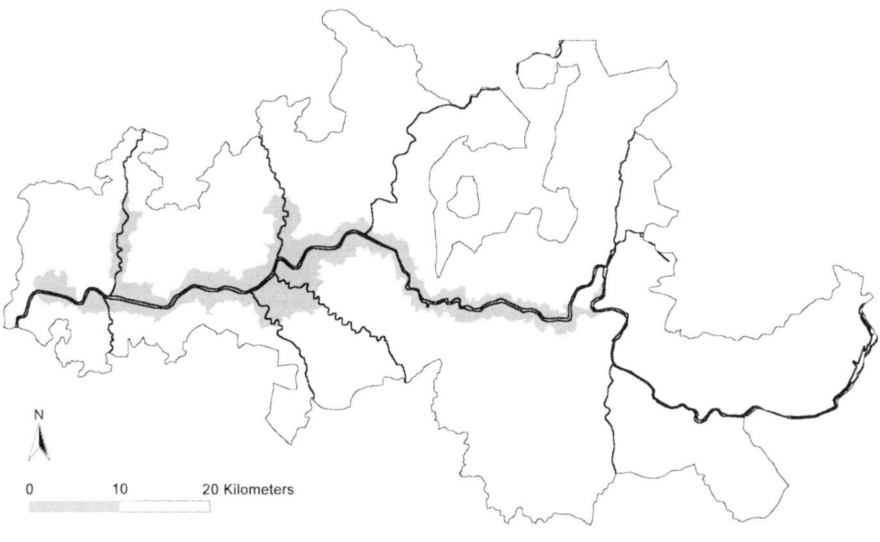

Figure 7-5. The Designated Area of Alto Douro

In the proposal for the designation of the Alto Douro these two mandatory criteria were applied to the authenticity/integrity and the statute of protection and also criteria iii, iv, and v. The authenticity/integrity of the cultural landscape of the Alto Douro were presented from the standpoint of the features that mark its unique nature, followed by a more detailed presentation of the most distinctive of these features: the antiquity of the demarcated region; the terraces; the cultural crossroads. The nature of the landscape is determined by wise management of limited land and water resources on extremely steep slopes. It is the outcome of permanent and intense observation, of local testing, of the profound knowledge of how to adapt the culture of the vine to such extremely unfavorable conditions. It is an expression of man's courage and determination. The proposal was submitted by Portugal in June 2000 and the approval arose in December of 2001 (Figure 5).

4. THE MANAGEMENT PROGRAM

The inscription of Alto Douro in the World Heritage List implied the preparation by Portugal of a management program. Throughout the past two decades, the Douro Demarcated Region has not only had a varied portfolio of instruments for development as it has also been able to obtain a varied and substantial set of funds for investing in infrastructures and production, all of which have helped the region increase its visibility and add momentum to its policies for presenting and promoting the region.

Although these policies and programs were not directly aimed at protecting the cultural landscape, they represented investments that contributed greatly to safeguarding the socio-economic force of the region and thus ensured the viability of the viticultural economy, the one that since time immemorial has been responsible for the existence of this cultural landscape. It is precisely because of those programs that new systems for planting vineyards such as *patamares* and vertical planting were developed

The management and protective strategy to be implemented is essentially based on the following planning and management tools and supporting structures:

- the Intermunicipal Plan for the Alto Douro Wine Region (PIOT), directed at conserving and improving the living cultural landscape of the Alto Douro, whose guidelines are to be handed on to the Municipal Master Plans, the prime local land management instruments;
- a development agency in the dependence of the thirteen municipalities with technical responsibilities to implement the PIOT and ensure the preservation of the landscape values;

- a non-governmental organization to promote Alto Douro and to maintain an ongoing dialogue between the public and the private agents acting in the region.

PIOT is a recent instrument within the Portuguese landscape planning legal framework. It is considered an instrument to foster integrated development among municipalities. It has a strategic character to identify alternative development scenarios that adjusts well to the objectives to safeguard a cultural landscape spread through a large number of municipalities.

One must remember the specific characteristics of the Alto Douro cultural landscape, namely its fragile nature, the existence of a product of excellence and of economic interest in a low populated area where the presence of people maintaining the agro-ecosystem is vital. So, the planning instrument had to be flexible in order to take into consideration these specific characteristics.

PIOT proved to be the appropriate instrument but had to be understood as a higher level of planning to be reverted to each of the local level instruments, the so-called Municipal Master Plans.

A set of objectives was attributed to PIOT (UTAD 2001):

- to define an inter-municipal strategy for the preservation and management of the cultural landscape;
- to integrate the existing local, regional and national plans and programs;
- to assess the transportation and infrastructures networks;
- to assess the economic activities and the impact upon the environmental, social and cultural dynamics.

A methodology including two levels was defined to prepare the plan based upon the objectives and an identification of the variables to be considered (Figure 6). The first level aimed at the elaboration of a master landscape plan, exploring the principles contained in the Municipal Master Plans and the existing land use regulations. The second level produced an action program and the recommendations for the revision of the Municipal Master Plans.

A key factor in the process was public participation. From the very beginning the planning team identified the major agents in the region, namely those related with the wine business in a broader sense as well as the tourism operators, and the local and regional authorities. A series of public meetings took place all over the region and during both levels of implementation of the working methodology.

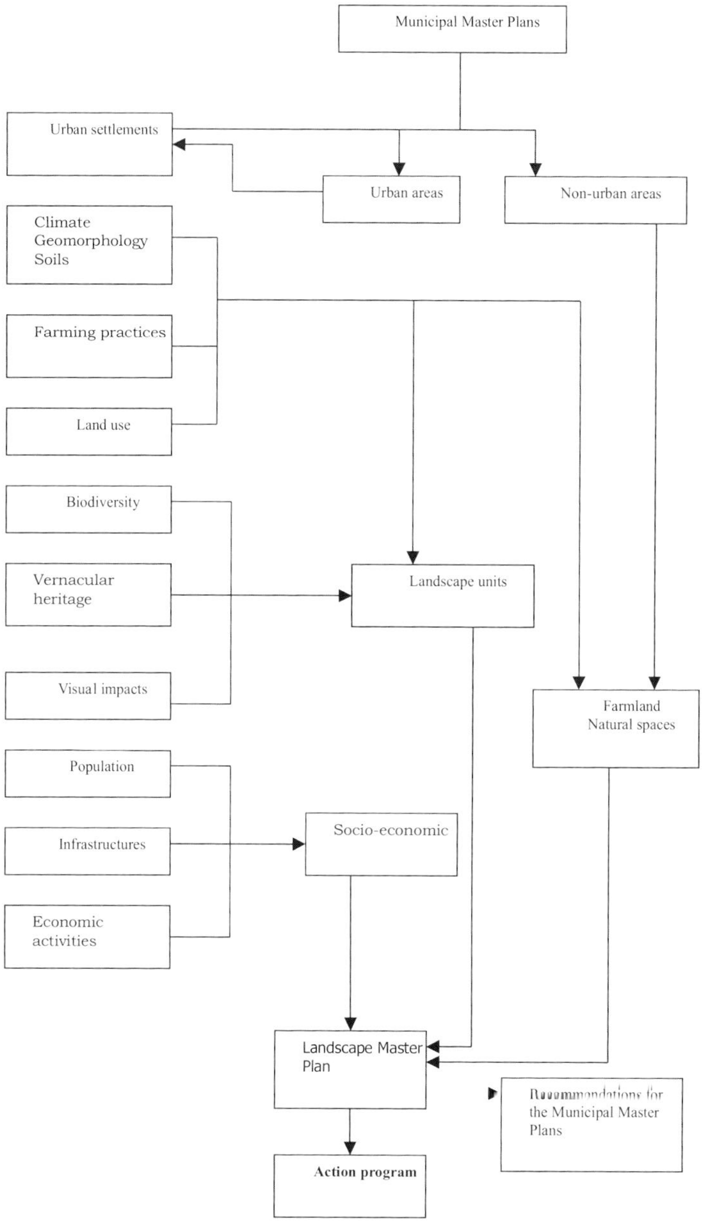

Figure 7-6. Methodology of the planning process

The diagnosis included the following variables: urban settlements, climate, geomorphology, soils, farming practices, land use (Figures 7 and 8), biodiversity, vernacular heritage, visual impacts, population, infrastructures and economic activities. The existing Portuguese legal framework calls for a land classification and qualification. Basically the classification implies the definition of urban and non-urban areas, a concept where construction rights prevail in the urban areas. The major concern of the plan (Figure 9) was indeed the non-urban areas, the vine areas and shrubbery areas. It was assumed that the management of urban areas as defined within the Municipal Master Plans was the responsibility of the local authorities and ruled according to the existing regulations. The non-urban areas where crops – wine, olive oil and almonds – prevail together with areas of shrubbery with a significant value in terms of biodiversity and landscape pattern were qualified as farmland or natural spaces. Both areas correspond approximately to 55% and 45% respectively of the total non-urban area. It was considered that given the characteristics of climate and geomorphology production forests were to be excluded from the Alto Douro. The socio-economic studies were integrated in the diagnosis as a complement to this spatial analysis supporting the understanding of people's dynamics in the region and the implications upon the landscape character.

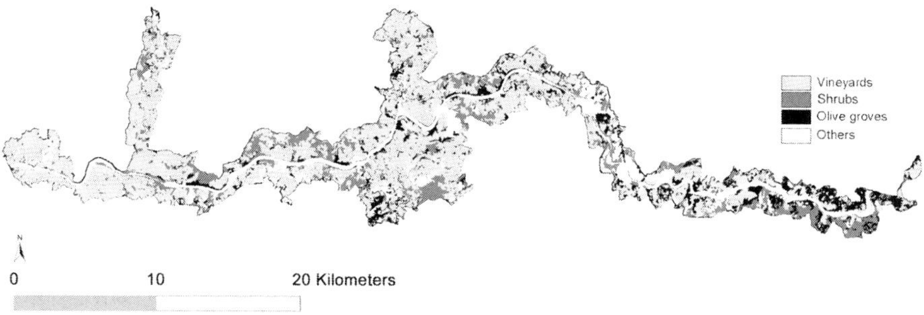

Figure 7-7. Land use of Alto Douro

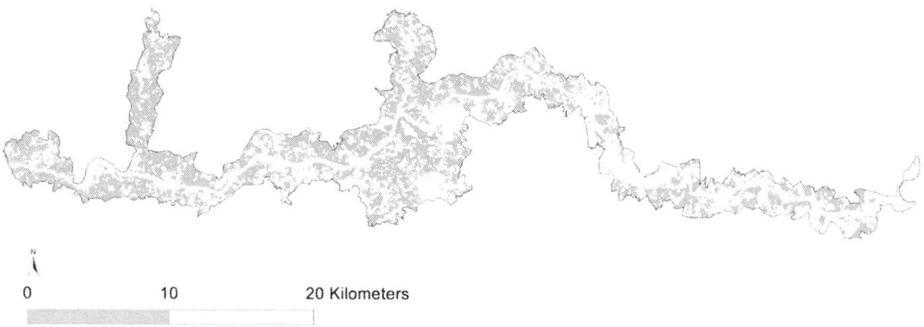

Figure 7-8. The Terraced Landscape

A major step in the diagnosis was the definition of landscape units once the character of the Alto Douro landscape was based on diversity. Together with the understanding of the differences in the landscape and the economic activities, the landscape units were of significance to the spatial qualification of farmland and natural spaces and, finally, to the preparation of the Landscape Master Plan.

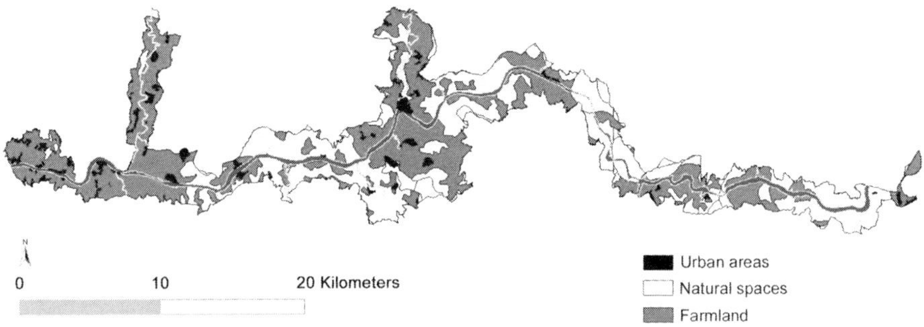

Figure 7-9 The Landscape Master Plan

At the second level, based upon this master plan, two other important steps were taken: the definition of the recommendations to be presented as guidelines for the revision of the Municipal Master Plans and an Action Program with four major objectives:

- contribute to the enhancement of the landscape and the vernacular heritage;
- contribute to the minimization of major impacts in the landscape;
- contribute to the environmental and life quality as the basis for sustainable development in the region;
- develop strategies to promote for the landscape values in Alto Douro as well in the Douro Demarcated Region.

The Program comprises five major actions within a time framework and investment plan: landscape planning and management, sustainable agriculture development, enhancement of landscape values, qualification of urban settlements and promotion of economic activities.

5. CONCLUSIONS

The Douro Demarcated Region is a most impressive cultural landscape that recently became inscribed in the World Heritage List. However, the inscription of Alto Douro was but a first step towards the recognition of the region's value and also a very significant tribute to the various generations that have created – and continue to create – this landscape. A second major step is the capacity to implement the briefly described management program to ensure its future in terms of the preservation of the unique character of the landscape within an economical and socially dynamic system. To achieve this target calls for an open-minded attitude to build strong links between public and private entities and a strong local authority that in accordance with the municipalities looks after the technical implementation of the plan on a daily basis.

The process to produce the intermunicipal plan counted on significant public participation. This should be assumed just as a first stage of an ongoing process so that the various peoples and institutions' interests in the Douro Demarcated Region find in this opportunity a way to achieve their goals. However it is important that it is understood that the future of the region depends on their capacity to continue to establish compromises and tradeoffs in order to find suitable and acceptable solutions for the good of everybody and of the landscape.

ACKNOWLEDGEMENTS

The authors would like to thank the coordinator of the Alto Douro candidacy and of the intermunicipal plan, Prof. Fernando Bianchi de Aguiar, as well as all the other team members.

REFERENCES

Andresen, T., Curado, M.J., & Marques, P.F. (2000). A candidatura da Paisagem Cultural do Alto Douro Vinhateiro para inscrição na Lista do Património Mundial. In *Arquitectura e Vida* (pp. 80-85). Lisboa.

Andresen, T. (1999). O Alto Douro Vinhateiro – Uma Paisagem Evolutiva Viva. *Actas do Seminário 'Estratégias de valorização de uma Paisagem Cultural: O Vale do Douro'*. (pp.13-18). Zamora: Fundação Rei Afonso Henriques.

Audrerie, D., Souchier, R. & Vilar, L., (1998). *Le patrimoine mondial. Que sais-je?* Paris: Presse Universitaries de France.

Bianchi Aguiar, F., Dias, J.M., Andresen, T., Curado, M.J., Marques, P.F., Pereira, G.M., et al. (2000). Candidatura do Alto Douro Vinhateiro a Património Mundial. Zamora: Fundação Rei Afonso Henriques.

Council of Europe, UNEP, European Centre for Nature Conservation. (1996). *The Pan-European Biological and Landscape Diversity Strategy*. The Netherlands.

Council of Europe (2000). *European Landscape Convention and Explanatory Report.* Strasbourg.

Rossler, M. (2000). The implementation of the World Heritage cultural landscapes categories. In M. Rössler & G. Saouma-Forero (Eds.) *The World Heritage Convention and cultural landscapes in Africa.* Report of the Expert Meeting (pp. 7-15). Tiwi.

UNESCO (1972). *Convention concerning the protection of the World cultural and natural heritage.* Paris.

UNESCO (1999). *Operational Guidelines for the Implementation of the World Heritage Convention.* Paris.

Universidade de Trás-os-Montes e Alto Douro (UTAD). (2001). *Plano Intermunicipal de Ordenamento do Território do Alto Douro Vinhateiro*: *Vol. 1-2*, (p. 199). Vila Real (unpublished document).

Chapter 8

CULTURAL AND HISTORICAL VALUES IN LANDSCAPE PLANNING: LOCALS' PERCEPTION

Helen Alumäe, Anu Printsmann & Hannes Palang
Institute of Geography, University of Tartu, Estonia

1. INTRODUCTION

Culture is the hidden hand of land use planning. Culture bounds the land in diverse ways. Culture marks the corners and edges of place; it selects which places will be sacred and which will be sacrificed; it yields maps of place and bestows place names; and it decides the aesthetics and ethics of the land (Geisler 2000). Cultural heritage has been discussed by many authors of different disciplines for many decades (e.g. Meinig 1979; Lowenthal 1985; Daniels & Cosgrove 1988; Graham et al. 2000; Claval 2002; Maaranen 2003). The relationships between land use and culture are of perennial interest in such disciplines as cultural anthropology, human geography, natural history, landscape architecture and human ecology. In recent years also the landscape ecologists have started to realize the importance of cultural heritage and its role in the landscape planning process and landscape management, and a large number of works have been published in this field of research (Nassauer 1997; Stenseke 1999; Luz 2000; Oreszczyn 2000; Cantwell & Adams 2003; Fry 2003).

During many decades, in West-Europe, but even more so in East-European countries, planning was the fun of few chosen experts, who, as excluded outsiders, made plans that often did not function in the actual landscape. "This may be compared", writes Cullen (cited in Relph 1976: 52), "to God creating the world as someone outside and above the thing created". Just as we can argue for God having made some mistakes, we can point at

H. Palang and G. Fry (eds.), Landscape Interfaces, 125–145.

numerous shortcomings of such outsideness, as noted by Relph (1976). In top-down planning, there is little room for public participation in defining cultural-historical heritage in a local area. Having realized this, there is nowadays a growing tendency to take lay people's opinions into account when planning landscapes in most countries of Europe. More and more, planning is based not only on expert judgments, but on the opinion of lay people, such as local inhabitants and different interest groups who will be most influenced by planned changes. Many authors from different disciplines have studied the role of cultural-historical values among people and have stressed their importance as an environmental quality (Schama 1995; Lowenthal 1985). However, very often the theory is good, but lacks support from the local population in practice. The lay people are overloaded with information and personal problems, and that prevents them from taking deep interest in planning process.

In this paper we focus on local peoples' perception of cultural and historical values in the landscape as revealed by results of delimiting valuable landscapes in four study areas of Estonia. Perception of visible and non-visible values in the landscape and the *genius loci* of places are discussed as well.

2. LANDSCAPE IN SCIENTIFIC DISCOURSE

Landscape has often been compared to palimpsest – meaning a medieval writing block where an original inscription would be erased and another written over it (Crang 1998; Espersen 1999; Marcucci 2000). Alike, culture inscribes itself on the landscape as the sum of erasures, accretions, anomalies and redundancies over time. As Johansson (1993) puts it, in decoding cultural heritage, every place provides more or less strong evidence of how the place has been used at different times. Traces of different eras force their way into the present. If we arrange these traces in patterns that belong to the same time, we find overlapping layers that shine through each other – they are somewhat transparent. The pattern of accumulated forms and change suggests a lot about the evolution of the landscape and the local culture. It implies that landscape has been shaped by and is shaping the people living there, thereby becoming a bank of cultural memories – some still in use, others as residues of past practices and knowledges. Above all specific landscape patterns and elements emphasize the link of people and land. However, while in the case of medieval palimpsest the overlaying layers are not connected in form or content, actual landscapes "are not fungible because each is a combination of physical, cultural, and locational features" (Marcucci 2000: 68). As suggested by Marcucci, a more accurate metaphor is to think of landscapes as legacies. Those existing today result from previous conditions and events in that particular locality. Also, a changing landscape is very much a function of historic conditions.

Landscape elements persisted from previous times till nowadays can be of singular or compound nature. An object is more valuable when it is not lying isolated in the landscape but is embedded within a cultural or historic context (Renes 1999). Landscapes and landscape elements may remain unintelligible to many of us because the social and cultural *context* is foreign to us, or because the context of the representation is unknown. This is the argument that the followers of the postmodern and cultural turn of landscapes studies are rightly emphasizing – landscapes as a way of seeing (Muir 2000; Widgren 2002).

According to Gestalt psychology, what we perceive can be described as a Gestalt, a whole that is more than the sum of its composing parts (Antrop 1997). Landscape is written in a multifaceted and symbolic language that can be interpreted differently by its observers. It refers to our perceivable environment and should be considered as holistic, relativistic and dynamic. Landscape identity is largely determined by holistic characteristics, and so is our perception and evaluation of landscape (Antrop 1998). Our perception is a complex process that analyses immediately and interactively the observation and links the results with our knowledge and past experience. This aspect of perception makes different people really 'see' different landscapes at the same spot and it determines the way they evaluate and appreciate the landscape, too (Antrop 2000). Therefore, everyone will see and read a story according to what one knows already and can recognize in the patterns and elements shaping the landscape.

All landscapes have a unique story. As Adam (cited in Ingold 2000: 189) suggests, for the native dweller, the landscape tells – or rather *is* – a story, a chronicle of dwelling. Antrop (2000) has claimed that the fascination of a landscape comes from the mysterious, age-old and indecipherable language it speaks. They all have a spirit linked to the land, a *genius loci*. However, Antrop and other authors have argued that the new landscapes – the monotonous and uniform modern landscapes – miss the spirit of the place, thereby reducing their attractiveness.

2.1 Historical meaning emphasizes the *genius loci*

According to Coeterier (2002), form is a combination of cognitive and affective aspects. Historically and culturally important sites have an existential value for people, at three levels: "they give or enhance place identity, personal identity and group identity". In our studies we focused on place identity landscape, which in our case could also be regarded as landscape identity. Historic sites especially, are the bearers of the place identity, or, *genius loci*. "It is what historic elements say, not about the genesis, but about the genius of the place" (Coeterier 2002: 120). Although areas of wilderness can possess a strong *genius* loci, the cultural heritage and the care of the local values are the main factors contributing to the genesis of *genius loci*. As Antrop (2000) suggests, the *genius loci* manifests itself in numerous signs in the landscape. Many elements express an age-old spiritual and symbolic meaning and have been shaped accordingly (Colquhoun 1997). Although many of these landmarks have general characteristics linking them to a common culture, details reveal their particular uniqueness and place in the landscape in which they are placed. The landscape might be the homeland of the resident, who is admiring the territory of his community. For the traveler, it might be a resting point with an overview of the journey so far and a look over the road ahead.

While *genius loci* is often comprised of the invisible heritage in the landscape – stories and legends that are known only to the local people – also the outsiders can 'sense' the spirit of the place when the landscape is full of stories associated with remarkable or outstanding landmarks. The *genius loci*, although almost never stated by any of the locals as a familiar concept, gives scope to one's imagination, and there is a strong feeling of respect for their ancestors' life and work. The preservation of their memory was considered important even for those who are not native to the certain place.

Distinguished features in the landscape attract stories – and if one of such objects should perish, legends attach themselves to the next distinguished feature. To take an Estonian example – if an old oak, which according to the legend was planted by a Swedish king, would die, the story would still live on – it would be linked to one of the nearby big trees, not specifically to oak. These stories have to have some kind of internal logics that usually explain the specialty of the object, e.g. holes in the stones that are evidence of devil's footprint (Hiiemäe 2001).

Historical meaning is a very interesting, but rarely used aspect. In some countries there is a tradition in valuing buildings or landscapes because of their association with famous historical persons or happenings. In some of these countries there is a lot of attention given to old battlefields, even when (as is usually the case) there is nothing to see. It is difficult to defend the preservation of these landscapes on the basis of their features and their present outlook. Instead the significance of these landscapes lies in the known, but now mainly invisible, history of its use through time.

Departing from the planning perspective, the delimiting of valuable cultural landscapes was one of the primary aims of the project. However, instead of trying to draw the impossible line between natural and cultural landscape, we support Tim Ingold's (2000) view that, in place of the naturalistic view of landscape as a neutral, external backdrop to human activities, and the culturalistic view that every landscape is a particular cognitive or symbolic view of place, we should adopt what he has called a 'dwelling perspective', according to which the landscape is constituted as an enduring record of – and a testimony to – the lives and works of past generations who have dwelt within it, and in so doing, have left there something of themselves (Ingold 2000). Also, in most cases there is no sense in dividing the landscape into either cultural or natural ones, because, as O'Rourke (2002) notes, landscapes simply cannot be meaningfully studied from an exclusively 'naturalist' or 'culturalist' perspective. Instead, we should understand cultural and natural landscapes as layers of the total landscape (Palang et al. 2000).

3. PLANNING PERSPECTIVE: MATERIALS AND METHODS

3.1 Methods

The project we discuss here was a part of a larger planning program initiated by the Estonian Ministry for the Environment, named *Delimiting Environmental conditions that direct settlement and land use*. The program consisted of two sub-projects, one focusing on ecological networks and the other on delimiting valuable landscapes. The valuable landscapes project combined three different approaches (see Palang et al. 2003 for more). These were following land use changes on old maps; expert opinions, and extensive interviews and questionnaires of different interest groups. The project was a part of county planning process, carried out following the same methodology in all Estonian counties (KKM 2001). In this paper we focus only on the results of the inquiries with local people and interest groups.

The project focused on five values (Figure 1). First, cultural and historical values that are sometimes hard to separate from each other. Under this category we were looking for areas that contain elements from different times; areas that have retained land use structure typical to one specific period (following the ideas of Cosgrove (1984)) about each social-economic formation creating its own landscape, Estonian landscapes could be divided into ancient, estate, farm, *kolkhoz*, and current postmodern landscapes; see Palang and Mander (2000) for more); and areas that are associated with famous personalities or are depicted on famous paintings, literature, film. Here we also included areas that comprise a number of singular elements that are under some kind of protection (nature or cultural heritage conservation). Second, aesthetical values were initially considered as the major objective of the project, but later, mainly due to subjectivity in assessing, their importance was reduced. Third, landscape's potential to offer recreational facilities were used in assessing recreational values. Fourth, as natural values we considered the abundance of natural diversity, rare species or communities. Finally, we also looked at identity value, i.e. the ability of landscape to allow local people to delimit their territory, to identify themselves with the landscape. For the local population, identity is one of the most important arguments (Renes 1999). Landscape identity also encompasses issues of *genius loci* and knowledge about local oral heritage, such as story telling. Above mentioned principles were used as guidelines to design the theme plans of valuable landscapes for the different county plans.

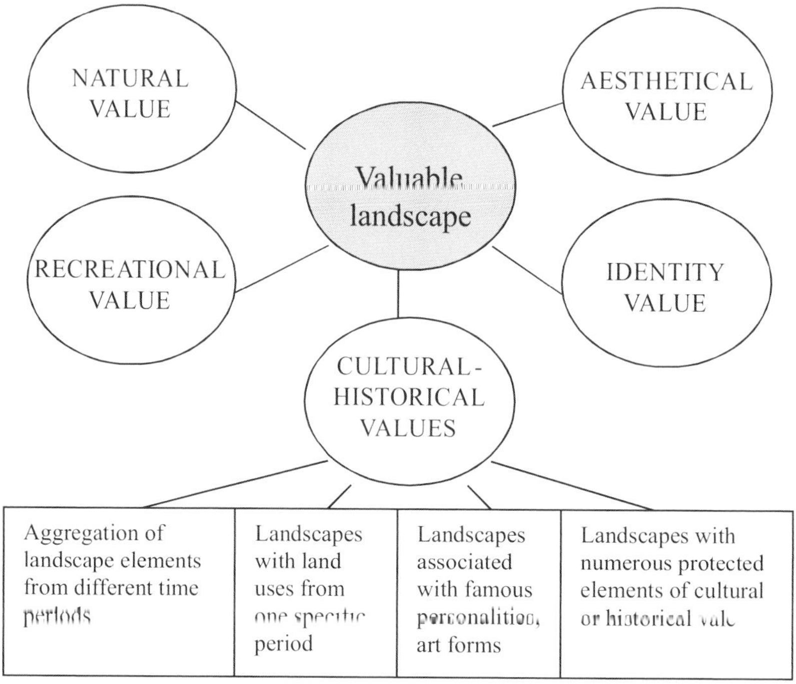

Figure 8-1. Values used for defining valuable landscapes (after Palang et al. 2003)

Local population played a very important role in delimiting valuable landscapes in the four study areas. In order to find out the preferences and value judgments of local people, questionnaires and interviews are most commonly used. Preference is given to open questions, which avoid prejudices that can be caused by pre-given answers. In the course of assessing landscapes in the case areas, we used questionnaires that differed a bit between counties. Questionnaires were distributed at public gatherings, at schools, local administrations and among local interest groups. In addition, we used open interviews with 1-2 to four persons at a time. Altogether we got feedback from about 1,000 people in the study areas. The participants were not chosen based on certain socioeconomic characteristics. We can argue that most of the people who filled in the questionnaires, have interest in their home landscape, since they voluntarily came to the meetings where questionnaires were distributed.

Although the aim of delimiting valuable landscapes is somewhat idealistic, the future of such landscapes is threatened by global processes, such as globalization. Furthermore, the preservation of past elements, technologies and mental heritage is important, but so far not helpful in making of living in the rural landscape. The modern technologies often put an end to traditional structures. Therefore, one of the main tasks of the project was how to support plans and activities that avoid the creation of such landscapes, via the use of management proposals.

3.2 Locals

In planning, the main participants are the experts, politicians, local population (with often important differences between farmers/landowners and others), and visitors. Although the latter group is not actively involved in the planning process, their views have to be taken into consideration in order to achieve a plan that would work in reality, because the increasing importance of global tourism can also be noted in Estonia, where there are hopes that tourism as an income source will largely replace agriculture. Each of these groups has their own values, and would make their own choices.

Peil adopts Heidegger's viewpoint about locals as people characterized by continuous dwelling in the landscape that has linked the people to the environment (Peil 2001). The local landscape can be characterized as "a region or area of which people have stories" (Peil 2001: 59). For Ingold (2000: 193), too, the landscape is the world, as it is known to those "who dwell therein, who inhabit its places and journey along the paths connecting them".

Moreover, the local people, although usually regarded as one clearly determined group of actors (see Palang et al 2003), is a very heterogeneous group. There are people who are deeply interested in the local history and the state of the cultural heritage in the region, while there are others who show disinterest and have very poor knowledge of the cultural heritage of the region. Those people, who have poorer knowledge about the place, are also less tightly connected to the place. Interestingly enough, it is often the newcomers who, though knowing fewer stories, show more respect for the traditional elements and landscapes. They often become the 'leaders' of the hamlets and the encouragers of landscape care in the area (see Gustavsson & Peterson this book for further discussion).

3.3 Case areas: location and history

The case studies were performed in five Estonian counties situated in the central – Viljandi and Jõgeva – and southern – Valga, Tartu and Põlva – part of the country (Figure 2) as part of the design of the theme plan of Valuable Landscapes for county plans in 1999-2001 (KKM 2001).

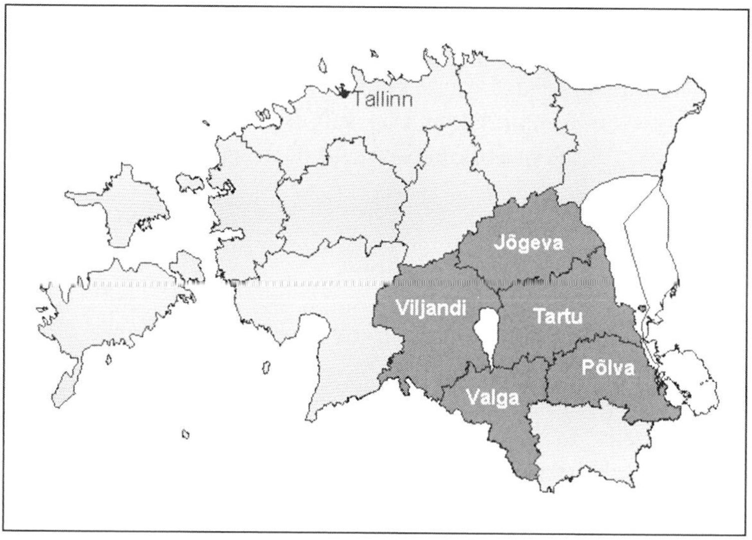

Figure 8-2. Location of the five counties

In order to understand the perception of cultural-historical values in Estonia, it is necessary to take a short look at the history of the study areas. In general, we can note that the history of the four counties has followed somewhat similar tracks. Before the christianization (13th century) Viljandi (at that time called Sakala) formed one of the mightiest and probably wealthiest regions of Estonia, with local chiefs staging fierce battles against the invading German crusaders. During the following centuries the Germans built several castles in the areas, such as Karksi and Viljandi, plus one in Helme in the neighboring Valga County. This is probably the reason for the local people of Viljandi County having the strongest feelings about castles and manor houses. The Viljandi area played also an important role in national awakening in the second half of the 19th century. During the Soviet times the area had one of the most industrialized agricultural enterprises of the country.

Valga and Põlva counties formed parts of the ancient Ugandi County that competed with Sakala in terms of might and wealth before the christianization, with Otepää being one of the most important strongholds. Jõgeva, however, was a loose collection of small parishes that later were also connected to the large Tartu County. From 15th century onwards this has formed the hinterland of the town of Tartu. The current Valga County was created in 1920, after Estonia had become independent, from parts of Viljandi, Tartu and Võru counties. Põlva and Jõgeva counties were created in 1949, the former contains parts of the Võru and Tartu counties. Before the 20th century these three counties used to be silent backwaters, with Otepää parish being a well-known exception as a tourist site from mid-19th century and some sort of spiritual center as well.

All the counties are dominated by agricultural land use, Põlva also by forestry. Currently the southern part of the country, bordering Latvia, suffers from rural decline, while in the central areas agriculture is doing relatively well. Jõgeva has had a relatively developed agricultural sector during the Soviet times. However, all the counties, especially those on the edge – Valga and Põlva – are threatened by continuous marginalization, characterized by fallow fields and abandoned farmsteads.

4. RESULTS

The results derived from the four case areas can be somewhat generalized to all areas, while some results were site-specific, but not necessarily county-specific as well. The regional differences researched within the study, are outlined in other articles (e.g. Palang et al. 2003). In this article we focus on more general issues regarding cultural and historical values in landscape assessment and planning.

The first, rather complicated issues were the concepts of *landscape* and *landscape values*. In order to get more information and not to scare people away, we avoided defining the term landscape. Since landscape was understood and perceived differently by different stakeholders and persons, the concept covered a very wide range of understandings of landscape. To most lay people, landscape consisted of single elements, most often man made elements, to which they attributed certain values. In the everyday language of lay people, the concept of landscape was almost unknown. The surroundings were not described as landscape. Instead, notions such as 'neighborhoods', 'nature' and 'home area' were most commonly used. So, when asked about beautiful landscape, people spoke about nature. When asked about valuable landscape, they mentioned architectural monuments. For example, when asked about valuable landscapes in Kolga-Jaani, a small town in Viljandi County, the local church was brought up as a local cultural-historical value. Cemeteries, schoolhouses, castles, manor houses and other architectural elements were valued highly as well. For lay people, landscape was an arena of everyday activities and they didn't think of it in terms of a holistic landscape with multiple values. For them, landscape was a place they dwell.

Another interesting aspect of local people's perception regarding cultural-historical, but also natural values, has been indicated by Stenseke (2000), and confirmed by our research: when speaking of landscape values in general or, more specifically, of cultural-historical values, local people tended to refer to features identified by the authorities who have been dealing with natural or cultural heritage in the area, showing little confidence in their own judgment. Many times in our research we found it important to underline that there were no right or wrong answers, and that we as planners were outsiders seeking to understand the local perspective of the place. This communication often needed diplomacy, and long hours of getting-to-know each other.

Although landscape researchers prefer to speak about landscape as a holistic unity, we, with the aim of explaining the local people' perception of cultural-historical values, found it important to make a difference between the visible and invisible in the landscape, as suggested by Widgren (1993). This made sense because it gives an idea of lay people's understanding of landscape. They spoke about visible features with ease; the invisible, which is hidden in stories, memories etc., was found to be more complicated, and even though the stories were there, the planners often lacked time to hear the stories. Lay people were not yet used to the idea of invisible, such as local stories, being valuable.

Generally, the results of the four case studies revealed that cultural-historical heritage had a typically unconscious, but still rather significant role in the formation of the landscape perception and values of the local people. The historical aspect could be described in terms of 'traditions' and 'roots'. As the studies in Estonia and also elsewhere (e.g. Stenseke 1999) have indicated, the local people valued signs showing that humans have lived in the place in ancient days. These signs could be represented either in visible or perceivable form. In this sense, the results gained from interviews and questionnaires largely overlapped with those of the experts, whose knowledge about the values of the areas was based on historical sources, such as old maps, and contemporary travel guides.

Typically, the local population put the highest cultural-historical value to estate complexes (Figure 3). Every county has at least 2-3 rather well preserved manor houses, which tended to top the list of valuable landscapes. Here, the meaning of landscape becomes even more limited than that of place – there are certain very limited localities, which were considered valuable.

Figure 8-3. Estonian countryside is characterized by manor houses dating back to 18th-19th centuries. Being distinctive cultural and historical values, manors like this in Mooste, Põlva County, were highly appreciated by the local people as well as experts. Photo: M. Saks

Perceivable, i.e. invisible values of landscape were mostly valued on a very intimate basis. Moreover, the influence of past top-down planning has made local people unsure about their own valuations – they did not believe that their personal preferences or values attributed to certain elements or landscapes would either interest the planner or affect the outcome of the planning project. However, we believe that our project had a positive impact on the local perception of local values in the sense that we made them feel that their opinion mattered. It was also interesting to see how some villages and farms took advantage of the project and used the project as an argument to stress the importance of landscape maintenance in their area.

5. DISCUSSION

5.1 Preservation needs and challenges

Widgren (1993) describes two basic approaches to landscape conservation. On the one hand, arguments for preservation may be based on the landscape itself and the older features still visible in it: the buildings and monuments, the characteristic type of field boundaries, the actual farming practices and the vegetation created and managed by these practices. These arguments are based mainly on the present outlook of the landscape. On the other hand, Widgren admits that there is also a general agreement that historical significance can be given to a landscape by the artistic or literary representations that have been inspired there or by the historical events that have taken place there. This argument supports the fact that knowledge of events or circumstances that can no longer be seen in the landscape, can, nevertheless, give meaning to a landscape. Our task as landscape planners was to take both sets of arguments into consideration.

The outcome of this planning exercise was not to create new protected areas, but rather to establish rules and examples of good practice for further management, so that the outstanding values indicated during the project could be taken care of and to become sustained. Since one of the main ideas of the plan of valuable landscapes was to conserve the landscapes that were considered valuable on one or more basis, the issue of landscape conservation has been rather central in discussing the qualities of landscape. Several scholars from different disciplines, such as landscape ecology, landscape design, tourism etc. (Norderhaug et al. 1996; Saltzman 1999), to name a few, have been involved in discussions on to what degree should we try to conserve certain landscapes, landscape elements or processes. It is clear that preserving landscapes as museums – what Relph (1976) has called museumization – is not a well-supported idea and should be handled with caution. There is a recognition that the conservation, maintenance, as well as effective utilization of at least part of the cultural heritage, in the form of existing historical landscape structures is not possible via professional bodies, state institutions etc. Success relies largely on the incorporation of general public. Thus the identification of the value judgments of the local people concerning cultural landscapes and cultural heritage in the landscape formed an essential part of the evaluation.

There are many activities, such as craft inventories and local history research that prove interest in cultural heritage at the local level. In Estonia, the local history research got a real kick after new independence in 1991. Many local village improvement programs were started in the rural areas. The best example of such initiative is *Kodukant* – the Movement of Estonian Villages and Small Towns. In 1997, Estonia's President Lennart Meri started the movement *For More Beautiful Homes*. This initiative was very much welcomed in the rural areas of Estonia and largely due to the scheme a lot of landscape conservation and improvement activities have already been implemented in numerous villages. Furthermore, there are people, mostly school children and elderly people, who have done local and genealogical research, collected pictures and interviewed local people. Many small villages or regional museums have been set up, often related to famous personalities that have lived or been born there or to novels written and films made at this certain place. Planners, too, have realized that paying attention to the cultural values of landscape at the local level can contribute to the maintenance of vitality in the rural area.

5.2 Landscape – valuable to whom?

Similarities and differences of lay people and experts stress the importance of the involvement of lay people in planning, since the criteria expressed by them are insufficiently covered by expert opinions. The results of our research supported the arguments made by Coeterier (2002), who in his study of lay people's evaluation of historic sites indicated that to experts, the value of historic elements or patterns is based on their information value, while for lay people, the main criterion is form. In our study, one of the main criteria for the first selection, i.e. the expert selection of valuable landscapes was well-preserved land use pattern. The local people seemed not to recognize that certain aspect of landscape heritage. One reason for the relatively low appreciation of field patterns of the past could be the fact that it is the everyday landscape, the working location. That is the landscape people are used to in their everyday activities and with few exceptions, they do not think of that in value categories.

Although results seemed to indicate that several of the old farming practices and related landscape patterns, such as wooded meadows were evaluated highly as well, it seemed to be caused by expert opinions. Stenseke (1999) shares that view, making an interesting point that there is major difference between recent history, meaning history mainly known by traded knowledge, and old history, meaning historical objects older than 200-300 years, of which most knowledge is generated by specialists: archaeologists and historians. A landscape or landscape element can gain value for the local people if it is highly appreciated by outside experts. We can observe the same tendency with tourist sites – many landscapes were now considered valuable on the local scale because these landscapes had tourist value (which was usually associated with particular historical or cultural values of the area). Sometimes people just knew that something was there, but they did not know what it was or why was it considered valuable. Recent history is more closely related to present inhabitants and their holdings.

Figure 8-4. The Viiralt Oak in Viljandi County – a noteworthy landscape element with strong cultural and historical background. The local people involved in the designation of valuable landscapes, however, did not consider the oak and the landscape as important

There was another example that vividly illustrated the landscapes of tourism: the so-called Viiralt oak located in Viljandi County (Figure 4). The old oak has been pictured by a famous Estonian artist, Eduard Viiralt, in his graphic sheet *The Landscape of Viljandimaa*, in 1943. The magnificent oak stands in its pride solitude in the middle of fields near the town of Viljandi, and can be well observed from the main road. It is definitely the symbol, the icon of the Viljandi County, being depicted on the cover of every tourist brochure. Curiously enough, the oak seemed to have no *genius loci*. It was never mentioned in any of the answers of the questionnaires. It is hard to find explanation to that. Is the sight of a single tree on a field too ordinary or perhaps it is just an icon artificially constructed by tourist agencies? The oak and the surrounding landscape are presented as 'the very landscape of Viljandimaa' for tourists. As such, the oak landscape seems to be valuable for tourist agencies, not for the local population.

There are quite numerous landscapes in the study areas that have cultural-historical importance from the literary point of view. One of the best-known literary places in Estonia is Palamuse, an ordinary village with a church and school, in the drumlin field of Jõgeva County. The scenes of one of Estonia's most famous books *Kevade*, written by Oskar Luts in 1912, were set up at Palamuse, so it is a place known for every Estonian. The book is so popular that people mix up the reality and fiction. Although Palamuse and its people were the prototypes for the characters in the book, people very often think of the stories as something that have really happened. As such, Palamuse forms the core of identity and touristic interest of Vooremaa (in English 'The Drumlin Field').

It is also interesting how Vooremaa, the region, which as a whole has perhaps the strongest *genius loci* of all places studied, is significant both naturally and culturally. The drumlin field, being naturally unique and outstanding also on the European scale, is related to more legends of Estonian national hero Kalevipoeg than any other areas in Estonia. The lakes between drumlins are the footsteps of Kalevipoeg, while other valleys are his beds. This area is valued both by locals and tourists.

5.3 How old should a trace be to become valuable?

When determining cultural and historical values of a place or landscape, the time aspect becomes very important. We tend to give higher appreciation to elements that date back from early history. According to our research, the historical objects and stories of 100-300 years ago seemed to possess the highest value among lay people.

History that dated back from more than a couple of centuries ago was most highly appreciated by experts, who drew their information from different literary sources which emphasized the importance of ancient and medieval elements. The oldest noteworthy cultural-historical elements in Estonian landscapes are the Estonians' strongholds, dating back from the 12th and 13th centuries. These are not that well preserved because they were built of wood. There is one especially noteworthy site of a stronghold in Viljandi County, but not in the other studied counties. The ancient strongholds of Estonia remind the case of battlefield landscapes of Britain: there is not so much to see, but there is the legend, the history, and the sense of past events. The events taken place there are directly linked to the fight for independence and therefore extremely important for the national identity of Estonians, although most of the battles in history have been lost.

Estonia is quite rich in the objects of cultural heritage that date back from the period of feudalism (13th-18th century) – German, Danish and Swedish reign periods – and Czarist-Russian (18th-19th century) period. However, most of these elements are small or in a rather poor condition. There have never been any magnificent, outstanding castles in Estonia, instead, there are plenty of small estates around that are very important landscape elements. Another type of visible architectural landscape elements is ancient strongholds. Both types of buildings actually represent the history of foreign power in Estonia. Estonia has only been independent for very short time periods in its history, and interestingly enough, the symbols of foreign power from the far past are valued quite highly, contrary to the symbols from the Soviet period. It seems that these symbols are the reminders of the hard rural life of our ancestors. Estate owners were mostly Baltic Germans, and surely they were disliked by the oppressed Estonians, or country-folk, as the nation called themselves. But the Germans have left strong landmarks here, including the field patterns. Following the landlords' way of life, the local farmers started to establish flower gardens around their houses as well. Although the landlords were hated and in many times, ridiculed, they symbolized the good life and the farmers, too, wanted to have a good life. And while actual well-being was not achievable, the appearance was.

On a large scale, the most recent history (in the Estonian case, the post World War II period) was not considered valuable. While the layers of old history and culture, i.e. that of pre-WW II period, were highly appreciated on the local as well as expert level, the traces of most recent history, the Soviet period, were not appreciated; moreover, this period was almost a 'taboo' topic. However, local people perceived landscapes of the recent past, most exemplified by neglected collective farm buildings that characterize our rural landscapes, differently. There were people, usually Soviet time collective farms workers, who longed for the gone period and to whom these buildings symbolize the good old Soviet times. To these people the Soviet farm buildings had a positive value. Others, usually older and younger people and those with higher education, conceived these elements as something negative that has to be torn down.

In planning, the so-called Soviet landscapes are the most difficult to evaluate. In our research, we asked people how they felt about the Soviet rural centers and of other elements of that time. It was obvious that people did not, yet, think of these landscapes in historical-cultural terms. The most common reaction was surprise: *why would we consider that to be valuable?* The settlement and land use patterns of the Soviet time were unnatural, unaesthetic, not harmonious and ecologically unfriendly (Figure 5). Therefore, the patterns were imposed on people and landscape, the only exceptions being former manors and churches that were rebuilt into collective farm centers or schools and gymnasiums, respectively. This exposition of foreign power can be compared to that of Christian crusaders building their churches into holy groves. Nevertheless, this dominating reaction was contradicted in Vastemõisa (Viljandi County), a typical Soviet-style collective farm center. The people of Vastemõisa were very proud of their Soviet heritage, explaining that it was *their heritage, their home* that was important to them. They specifically asked us to include the Vastemõisa settlement in the list of valuable landscapes. Here it is intriguing to add that Vastemõisa, meaning 'New estate' in Estonian, has a rather rich cultural-historical heritage including the German estate in the neighborhood. Quite unexpectedly, unlike in most other case areas, the Soviet-period village center was attributed a higher value than older, and perhaps more commonly appreciated elements in the area.

Figure 8-5. The Soviet layer in Estonian rural landscape. The extensive farms of concrete blocks were mostly regarded as being ugly or, being sad remnants of the 'good old Soviet times', but usually this most recent time layer was not regarded as culturally or historically valuable. Photo: H. Palang

It can be argued that the Soviet time was not appreciated because it symbolized the foreign reign, which was so much hated. However, the same could be said for the bulk of cultural-historical elements dating back from 13th-19th century: most of the important symbols of today's Estonian landscapes represent the power of feudalists who reigned the country – estates and churches are but a few examples of those. So, we can argue that it is not so much about the foreign power, but the time distance that separates us from that. The Soviet time is still painful to remember, while the hardships of the German time have been forgotten. Not enough time has passed since the end of the Soviet reign – the traces are not yet old enough to be regarded as valuable.

6. CONCLUSIONS

The case studies affirm the argument made by Fry (2000) that people are emotionally attached to landscapes, sometimes to several different landscapes. However, Fry sees dangers in many of the current initiatives to neutralize landscapes into physical building blocks of the countryside. These will fail to capture their symbolic value and hence fail to understand why landscapes are so important to people. As suggested by Stenseke (1999), pointing out features of value on an inhabitant's land will inspire pride and make him/her more careful with introducing sharp landscape changes. This is very important as a planning strategy. Planning is a useful tool in getting local people interested and involved in their landscapes. Often people in rural areas lacked confidence in taking care of their landscapes, and we feel that our planning exercise has, as communicative planning showing interest in the local histories and values, contributed to the rise of local care of cultural and historical values, which was an indirect, but very important aim of the whole project.

In most cases cultural and historical values in landscape were presented as visible elements, not as a whole. The exceptions were landscapes, which were characterized by a strong identity – a *genius loci*. Our research indicated that for the local people, too, an object was more valuable when it is not lying isolated in the landscape but is embedded within a cultural or historic context. The whole is more valuable than the sum of its parts. It is the whole that makes up the *genius loci*, which was usually appreciated by the local population as well as outsiders, such as experts. This has been well proven by the case studies, which indicated that the most valuable landscapes were those, which comprised visible historic elements as well as the perceivable cultural meaning of places. However, not all history was equally appreciated; the most value was put to the traces of history dating back to 80-400 years ago (before the Soviet time).

The geographical differences in perceiving cultural and historical values in landscape were not well documented. The noted differences were site-specific, not county-specific. In rural landscape, people were much influenced by outstanding individuals who had lived or were living in their area. Also, written records of one's home area seemed to influence one's views of the local heritage. Moreover, emotional attachment to landscape is important as well.

ACKNOWLEDGEMENTS

The authors wish to thank Age Koppelmaa-Sild, Mari-Liis Otsing and Piret Pungas for helping us in gathering the materials; the people from county governments – Tiia Kallas, Vello Lepik, Marika Saks and Tarmo Möller – for their encouragement. Our sincere gratitude to Kristiina Hellström and Kalev Sepp for their input in developing the methods of the project.

REFERENCES

Antrop, M. (1997). The concept of traditional landscapes as a base for landscape evaluation and planning. The example of Flanders Region. *Landscape and Urban Planning, 38*, 105-117.

Antrop, M. (1998). Landscape change: Plan or chaos? *Landscape and Urban Planning, 41*, 155-161.

Antrop, M. (2000). Where are the Genii Loci? In B. Pedroli (Ed.), *Landscape – Our Home. Lebensraum Landschaft* (pp. 29-34).

Cantwell, M., & Adams, C.W. (2003). An Aboriginal Planning Initiative: Sacred Knowledge and Landscape Suitability Analysis. *This volume*.

Claval, P. (2002). *The language of the rural landscape*. Presentation at 20th Session of the Permanent European Conference on the Study of the Rural Landscape, Tartu-Otepää, Estonia.

Coeterier, J.F. (2002). Lay people's evaluation of historic sites. *Landscape and Urban Planning, 59*, 111-123.

Colquhoun, M. (1997). An exploration into the use of Goethean science as a methodology for landscape assessment: the Pishwanton Project. *Agriculture, Ecosystems & Environment, 63*, 145-157.

Cosgrove, D. (1984). *Social formation and symbolic landscape*. Madison: University of Wisconsin Press.

Crang, M. (1998). *Cultural geography*. London: Routledge.

Daniels, S., & Cosgrove, D. (1988). *The iconography of the landscape: essays on the symbolic representation, design and use of the past environments*. Cambridge: Cambridge University Press.

Espersen, S. (1999). The Framer as a Steward of the Cultural Landscape. In G. Setten, T. Semb, & R. Torvik (Eds.), *Shaping the Land, Vol. 3. The future of the past*. (pp. 596-605). Proceedings of the Permanent European Conference for the Study of the Rural Landscape, 18th session in Röros and Trondheim, September 7 – 11, 1998.

Fry, G. (2000). The landscape character of Norway – landscape values today and tomorrow. In B. Pedroli (Ed.), *Landscape – Our Home. Lebensraum Landschaft* (pp. 93-100).

Fry, G.L.A., Jerpasen, G., Bakkestuen, V., Skar, B., & Erikstad, L. (2003). Locating cultural heritage sites in the landscape – a hierarchical approach based on landscape indicators. *Landscape and Urban Planning*. In press.

Geisler, C.C. (2000). Estates of mind: Culture's many paths to land. *Society & Natural Resources, 13*, 51-61.

Graham, B., Ashworth, G.J., & Tunbridge, J.E. (2000). *A Geography of Heritage. Power, Culture and Economy*. London: Arnold.

Hiiemäe, M. (2001). Maastik ja kohapärimus. In H. Palang, & H. Sooväli (Eds.), *Maastik: loodus ja kultuur. Maastikukäsitlusi Eestis* (pp. 86-95). Tartu: Publicationes Instituti Geographici Universitatis Tartuensis 91.

Ingold, T. (2000). *The Perception of the Environment*. London: Routledge.

Johansson, B.O.H. (1993). Mindscapes, cultural heritage and rural landscapes. *The future of rural landscapes. Bebyggelsehistorisk tidskrift 26*, 34-42.

Keskkonnaministeerium (KKM) (2001). *Väärtuslike maastike määratlemine. Metoodika ja kogemused Viljandi maakonnas.* Tallinn: Keskkonnaministeerium.

Lowenthal, D. (1985). *The past is a foreign country*. Cambridge: Cambridge University Press.

Maaranen, P. (2003). *Landscape archaeology and management of ancient cultural heritage sites: some notes based on Finnish experiences. This volume.*

Marcucci, D.J. (2000). Landscape history as a planning tool. *Landscape and Urban Planning, 49*, 67-81.

Meinig, D. (1979). *The interpretation of ordinary landscapes: geographical essays*. Oxford: Oxford University Press.

Muir, R. (2000). *The New Reading the Landscape: Fieldwork in Landscape History*. Exeter: University of Exeter Press.

Nassauer, J.I. (1997). *Placing nature. Culture and landscape ecology*. Island Press.

Norderhaug, A., Austad, I., & Hauge, L. (1996). Holistic studies of Nordic land use changes: a focus on the traditional western Norwegian farm as a biological and cultural system. In R.H.G. Jongman (Ed.), *Ecological and landscape consequences of landuse change in Europe* (pp. 130-142). Proceedings of the first ECNC seminar on land use change and its ecological consequences, 16 –18 February 1995, Tilburg, Netherlands.

O'Rourke, E. (2003). *Socio-natural interaction and landscape dynamics in the Burren, Ireland.* Presentation at 20th Session of the Permanent European Conference on the Study of the Rural Landscape, Tartu-Otepää, Estonia.

Palang, H., Alumäe, H., Printsmann, A., & Sepp, K. (2003). Landscape values and context in planning: an Estonian model. In J. Brandt, & H. Vejre (Eds.), *Multifunctional landscapes.* WIT Press.

Palang, H., Alumäe, H., & Mander, Ü. (2000). Holistic aspects in landscape development: a scenario approach. *Landscape and Urban Planning, 50,* 85-94.

Palang, H., & Mander, Ü. (2000). Maastiku muutused Eestis. In Frey, T. (Eds.) *Kaasaegse ökoloogia probleemid VIII: Loodusteaduslikud ülevaated Eesti Maa Päeval* (pp. 169-179). Tartu.

Peil, T. (2001). Maastike keskel. In H. Palang, & H. Sooväli (Eds.) *Maastik: loodus ja kultuur. Maastikukäsitlusi Eestis* (pp. 57-66). Tartu: Publicationes Instituti Geographici Universitatis Tartuensis 91.

Relph, E. (1976). *Place and placelessness*. London: Pion.

Renes, J. (1999). Evaluating historic landscapes. In G. Setten, T. Semb, & R. Torvik (Eds.) *Shaping the Land: Vol. 3. The future of the past.* (pp. 641-650). *Proceedings of the Permanent European Conference for the Study of the Rural Landscape*, 18th session in Røros and Trondheim, Norway, September 7th–11th 1998.

Saltzman, K. (1999). Ideals and practice in the care of landscape. A conflict and causes. In G. Setten, T. Semb, & R. Torvik (Eds.), *Shaping the Land: Vol. 3. The future of the past.* (pp. 663-675). *Proceedings of the Permanent European Conference for the Study of the Rural Landscape.* 18th session in Röros and Trondheim, Norway, September 7-11, 1998.

Schama, S. (1995). *Landscape and memory*. London: Harper Collins.

Stenseke, M. (1999). Landscape perspectives on landscape values. In G. Setten, T. Semb, & R. Torvik (Eds.) *Shaping the Land: Vol. 3. The future of the past* (pp. 543-555).

Proceedings of the Permanent European Conference for the Study of the Rural Landscape. 18th session in Røros and Trondheim, Norway, September 7-11, 1998.

Stenseke, M. (2000). Landscape values: the prospects for communicative planning in landscape management in Sweden. *In Reshaping of Rural Ecologies, Economies and Communities. Conference Proceedings, Commission on the Sustainability of Rural Systems.* International Geographical Union.

Widgren, M. (1993). Landscapes as museums – Landscapes as archives. *The future of rural landscapes. Bebyggelsehistorisk tidskrift 26*, 65-70.

Chapter 9

STAKEHOLDER LANDSCAPES AND GIS

Institutional Visions of Landscape and Sustainability in the Management of the Sherwood Natural Area, UK

Robert Fish, Roy Haines-Young & Jorge Rubiano
School of Geography, University of Nottingham, United Kingdom

1. INTRODUCTION

Writing over a generation ago in her celebrated work on the future of Britain's landscapes Nan Fairbrother makes the following comment:

> "Landscape is the battleground where land use works out its own salvation. Economic, social, transport and other problems cannot be solved in isolation but must first be combined and related in a single situation, and landscape is the melting point where all uses – each self intent and often conflicting – meet and mould each other and reach a realistic balance. It is why total environment is most usefully approached as landscape, which translates this wide range of abstracts (land use, population densities, settlement patterns, traffic flow, local site conditions and so on) into physical reality (it is also, for humans beings who are naturally visual, a visual statement of a complicated situation)" (Fairbrother 1970: 291).

Fairbrother's comment is prescient with regards to the evolving terms of the sustainability agenda. In expressing the view that seemingly different human concerns and priorities do not exist as mutually exclusive categories, but rather are shaped and influenced in relation to each other, she rehearses familiar premises of mainstream approaches to environmental management: on the one hand that of integrative thinking; on the other, the possibility of striking 'realistic balances'. More specifically, Fairbrother's *modus operandi* is one of landscapes, a framework for action that arguably captures a key

H. Palang and G. Fry (eds.), Landscape Interfaces, 147–161.

element of many contemporary formulations of managing for sustainability. As Paul Selmon has recently explained:

> "The concept of landscape is one which contributes to our ability to frame integrative policies: it fuses the patterns and processes of rivers, soils, rocks vegetation, animals and people, and thus acts as a basis for addressing complex sustainability issues. Landscape is thus starting to underpin our approach to the use of natural capital more generally, with planning and management units based more on bioregional units and less on administrative boundaries" (Selmon 2000: 99).

Though ideas of landscape are not reducible to the strictly visual, as Selmon himself notes, it is nonetheless fair to claim that the possibility of framing such integrative policies rests in part on the premise that, in moving beyond abstract, seemingly separate categories – such as 'economy', 'culture' and 'environment' – into the actual and material, a landscape perspective is irredeemingly also a visual perspective, simplifying and clarifying the complexities at stake in managing for sustainability: "a visual statement of a complicated situation" as Fairbrother was to put it. Thinking in landscape terms, then, is thought to be a useful and succinct way of making sense, and ultimately exploring together, some of the material outcomes of how we think and act towards the material world around us.

And yet, while attempts to marry together seemingly disparate concerns at landscape level have been an emerging feature of policy agendas for sustainability, both as a framework for planning and action, and a means by which progress on sustainability issues can be judged, monitored and reported, we would do well to heed the words of Fairbrother in describing landscape in the combative terms of a 'battleground'. Indeed, if a landscape context to management is a useful and arguably necessary way of grasping the interplay and ultimately integration of concerns too frequently constituted as separate, it is also a construct by which different groups express values, attitudes and terms of action in often competing ways. Put simply, the problem context for such an agenda is how to manage and overcome the many different visions of landscape people hold. In this chapter we wish to highlight some of the challenges raised in attempting to build consensus out of different visions of landscape and sustainability by drawing on the insights of a research project exploring issues of landscape ecology in the Sherwood Natural Area of Nottinghamshire, UK.

In particular, in this chapter we wish to explore the possibility of developing a decision support tool that can integrate and compare different institutional stakeholder visions of this area's character and potential. We begin this discussion by outlining a contemporary context to environmental management that emphasizes the need for wider stakeholder involvement within decision-making processes. We go on to highlight how this agenda of stakeholder governance, whilst challenging the exclusive authority of 'expert' knowledge within decision-making, can profit from new types of information handling tools and methodologies that characterize different stakeholder opinion. By introducing some of the landscape characteristics that define the Sherwood Natural Area, we go onto describe an approach that can potentially start to grasp the multiple views that different groups hold about this area. We show how this approach relies on a mixture of qualitative and quantitative data analysis taking its shape through the use of Geographical Information Systems (GIS). We conclude the discussion by arguing that, while such an approach is not an end in itself, it has an important place within the new frameworks of deliberation and consensus building currently being advocated by the stakeholder agenda.

2. THE STAKEHOLDER AGENDA IN ENVIRONMENTAL DECISION MAKING

We suggested above that the problem context for issues of sustainability and landscape is how to manage and overcome different visions of landscape. In one sense, this problem appears to be accentuated by the recent emphasizes of the policy making community more generally, since it is an article of faith within recent theorizations of, and approaches to, environmental management that decision making processes should actively engage with wider stakeholder knowledges (Burgess 2000; Glicken 2000). The background to this trajectory of argument has been explained in detail elsewhere (e.g. Irwin 1995), but broadly put, it rests on the idea that decision making processes have historically been predicated on the wisdom of science, which, disengaged from a situated contexts, tendered to steamroller over wider stakeholder understandings of environmental concern, sometimes with undesirable and avoidable outcomes. The implication was that such stakeholders were deemed rather ignorant in relation to expert knowledge, and therefore needed to be educated into the ways and logics of scientific reasoning. Hence, as Irwin (1995) notes in terms of the UK, the historical prevalence of programmes emphasizing the 'public understanding of science'.

However, as this argument for stakeholder governance has been gradually articulated, so too have the pragmatics of handling difference begun to emerge. Engaging with wider sources of judgement, meaning and valuation has been understruck by methodologies emphasising ideas of 'deliberation' and 'consensus building' (Bloomfield et al. 2001). In this sense, the stakeholder agenda within environmental management is actively seeking to overcome the complexities it appears at first to accentuate. It develops the idea that multiple and competing demands surrounding management issues can actually be overcome. It argues that a more productive and creative interaction between different stakeholder groups – one in which priorities and strategies are socially determined and agreed upon in a much broader sense – can emerge.

In landscape terms, the implication of this stakeholder agenda is to effectively reverse Fairbrother's proposition of a 'battleground' landscape. Expressed in terms of deliberation, inclusion and participation, decision-making processes are the site of more, rather than less, consensual landscape forms (Burgess 2000). One of the outcomes of this methodological emphasis has been to bring into sharper focus the need to develop ways of representing possible and alternative futures and their consequences for sustainability. As a result, though the stakeholder agenda appears, in one sense, to downgrade the privileged status of expert knowledge in environmental decision-making (i.e. it is merely now a 'stake' to be deliberated), recent advances in visioning techniques become one way that the goals of this agenda may actually be realized. In particular, in this discussion we wish to argue that in GIS we see the possibility for developing decision support tools that can provide a useful input into methodologies where different individuals and groups articulate and discuss visions and possible scenarios for landscape management.

GIS have been seen as key decision making tools for environmental planning (e.g. Kangas et al. 2000; Appleton et al. 2002; Hakley 2003) though have been thought by some to be complicit with the very processes the stakeholder agenda is attempting to critique: that is promoting the views of the expert and excluding key stakeholders and their views of what constitutes relevant knowledge (see Craig et al. 2002). In this discussion we wish to demonstrate the value of GIS in representing different and sometimes opposing visions that stakeholders might hold about the world. This trajectory of work emphasizes the role of systematic methodologies in creating a platform for more deliberative, discursive decision making within stakeholder governance. In the remainder of this discussion we outline how such a platform was created in the context of our case study area. By way of an introduction to this process, consider some background information on the location in question and how we came to focus upon it.

3. THE CASE STUDY CONTEXT: THE SHERWOOD NATURAL AREA

Figure 9-1. The Sherwood Natural Area

Our study takes as one of its analytical starting points the idea of the 'Sherwood Natural Area', an area located in the north west of the county of Nottinghamshire UK and one of 120 similar such designations comprising the Natural Areas framework devised by the quasi-governmental organization English Nature in the mid 1990s (see Figure 1). The natural area framework is thought to differentiate the country into characteristic landscapes comprising a mix of wildlife and natural features, uses of land and wider symbols of cultural history. As an example of one such designation, the Sherwood area is an interesting example of how cultural meanings do not square well with material realities. While a key element of its cultural meaning is focused around the idea of 'Sherwood Forest', driven in a significant sense by its association with Robin Hood Mythology, in reality, the area of remaining mature oak woodland and associated heath communities is small. Recent estimates suggest these components together comprise approximately 7% of a landscape dominated by farmland (50%), coniferous and broad-leaved plantations (16%) and urban areas (13%). Other important symbols of cultural heritage such as coalmines are small (3%) and since the late 1980s most are now derelict (NCC 1992). Faced with such a complex picture, some people say we should try and recreate a forested landscape in Sherwood, but others living in the area are less interested. Even amongst those who agree, there is disagreement about how it should be done and what form the forest should take. Should the planting be mainly directed towards sustaining or enhancing biodiversity? Alternatively, should it be directed mainly towards supporting recreation and visual amenity, an important element underpinning the economic viability of the area.

4. METHODOLOGY FOR ASSESSING STAKEHOLDER VISIONS IN THE SHERWOOD NATURAL AREA

In a recent discussion of envisioning the future Costanza (2000) laments that: "recent work with business and communities indicates that creating a shared vision is the most effective engine for change in the desired direction, yet most effort in 'futures modeling' has focused on extrapolating past trends rather than envisioning alternative futures". Writers such as Pinter et al. (2000) suggest that thinking about the future may reduce the risk of unpleasant surprises and broaden the perceptions of both the general public and policy makers in at least four areas: (i) consequence assessment – assessing the implications of present actions, decisions, policies; (ii) early warning and guidance – detecting and avoiding problems before they occur; (iii) proactive strategy formulation – considering present implications of possible future events; and finally (iv) normative scenarios – envisioning aspects of possible and alternative futures. The work we describe here is one methodological component by which each of these aspects can begin to be broached. In particular, we outline a methodological process that sought to define and compare institutional stakeholder visions of the Sherwood natural area through a GIS, and one that can be then be validated by the stakeholders concerned. We shall explain in the conclusion how we consider this to be a necessary first step in realizing the potential of a much broader deliberative process among stakeholders. It comprised four key steps.

First, the identification of key institutional stakeholders who are formulating policy visions for the area, and the representation of these visions in spatial terms. This stage involved interviews with representatives in which participants identified locations of ideal places or preferred sites that corresponded with their visions, together with the landscape variables that defined them. Second, the data collected were used as an input for the construction of geographical models of stakeholder 'vision spaces', that is, areas where it is highly probable to find places with similar landscape characteristics to the initial points identified by the stakeholders. These individual vision spaces were then overlayed to highlight the different extents to which stakeholder visions coincided. Fourth and finally, the composition of the landscape embedded inside these individual and shared vision spaces were examined.

5. LANDSCAPE GOALS AND PREFERENCES OF INSTITUTIONAL STAKEHOLDERS

As explained above, the first substantive stage of this research involved an interview analysis of the different goals, assumptions and processes underpinning institutional stakeholder visions of the Sherwood Natural Area, particularly as these related to identifying particular locations for these visions and their associated landscape characteristics. In total, seven organizations of a public or quasi-public status involved in partnerships specifically devoted to the management of Sherwood were considered by the study: Basetlaw District Council (BDC); English Nature (EN); the National Farmers Union (NFU); Newark and Sherwood Forest District Council (NSDC); Nottinghamshire County Council (NCC); Nottinghamshire Wildlife Trust (NWT) and Sherwood Forest Trust (SFT). This sample was by no means exhaustive of institutional interests impinging upon the future of Sherwood, but for the purposes of our work, represented an indicative mix of policy scales – national, regional and local – that were bringing to bear ostensibly quite visions of management upon the area. The results we describe here also include an eighth insight; those derived from a pilot interview with an academic from the University of Nottingham (UNOTT), whose expertise lies in the field of landscape ecology and whose knowledge of the study area was extensive.

The interview process followed a similar structure in all cases. The project was initially introduced, and then from this basis, participants were shown a poster of the area annotated variously with: aerial photography of different landscape types; labels of key places, institutional boundaries such as those of Sherwood Natural Area devised by English Nature, including within this SSSI designations1, and boundaries of wards. It was against this context that respondents were asked to highlight on the map those locations that, for their organization, represented or fulfilled their ideal landscape conditions for the Sherwood area. This process of identifying places as expressions of stakeholder preferences was easily understood by the respondents, although re-working institutional visions into an explicitly spatial form was also felt to be a challenging exercise. Respondents were

[1] 'Sites of Special Scientific Interest' are an English Nature designation for landscapes considered to be of significant wildlife and/or geological interest.

then asked to justify their choice of locations (Table 1). A common feature of these justifications was reference to the presence of heathland, combined with a strong concern for mosaics of ancient broadleaved woodland and acid grassland, as well landscapes where there was perceived to be a notable wildlife interest. Other explicit references to landscape features included justifications based on the presence of broadleaved and coniferous plantations, wetlands and meadows, and signifiers of recent industrial history, such as evidence of coal mining and gravel extraction. Selections were also based on more contextual landscape processes or qualities such as issues of accessibility and the recreational functions of particular locations, as well as landscape designations, such as SSSIs, and symbols of historical ownership, such as the Estates of the Dukeries. Such justifications, however, were not mutually exclusive: concerns for locations with SSSI status underpinned concerns for locations with a special wildlife interest; symbols of history such as the Estate of Dukeries were seen as complementary to maintaining characteristic landscape features, such as the remaining ancient woodland; while the presence of farming was closely allied with the creation of employment.

The important issue to note here is that such justifications were the basis from which more comprehensive spatial visions of the area for individual stakeholders could be produced and then evaluated in relation to each other. In particular, the initial starting point for this work involved generating a geographical information database covering each of these variables and then processing them as geographical layers for the natural area more broadly. The creation of the database was carried out in several stages. First, a preliminary search for geographical information was made in relevant organizations of Great Britain and Nottinghamshire. A key source of information consulted in this respect was the 1992 Nottinghamshire Land Cover Map, which encompasses land cover classes directly comparable with the themes at work in the stakeholder visions, such as ‘acid grass and heathland’, ‘coniferous’ and ‘broadleaved plantations’, ‘farming’, and ‘inland water’. Other sources included: the Ordnance Survey, where information covering accessibility networks of roads and recreational sites could be determined; the British Geological Survey, for identifying river corridors; the 1991 census for England for employment data; and English Nature for the location of designations such as SSSI’s. Such data came in a variety of formats and varied in extent, quality and resolution. The second task therefore consisted of standardizing and converting files into a common format. ArcView 3.2a and ArcInfo were interactively used to convert everything into arc, shape and grid formats. Layers of GIS data that best corresponded to each of the reasons outlined in Table 1 were then allocated to each stakeholder institution.

Table 9-1. Key reasons for selected sites by institutional stakeholders

Key reasons for choice of selected sites	NCC	UNOTT	SFT	BDC	NFU	NWT	NSDC	EN
Accessibility			•					•
Ancient broadleaved woodland	•	•	•	•		•		•
Broadleaved Plantations			•	•	•	•		•
Coal mine sites & gravel extraction		•				•	•	
Coniferous plantations	•			•		•		
Dense forest	•				•			
Estates of the Dukeries	•	•						•
Farming					•			•
Good grazing management			•					•
Golf courses			•	•				•
Heathland and acid grassland	•	•	•	•	•	•	•	•
Historic gardens				•		•	•	
Lakes							•	
Parkland						•		
Provision of employment					•			
Provision of housing	•							
Recreational function			•	•	•	•	•	•
River Corridors				•		•	•	
SSSIs designations						•	•	•
Wetland and meadows		•			•	•		•
Wildlife interest	•	•	•			•	•	•

The outcome of this stage was to generate a series of stakeholder vision spaces, a series of maps of the Sherwood Natural Area where it is highly probable to find landscapes similar to those expressed in the interviews. Significantly for the analysis, such individual vision spaces were then overlaid to identify the locations where stakeholder visions come together to varying degrees. These levels of agreement between the stakeholder visions space are depicted in Figure 2. At one extreme, in the areas depicted in white, the map displays locations in Sherwood which are essentially 'anti-visions': that is, zones which fail to have congruence with a single stakeholder vision of Sherwood. The progressively darker zones represent different extents to which stakeholders visions are at work in the area: 'Agreement level 1', for instance, denotes the presence of a single stakeholder vision, while the darkest area, 'agreement level 8', reflect areas where all eight stakeholder visions are at work.

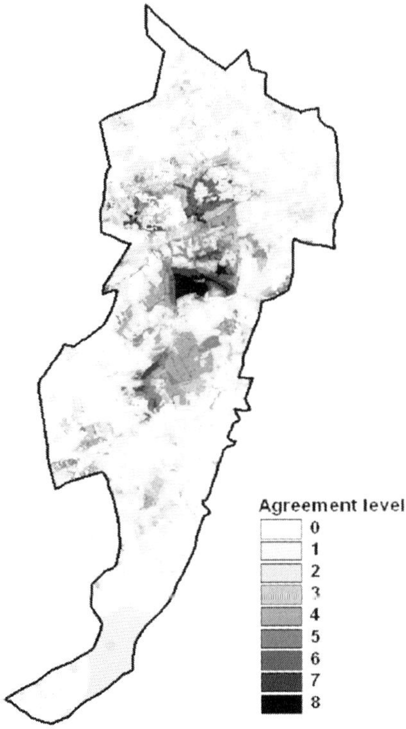

Figure 9-2. Levels of agreement between the eight stakeholder vision spaces

6. LANDSCAPE PATTERN ANALYSIS OF STAKEHOLDER VISIONS

It was against this context that a compositional analysis of the landscapes was carried out. By landscape composition we refer to the types of land cover found in the areas delimited by the boundaries of each vision space and agreement level. This analysis was carried out using the NCC land cover classes noted in section 5.

In terms of landscape composition some interesting similarities and differences are found between stakeholder visions. In Figure 3, for instance, we depict the relative landscape land cover inside each vision space plus the proportion of land cover types for the whole extent of Sherwood Natural Area. What is initially striking here is the congruence between the vision spaces of the Sherwood Forest Trust and the National Farming Union where coniferous and broadleaved plantations and farming comprise more than 60% of the total land cover. So too are there similarities between the visions of the University of Nottingham, Basetlaw District Council and Nottinghamshire County Council where acid grass and heathland, and coniferous and broadleaved plantations again constitute over 60% of the total land cover of the vision spaces. The remaining stakeholders, however, were quite dissimilar. English Nature had the highest proportion in heathland and broadleaved plantations at approximately 70%. A more balanced composition, but one with a large proportion of broadleaved plantations still, was found in the case of the Nottinghamshire Wildlife Trust, which was also second among stakeholders' visions in the area given over to Heathland. The most balanced distribution of landcover types assigned to a particular stakeholder vision was found in the case of the Newark and Sherwood Forest District Council.

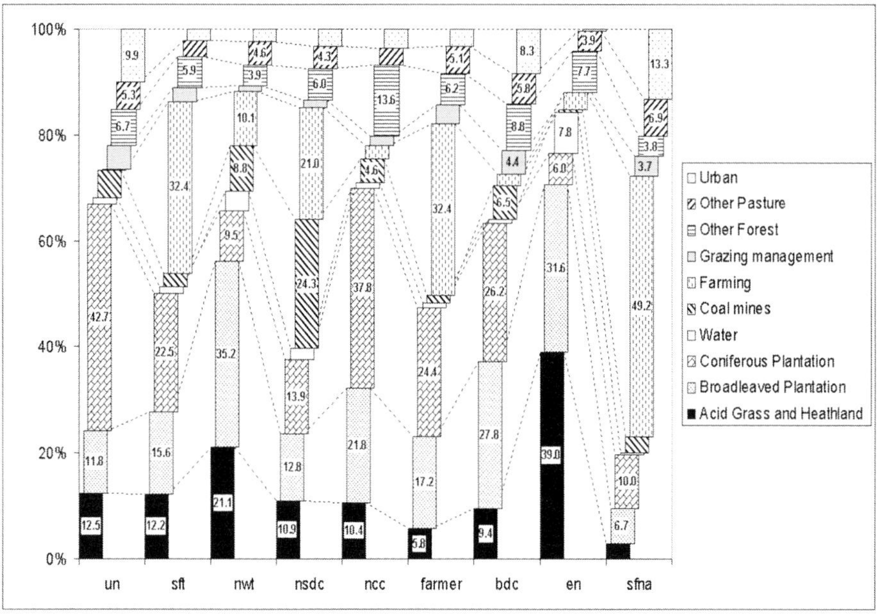

Figure 9-3. Relative land cover composition inside the eight vision spaces

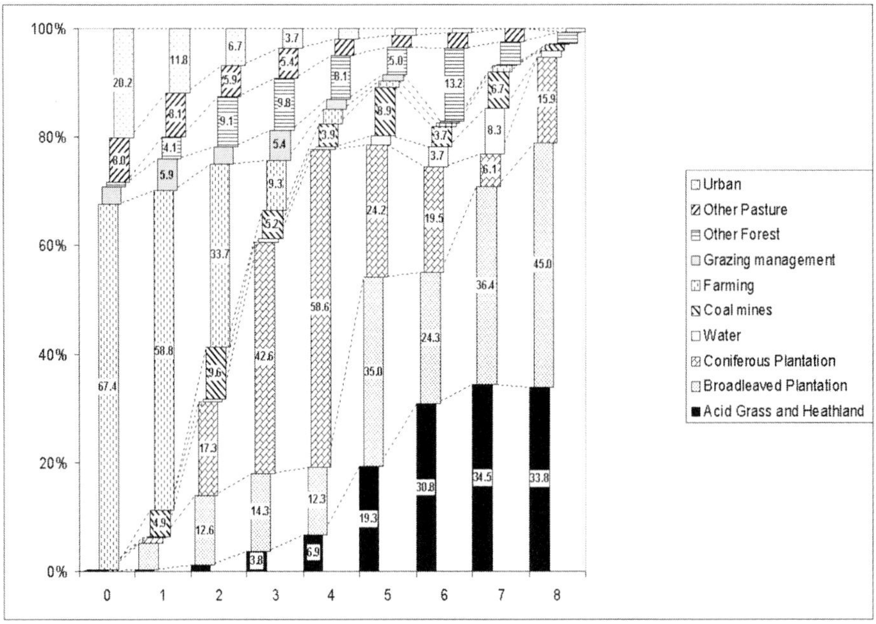

Figure 9-4. Land cover composition in different agreement zones

Figure 4 shows a more complex analysis of the stakeholder visions. It represents the proportions of land cover types among different degrees of agreement among stakeholder visions, depicted above in Figure 2. The proportion of land cover types displayed in column '0' represents the landscape composition of areas outside of all stakeholder vision spaces. These areas were predominantly represented by the land classes of farming (67.4%) urban (20.2%) and pasture areas (8%). What is interesting to add here is way these proportions are actually rather similar to the area proportions for the Sherwood Area as a whole. That is to say, the proportions of land cover that make up Sherwood are not, paradoxically, found in any of the individual stakeholder vision spaces. It is in column '8', however, where we find a landscape composition upon which all stakeholder visions come together, and it is principally composed of broadleaved plantation (45%), heathland (33%) and coniferous (15.9%) land classes. In the intermediate columns the proportion of land cover types fluctuated. When approaching total agreement, the trend is of an increase in heathland; broadleaved plantations and water cover types. There is also a decrease in farming, urban and pasture areas. Coniferous plantations and coal mine sites had a low proportion towards both extremes, but were relatively high towards the middle.

7. CONCLUSION – A SCIENCE OF STAKEHOLDERS?

It was noted in the introductionary sections to this chapter how the terms of the environmental agenda are being gradually recast around the idea of 'stakeholder governance'. This agenda has made a strong case for the active inclusion of wider social knowledges within environmental decision making processes, ones historically downgraded *vis a vis* the expert. However, we wish to argue in light of the case study, that such an agenda must be sensitive to the empowering role that aspects of scientific thinking and approach, such as GIS, can bring to discussions of environmental management. In particular, we wish to argue that ideas of stakeholder participation do not preclude science. Rather, they change the terms on which science participates. In one sense, as we argued earlier, science now simply becomes a stakeholder in this participatory sphere. That is to say, scientific knowledge becomes one line of reasoning in a process that emphasizes deliberation between different stakeholder knowledges. In terms of the discussion in this chapter, for instance, we might highlight the role of English Nature – a quasi-government scientific organization – in defining a particular vision of the Sherwood Natural Area as indicative of this role. We might succinctly describe this role for scientific knowledge using the label 'science *as* stakeholder' – a discrete and clearly 'interested' knowledge within the wider stakeholder sphere.

Alternatively, we might consider science as a source of information and methodologies that different stakeholder groups choose to draw upon for articulating and clarifying their visions. Science in this sense arguably takes on qualitatively different roles. It becomes the means by which stakeholders express particular worldviews and submit them to the wider deliberative sphere. That is to say, GIS is part of a process by which particular management issue is envisioned. In contradistinction to the idea of 'science *as* stakeholder', we might describe such a role for science with the idea of a 'science *of* stakeholders'; a platform for stakeholder deliberation, that seeks to facilitate the process of integrating, comparing and contrasting stakeholder visions. The example we develop here – that of the Sherwood Natural Areas – clearly falls in to this category, and has, we would argue, a logic with wider purchase on work attempting to genuinely 'spatialize' the dilemmas at stake in imagining and managing future landscapes. Indeed, there is nothing peculiar about Sherwood in being a site for different and contested visions of landscape, and nothing peculiar about Sherwood in being amenable to the science of stakeholders we have envisaged here.

Such a science is not the final word on the decision making process. It is a series of inputs that allow this process to learn more about a particular management problem, and perhaps even transform the very problems that it believes it is dealing with. It is part of a trajectory of work that is best described as a 'soft systems' approach: one that recognizes that problem solving is essentially an iterative process; that we discover more about a problem as we attempt to solve it; that problems are continually refined as we discover more about the issues and complexities at stake.

One of the implications of this argument is that a science of Sherwood stakeholders is only meaningful insofar as it should be validated, contradicted and potentially altered by the very institutional stakeholders who provided the terms on which it was defined: how do the NFU react to the representations of the vision? how do they react to the views of other institutional visions? what are the implications or trade-offs of pursuing a EN vision at the expense of others?; how can we rework stakeholder visions in light of their stakeholder reactions? How to these visions stand in relation to wider lay knowledges?; and so forth. The particular science we are proposing, in short, is one designed to bring into sharper focus the nature and uncertainties of stakeholder visions. The particular GIS insight we provide here must therefore be seen as a 'first draft' of a deliberative process by which different social constituencies, institutional or otherwise, can begin to examine and compare visions in an iterative and perhaps ultimately consensual fashion.

ACKNOWLEDGEMENTS

We would like to extend our thanks to all the individuals and institutions that participated in this research, and to the UK Forestry Commission for helping to fund this research.

REFERENCES

Appleton, K., Lovett, A., Sunnenberg, & Dcokerty, T. (2002). Rural Landscape Visualisation from GIS databases: a comparison of approaches, options and problems. *Computers, Environment and Urban Systems 26*, 141-162.

Bloomfield, D, Collins, F., Fry, C., & Munton, R. (2001). Deliberation and inclusion: vehicles for increasing trust in UK public governance? *Environment and Planning C 19*, 501-513.

Burgess, J. (2000). Situating knowledges, sharing values and reaching collective decisions. In I. Cook, D. Crouch, Naylor, & J. Ryan (Eds.) *Cultural turns, geographical worlds* (pp. 273-87). London: Prentice Hall.

Costanza, R. (2000). Visions of alternative (unpredictable) futures and their use in policy analysis. *Conservation Ecology, 4* (1), 5. http://www.consecol.org/vol4/iss1

Craig, W., Harris, T., & Winer, D. (Eds.). (2002). *Community participation and Geographic Information Systems, 22*, 84-89.

Hakley, M. (2003). Public access to environmental information: past, present and future. *Computers, Environment and Urban Systems, 27,* 163-180.

Kangas, J., Store, R., Leskinen P., & Mehtätalo, L. (2000). Improving the quality of landscape ecological forest planning by utilising advanced decision-support tools. *Forest Ecology and Management, 132*, 157-171.

Fairbrother, Nan. (1970). *New lives, new landscapes.* London: Architectural Press.

Glicken, J. (2000). Getting stakeholder participation 'right': a discussion of participatory processes and possible pitfalls. *Environmental Science & Policy, 3*, 305-310.

Irwin, A. (1995). *Citizen science: a study of people, expertise and sustainable development.* London: Routledge.

Pinter, K., Zahedi, K., & Cressman, D.R. (2000). *Capacity Building for Integrated Environmental Assessment and Reporting.* International Institute for Sustainable Development and The United Nations Environment Programme. http://www.iisd.ca/measure/iear.htm

Nottinghamshire County Council. (1992). *Land cover digital maps.* Nottingham: Nottinghamshire County Council.

Chapter 10

AN ABORIGINAL PLANNING INITIATIVE[1]

Sacred Knowledge and Landscape Suitability Analysis

Margot Cantwell & Chad W. Adams
Environmental Design and Management Ltd., Canada

1. INTRODUCTION

Located at the mouth of the Richibucto River in eastern New Brunswick, Big Cove is home to approximately 2,000 of Canada's First Nations peoples (Figure 1). The Mi'kmaq Nation inhabited the Richibucto River area long before the arrival of French missionaries and settlers in 1600. A close friendship of mutual respect formed between the Mi'kmaq and the Acadian people, the fate of these First Nations people changed dramatically following the British invasion of 1713. With the British came many peace treaties frequently adjusted according to British needs. Following the American Revolution, land grants were given to loyal British subjects with no regard to those who occupied this land, Mi'kmaq or Acadian. Threats from the First Nation peoples, verbal and physical, forced the British to grant the Mi'kmaq land for their exclusive use to maintain the peace treaties. In 1802, a land grant of 51,200 acres was given to the Mi'kmaq community (Augustine personal communication). The acreage of this reservation dwindled over the next century to 2,222 acres, as parcels were taken by the British to fulfil their own needs. Recent years have seen this property expand to approximately 5,300 acres, but it remains too small and marginal for the needs of the community.

[1] This case study is part of a soon to be published book on landscape suitability analysis. All rights remain with the publisher of that book.

H. Palang and G. Fry (eds.), Landscape Interfaces, 163–184.

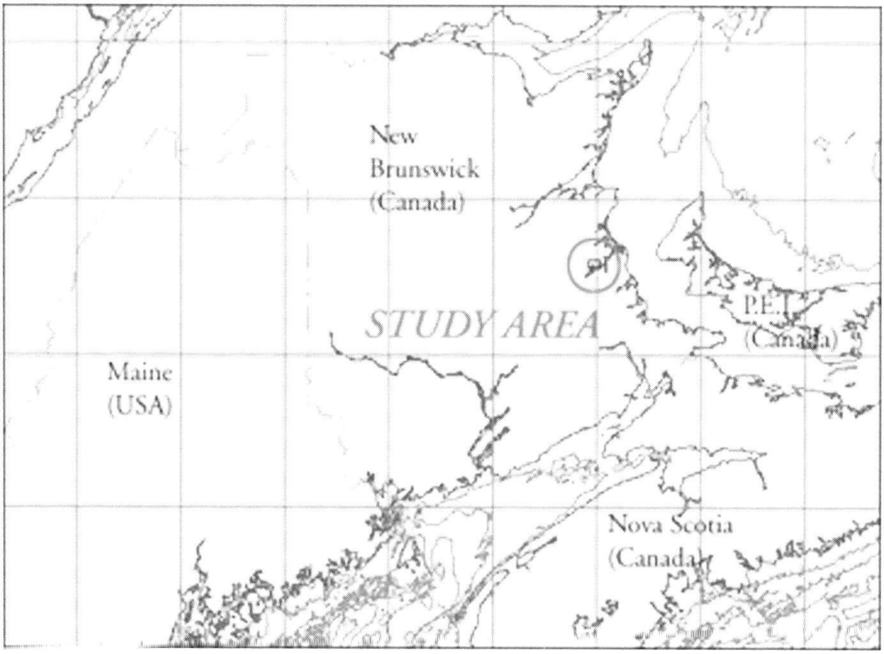

Figure 10-1. Key plan

The population of the Big Cove First Nations Community is very young, with over 60% of residents under the age of 30 years. These young people are at increased risk for substance abuse, depression, and suicide due to generations of economic hardship, lack of employment, and restrictions to traditional hunting and gathering practices. In 1999, the community suffered the loss of many young people to suicide, reflecting deep-rooted and long-standing social problems (Stephen Augustine, personal communication). Unemployment is also a significant problem for the Big Cove community, with rates as high as 75-90%, exacerbating the low self-esteem experienced by native youth. The Mi'kmaq leaders recognise the need for support, from the community and outside the reserve, to sustain their culture, provide a nurturing community, and maintain a meaningful and valued role in Canadian society.

During a Band Council meeting in 1999, a resolution was passed expressing a commitment to prepare a comprehensive community plan that would address 'development in an integrated and coordinated' way (Stephen Augustine, personal communication). This initiative was supported and sponsored by the Canadian Executive Service Organization (CESO).

The most pressing material need for the community is new housing to provide space for young people, separated from their parents, to help develop individual identity and boost self esteem. More than 200 units are required. Land space and quality restrict expansion possibilities. Since the community is largely surrounded by low, wet ground, construction of housing has been expensive and unplanned, often resulting in flooded basements. In addition, the community honours many areas of sacred value, demanding respect for these spaces and forbidding housing development in these areas.

The purpose of this project was to identify the best locations in the landscape to build new homes and services. Environmental Design and Management Limited (EDM) of Halifax, Nova Scotia, Canada, was hired by the community to assess the residential suitability of their land. Cultural valuation of the land was one of the primary goals of the community planning process. A secondary objective was to compare the cultural data supplied by the community to the science-based results of the landscape suitability analysis method.

The First Nations Peoples have traditionally used a paradigm of observation and interaction to understand the environment. This knowledge has been orally passed down through generations. In contrast, the scientific community gains knowledge by developing and quantifiably testing measurements conducted in a small system that are then generalised to larger systems. This knowledge is passed down through the written word. The comparison of the results of these two methods would help translate the First Nations' view of land use to a scientific base, readily understood by the government and public agencies. These groups have the greatest potential impact on the community.

2. PROJECT METHOD

The landscape evaluation incorporated a suitability analysis method. This method employs sophisticated Geographic Information Systems (GIS) assessment, based on models of construction cost, road suitability, erosion, water quality protection, residential development, on-site servicing, central servicing, visibility, species richness, and habitat protection. Each of these models requires digital data inputs and substantive fieldwork to create unique map layers and calibrate results. The EDM suitability analysis approach appealed to the steering committee for the planning project since it could incorporate maps of places of cultural significance (spiritual, traditional, or medicinal plant value). Furthermore, a map of culturally important areas could be used as a model input without disclosure of the specific locations or meanings to non-community members.

Background and community information, including history of the community, existing land use, population and housing characteristics, community servicing requirements and cultural heritage issues was gathered. The data were assessed according to five themes, including: land use goals and objectives, a biophysical and community inventory, cultural landscape values, and landscape suitability analysis and a resultant housing opportunity plan.

2.1 Land use goals and objectives

Steering committee and community members were interviewed to identify issues and establish goals and objectives for community land uses. A review of demographic characteristics and servicing needs was conducted. Constructed around a central core including the band hall, church, arena, playground and health care facility, the residential areas consist primarily of Canada Mortgage and Housing Corporation (CMHC)-style bungalows set on small lots of 40 or 50-foot width. Newer housing is set on larger lots (100 feet wide or greater), an arrangement much preferred by residents interviewed. Several issues were identified as being important to the community in terms of housing and service development. These included protection of culturally significant landscapes and water quality as well as dry basements, low building cost, cost-effective infrastructure, active recreation areas, and space for large footprint buildings.

2.2 Biophysical and community inventory

Natural Resources Canada, Agriculture Canada, and New Brunswick Natural Resources provided much of the information concerning physical aspects of the landscape. The land owned by the Mi'kmaq Nation covers approximately 2,200 ha on the shore of the Richibucto River in eastern New Brunswick (Figure 2). The topography is generally flat, with most of the land undeveloped and forested. The community is situated along the north shore of the river, covering approximately 500 acres. With a mean annual temperature of 5°C, the region's monthly average temperatures range from –9°C in January to 19°C in July. The predominant soils in the area are poorly drained and less than 1-2 m in depth, covering sandstone and siltstone bedrock. Aside from the Richibucto River, the dominant surface water feature, no other significant ponds or lakes exist. From the river, the land rises gradually toward the northwest, producing a southeast aspect to the community centre. The only significant relief occurs along the banks of the Richibucto River and the small tributaries that are incised into the land due to the eroded sandstone substrate.

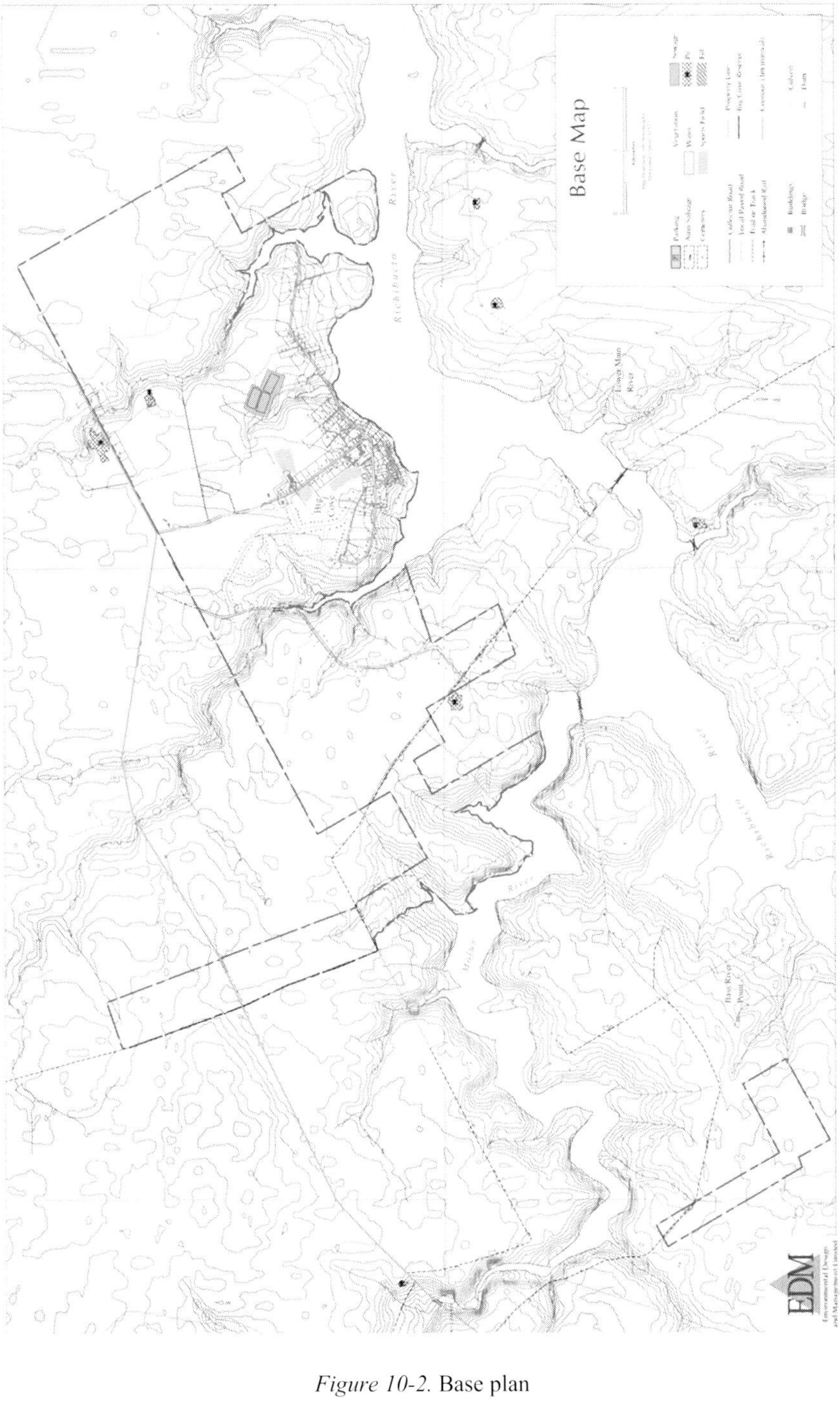

Figure 10-2. Base plan

Mixed hardwood and softwood species make up the majority of forest stands, although some small pure stands of hardwoods, pine, and cedar occur. The dominant hardwood species are trembling aspen (*Populus tremuloides*) and large-toothed aspen (*P. grandidentata*), while balsam fir (*Abies balsamea*) and red spruce (*Picea rubens*) are the most abundant softwood species. Since aspens are a relatively short-lived species, the dominance of this species indicates a relatively young forest that has been modified by forest fires and logging activities. The abundance of balsam fir in the understory indicates that the aspen will eventually be replaced by this species as the forest matures. Other species, such as red maple (*Acer rubrum*), black spruce (*Picea mariana*), eastern white pine (*Pinus strobes*), and grey birch (*Betula populifolia*), occurred widely but in low density. The age structure of the trees varied significantly from less than 10 to more than 80 years old. More than 155 species of plants were identified during a vegetation survey. Of these plants, 50 species are known to have medicinal value by the people of this community.

2.3 Cultural landscape values

Three different approaches were taken to identify landscape areas that require protection. First, workshops were held with elders and children of the community to capture past and future visions of the ideal First Nations community. Elders were asked to draw their world as they remembered it at five years of age, while the children drew their vision of the future. A distillation of values and visions was created from the verbal descriptions of these drawings. It included such ideals as 'respect for all', a 'strong sense of community', family and elder strength, preservation of land and environment and a 'sense of identity'.

Community members with expert knowledge of the land did field mapping of significant cultural areas. These maps included not only the significant area itself, but also included the appropriate buffer zone necessary for protection of its integrity and function. Base maps were provided to the community members, but no attempt was made to influence the definition of important areas.

Stephen Augustine, an anthropologist and native of Big Cove, was consulted to further understand the role of sacred knowledge. Mr. Augustine shared the story of Glooscap, as most of the rituals and landscape beliefs of the Mi'kmaq are described in this story of creation. The belief system of the Mi'kmaq people dictates that they may take only what they need from the landscape, and that they are: "equal to, not in dominion over, other flora and fauna". Culturally significant landscapes are not viewed as places of the past, but as alive and sacred.

Based on the information outlined above, three suitability models were developed to describe cultural landscape values within a suitability context. These models were reviewed and confirmed by the community in terms of accuracy and importance. River visibility was deemed an important criterion for this project, as the Richibucto River is a central connection to life in their cultural belief. The tradition of the people is to live near the river's edge, therefore proximity and visibility of the river is essential. Water quality protection is also essential in the planning process since water is considered sacred, 'the blood that connects all life'. Finally, landscape richness and the protection of all flora and fauna were deemed essential.

2.4 Landscape suitability analysis

Landscape suitability analysis is a method of finding the best areas in the landscape for a given use. Layers of input information are modelled or ranked to produce a map showing the most suitable locations. The input layers are composed of mapped information from various sources. These data, such as topography or forest inventory, are given values based on a ranked set of criteria or an equation.

A sample suitability model is presented in Figure 3, to illustrate the process. The sample is based on the water quality protection model written by EDM, and published by Environment Canada, 1996 (joint copyright, EDM Environmental Design and Management Ltd. and the Government of Newfoundland and Labrador). In suitability analysis, the values assigned to the data sets are derived by one of two methods. One method is physical modelling. Because the analysis is completed on a computer, physical, scientific and engineering values for each polygon can be easily calculated using the appropriate formulae. In the figure, the symbol *f* indicates where an equation is used to combine and/or value the data layers. For example, the function that precedes the erosion map is the Universal Soil Loss Equation (USLE). The second method is rule-based analysis. This method involves the development of a set of rules that facilitate locating the most suitable areas. In the figure, arrows without the *f* symbol indicate ranked criteria. For example, the K-value input map has values derived by ranking the soils map according to soil erodibility. The result of the analysis is a new map layer, called a suitability map, with all landscape areas valued according to the formula or ranking.

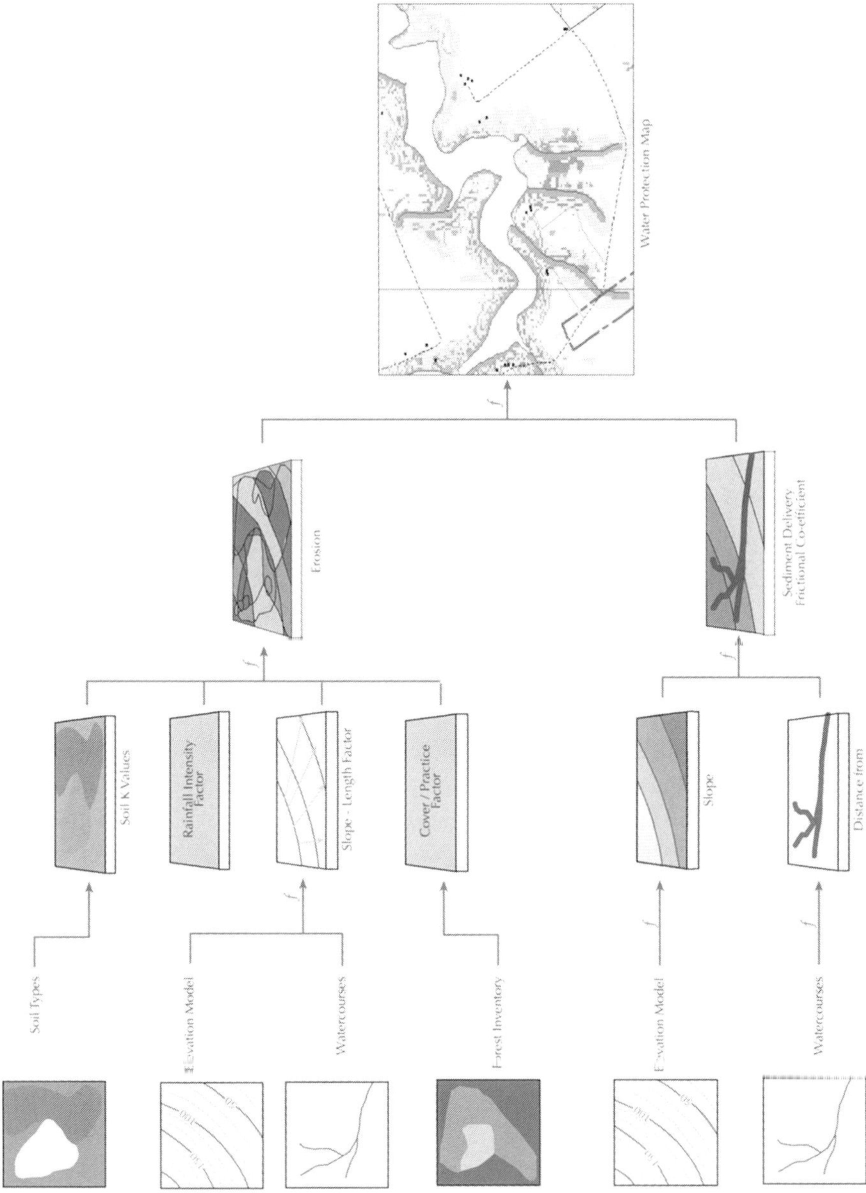

Figure 10-3. Sample suitability analysis – water quality protection

Each suitability map is created in a similar manner. Individual suitability maps do not integrate considerations of landscape and community preferences. Instead, each suitability map provides a clear view of the landscape from the perspective of one use or point of view. Therefore, each suitability map can also represent the effect of individual policies on the landscape, including intended and unintended effects.

The input layers used in this study are composed of mapped information from various sources. Issues raised by community members were represented by different suitability models used as inputs for the analysis (Table 1).

Table 10-1. Input layers and model results

COMMUNITY DESIRE	SUITABILITY MODEL
Dry basements	Building construction cost
Low cost to build	Building construction cost
Cost effective infrastructure	Road suitability
Cost effective infrastructure	On-site servicing
Cost effective infrastructure	Central servicing
Provide active recreation areas	Large flat areas
Provide large footprint buildings	Large flat areas
Protect culturally significant landscapes	Culturally significant areas
Protect water quality	Water quality protection

For example, the suitability model of building construction cost focuses on the issues of dry basements and low-cost housing. Similarly, the goal of cost-effective infrastructure was dealt with by input models of road suitability, on-site sewer servicing, and central sewer servicing. Information gathered from the biophysical and community inventory was used to create different models. The information included property data and boundaries, soil mapping, topography data, forestry and vegetation data, geology, service infrastructure, habitat data, and culturally significant areas. The different models were used to produce raw input data layers. The maps resulting from these individual suitability analyses included (listed in order of priority identified by the community):

- culturally significant areas – areas rich in medicinal plants or that are otherwise culturally significant were mapped and ranked. Pure stands of hardwoods were also considered 'significant', as their rarity holds both cultural and ecological importance;
- river visibility – this model identifies the areas from where the river can be seen, a factor revealed as 'essential' through community consultation;
- water protection – this model represents the sensitivity of the land to development based upon protection of the hydrology of the region, with sedimentation considered the greatest threat. The erodibility of the landscape is modelled using the Universal Soil Loss Equation (USLE) and a further analysis of sediment contribution from the tributaries is included (Snell 1984). This model assumes that the vegetation has been removed from the land, simulating a built condition. New Brunswick's legally defined 30 m buffer is shown around water bodies. Furthermore,

it was recommended that a hydro-geologist be consulted about water supply protection, before planning;

- species richness – the species modeling was based upon the New Brunswick Forestry Database containing information on the number of vertebrate species that potentially live in a specific vegetation class. The model includes mammals, birds, reptiles and amphibians;
- construction cost – this cost model considers the drainage, slope, and erosion potential of the soil. The best and most cost-efficient building sites are those that are well drained and gently sloped (2-8%), with slow erosion;
- road suitability – also based on the slope and drainage of the land, the road suitability cost model assumes that cost-efficiency is maximised when roads are built on soils that are at least 'moderately-drained' and on slopes of 2-6%. Flat areas and slopes >12% increase cost significantly;
- servicing suitability – this model is based upon existing sewer pipes and lift stations. Land uphill from the existing serviced area is considered within the servicing radius. Suitability for on-site sewage systems was modelled simultaneously since the existing infrastructure is performing at full capacity. The model attempts to determine the best system for a particular area. For on-site system construction, a slope of >2% and an overburden depth (soil above bedrock) of >1.5 m is necessary for proper drainage and filtration;
- large flat areas – this model determines areas that could have potential for large-scale development, such as commercial sites. Size (>5 ha), slope (2-6%), and drainage (better than poor) are the critical variables required in this model.

The maps do not yet represent a plan of where to build, but instead, illustrate the possible problems associated with particular areas. The darkest areas indicate the areas rated higher for the subject of analysis. The lightest areas indicate the areas scoring higher for the subject of analysis.

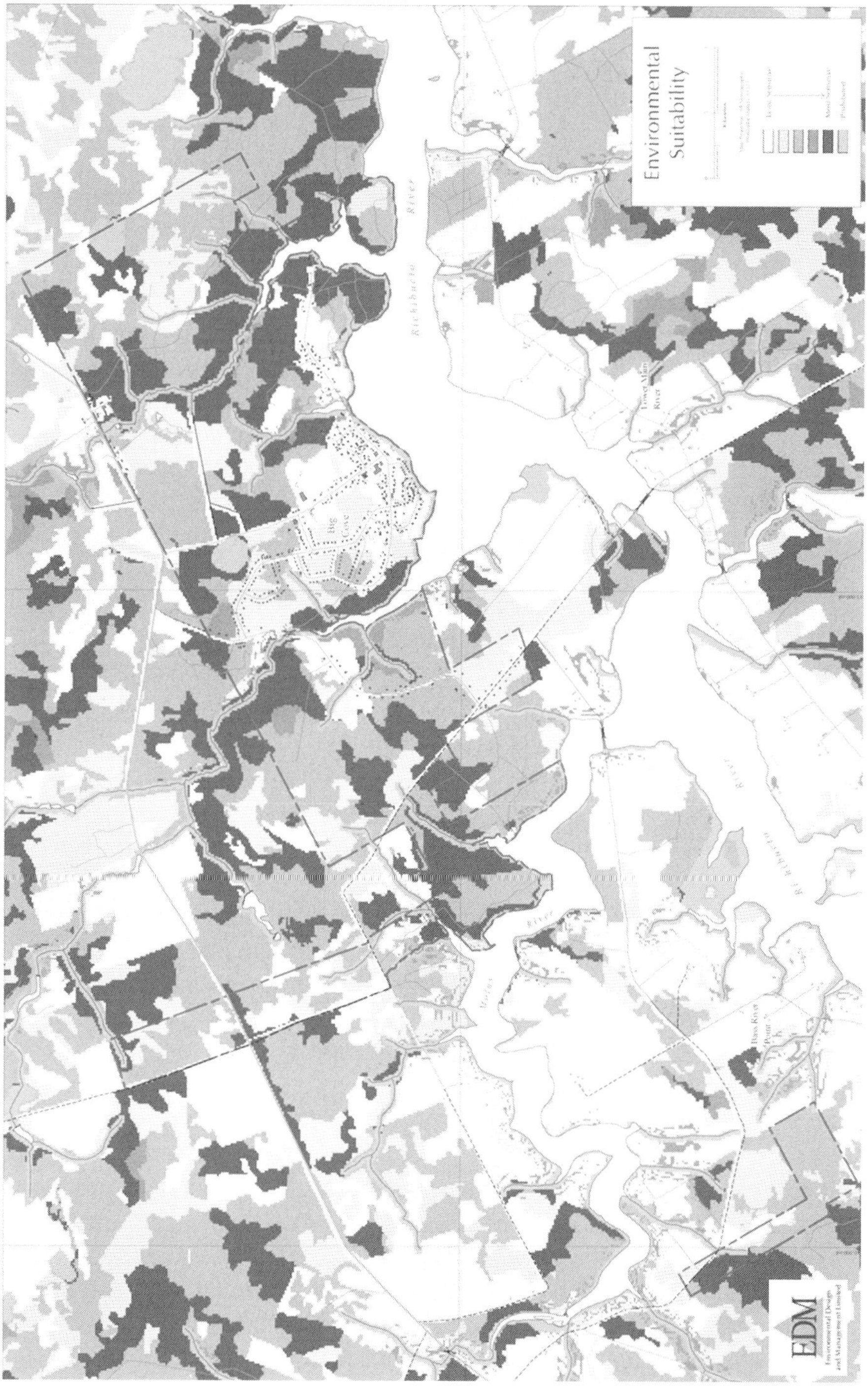

Figure 10-4. Environmental suitability

3. RESULTS

3.1 Residential suitability

A composite of the individual suitability maps was created based on values ranked by the community's steering committee members. Their decisions on the relative importance of the data layers were applied as weighting factors when the layers were added together. The results indicated a near complete lack of suitable construction sites. After further consultation with the community regarding the implications of ranking and compositing of individual suitability analyses, some compromises were reached. The most critical value remained 'culturally significant areas', with no compromise possible. River visibility, water protection, and species richness were identified as essential values, of nearly equal weight. Important values for consideration included building construction cost, road suitability, and servicing suitability, at near equal weights. Large flat areas were identified as being least important to the community.

Although river visibility was initially deemed an essential value in ranking of the criteria, it was not weighted as heavily in the final plan since it restricted the model to the point that nearly all of the available area was unsuitable. During presentations to the community members, it was decided that this value had to be compromised due to space limitations, an unfortunate impingement on the culture of the Mi'kmaq Nation.

The final residential suitability map revealed few areas within the community boundary suitable for building homes in a manner defined by community values. If building is to take place within the boundary, areas without a river view must be considered. Of the undeveloped land within the boundary, 80% is low, flat, and wet, characteristics suggestive of expensive building costs and poor housing conditions. While there are areas outside the boundary that may be suitable for building, these areas have not been extensively surveyed for culturally significant areas.

3.2 Landscape ecological open space plan

The value placed on the health of the 'landscape as a living system' prompted EDM to conduct a thorough analysis of the ecological functioning of this landscape. Combining the maps of culturally significant areas, species richness, and water protection created an environmental suitability map (Figure 4). This map provided a means of examining landscape sensitivity to residential development. Interpreting this map from a landscape ecological perspective led to the development of a Landscape ecological open space plan (Figure 5). The following landscape components were deemed necessary to preserve ecological integrity:

1. a minimum of 60% of the total landscape should be maintained as open space. Of this, 40-50% must be in large forested patches, 20-30% in small to medium forested patches, 20-30% in vegetated corridors and connections, and 5-15% within developed areas;
2. large undisturbed patches should include a diversity of landscape types, conditions, elevations (uplands/lowlands), and age classes;
3. essential recharge areas should be maintained;
4. vegetated corridor width should be at least 100 m to provide adequate interior habitat and species movement areas. There should be a diversity of widths and edge conditions;
5. the headwaters and highest order stream channels should be protected;
6. edge conditions should include a diversity of curvilinearity and grain size, and the shape of these edges should be maintained to encourage movement of 'good' flows in the landscape and to discourage 'bad' flows (i.e. invasive species);
7. essential (and rare) habitat types should be protected. At minimum, one of every existing or indigenous landscape type should be represented. In this landscape, mixed-wood stands and hardwood stands are particularly important for diversity;
8. the overall arrangement of this landscape should be one of connected diversity with some isolated patches and unique features. While the suitability analysis indicated a high incidence of these spatial features, ground-truthing and other analyses are necessary to complete the plan.

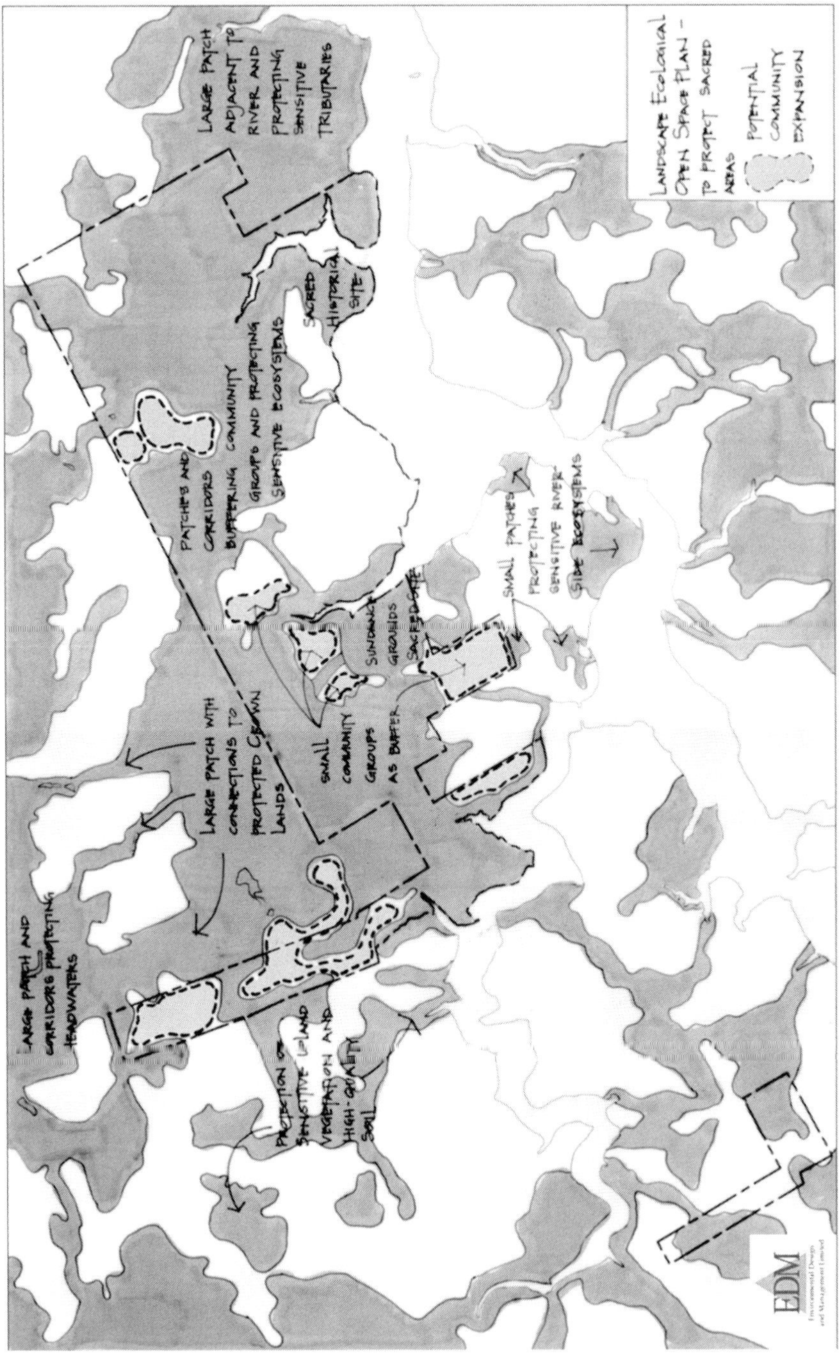

Figure 10-5. Landscape ecological open space plan

4. THE CORRELATION OF SCIENTIFIC AND CULTURAL ASSESSMENT

The scientific assessment relied heavily on the forest inventory database maintained by the provincial government in New Brunswick. This GIS-based inventory delineates common forest stands and provides descriptors of such parameters as land use, species composition by canopy layer, age structure, canopy closure, drainage, and soil type. Digital base mapping at a source scale of 1:10,000 that included hydrography and hypsography was used to describe drainage and slopes.

These inputs were assessed using standard engineering and scientific principles to define sensitivity to residential development. Areas with moderate to steep slopes on unstable soils, margins of watercourses, old growth forest, unique forest stands, and wetlands were identified as sensitive to development.

Areas having a high potential to impact water quality through sedimentation were assessed using the Universal Soil Loss Equation (USLE). This model incorporates soil characteristics, slope calculations, forest cover type, and distance from watercourses in the identification of areas that may contribute to degraded water quality.

The habitat quality for 157 species of wildlife was quantified using habitat models developed by the Provincial Department of Natural Resources. The study area contains suitable habitat for 74 of these species, and these 'potential habitats' were used to characterise vertebrate species richness. Species diversity often increases in disturbed systems where intolerant species are replaced by tolerant species. This also occurs naturally along ecotones of undisturbed systems. These factors would indicate that using species richness alone is insufficient to identify areas requiring protection from development, since it would be biased towards disturbed sites. However, the model used to generate species richness is based on vertebrate species that occur in normal successional stages of forest growth only (including regeneration on clearcuts and burned areas). For each species, the model identified habitat requirements including scale of interaction (home range size and body size), microhabitat, forest cover type, and forest successional stage. Thus, the models generally exclude species found exclusively in disturbed sites, with the exception of clearcut and burn areas.

An analysis of the final maps determined that clearcuts, fields, and other non-forested areas had the lowest species richness, even though forest species use non-forested habitats for a portion of their range. Based on this assessment, a high proportion of vertebrate forest species in a particular area is an appropriate indicator of protection requirements since removal of the forest would result in habitat loss for these species. Finally, the map of water quality protection was combined with species richness results to define areas of environmental suitability. This environmental suitability map was used in the comparison of scientific and cultural data.

To access the cultural data, representatives from the local community were responsible for collecting information regarding culturally important resources and areas. Interviews were conducted and oral history gathered from the community to identify these areas on field maps. Big Cove's cultural knowledge has been held and passed down orally through close association with the environment. In addition to areas of spiritual significance, areas important for collection of traditional medicinal plants were also identified. The final map (Figure 6) used for the comparative analysis was weighting (by community consensus) of cultural data, illustrating significant and not significant areas.

The scientific assessment and the cultural assessment each produced a map depicting areas of importance. Results of a comparison between the ecologically important areas and the culturally important areas revealed an interesting convergence despite the two different methods. Important areas from the cultural perspective overlapped frequently with areas identified as having high ecological value. The data were compared using a GIS overlay technique, and a crosstab was created showing the area overlap of each category. An agreement of 82.8% was found between the two maps. An alternate comparison was conducted based on a polygon count using the forest inventory boundaries. This assessment agrees with the area analysis result in that 79 of the 91 polygons (86.8%) classified in the cultural assessment agreed with the results of the scientific method (Figure 7).

This convergence was also observed in the final residential suitability map (Figure 8). By removing the cultural data from the residential suitability, it was possible to assess whether the culturally significant areas were also areas identified as being unsuitable for development. Results indicated that 83.5% of the important cultural areas occur in sites identified as being unsuitable for residential development.

Some areas of significance identified by the community that were not corroborated by the 'scientific' model. These areas treed bogs (4 areas) and an agricultural field (1 site). This points to a limitation in the forest model in that these sites were not classed as wetlands, but merely as 'non-productive' sites within the forest inventory database. The cultural analysis has correctly identified these sites as having environmental sensitivity.

The areas identified as having cultural significance encompass a variety of habitat types. As well as the treed bogs and agricultural field, wetlands, four hardwood stand types, two softwood stand types, and five mixed wood types were included.

Figure 10-6. Culturally significant areas

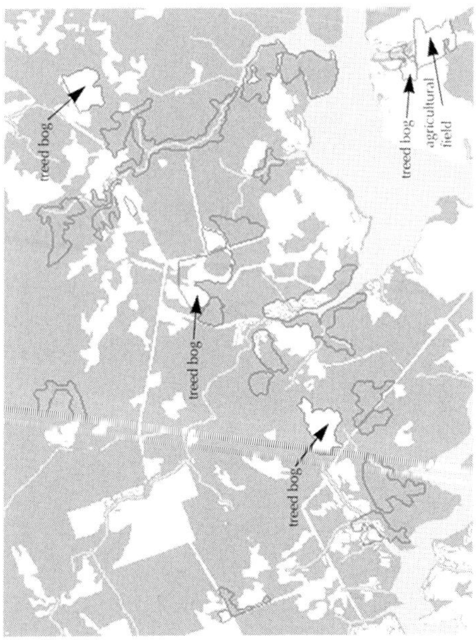

Habitat Types with Cultural Significance		
Category	Stand Type	Area (km^2)
Non-forested	Agricultural Field	0.13
Non-forested	Non-productive	0.44
Non-forested	Wetland	0.04
Softwood	Spruce/Balsam fir	0.93
Softwood	Stunted Softwood	0.10
Hardwood	Intolerant/Tolerant Hardwood	0.19
Hardwood	Tolerant/Intolerant Hardwood	0.09
Hardwood	Intolerant Hardwood	1.99
Hardwood	Tolerant Hardwood	0.17
Mixed Wood	Intolerant Hardwood/Spruce	0.07
Mixed Wood	Spruce/Intolerant Hardwood	0.03
Mixed Wood	Tolerant Hardwood/Balsam fir	0.06
Mixed Wood	Tolerant Hardwood/Spruce	0.07
Mixed Wood	Spruce/Tolerant Hardwood	0.45

Note : Mixed species groups such as "Intolerant/Tolerant Hardwood" and Tolerant/Intolerant Hardwood" are differentiated by abundance. For "Intolerant/Tolerant Hardwood", Intolerant species are more abundant than Tolerant species.

Assessment of Culturally Significant Areas	
SCIENTIFIC METHOD	CULTURALLY SIGNIFICANT AREA
Not Significant	0.82 km^2
Significant	3.94 km^2

Figure 10-7. Data correlation

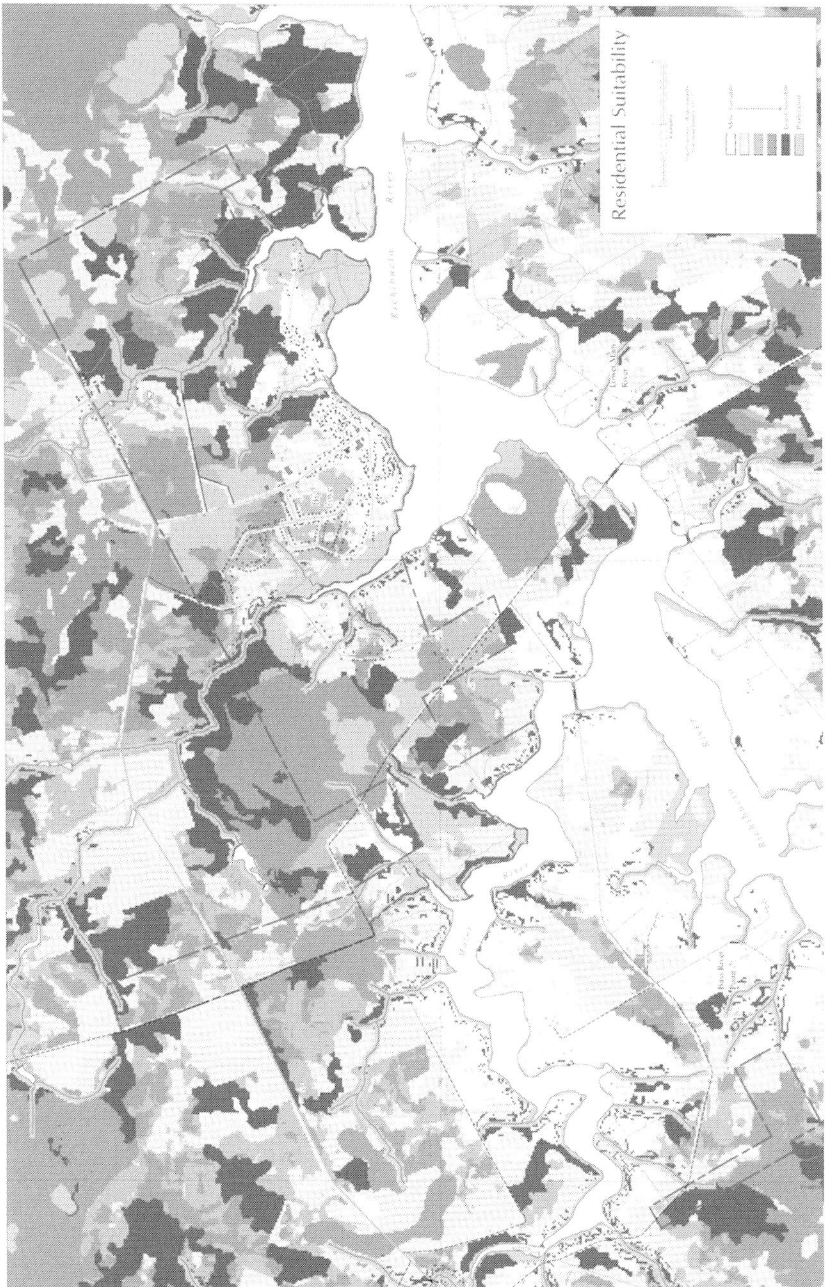

Figure 10-8. Residential suitability

5. DISCUSSION AND CONCLUSIONS

The suitability analysis and community planning work yielded a viable residential siting plan for Big Cove. It also paved the way for important conclusions with relevance in a broader context.

The final section describes some analytical challenges and recommendations for improvement.

5.1 Discussion

While results indicate that few areas within the community, as defined by community values, several concerns and questions exist in the analysis method. Since most of the models used were a combination of analytical and predicted information, it is possible that the input data were not precise enough to make accurate assessments. In the model for water protection, the most important factor was slope or topography. The New Brunswick 1:10,000 mapping has a 5 m contour interval precision. This was interpolated to 3 m, which induces error but helps with definition. Nevertheless, the topography used was insufficient in describing the landscape subtleties and was unable to act as reliable input data for the water protection model. Similarly, the river visibility model relies almost exclusively on topography. Without a precise topographic map, there may have been more or less areas available for development.

The species richness model developed by EDM for this project relies upon the presence of mast (woody debris, food and cover) on the forest floor. This forest component is critical for fauna, and it was not mapped in the forest inventory, thus assumptions were made based upon stand composition. Without substantial fieldwork to verify this correlation, data error may have been introduced.

The significant correlation between culturally significant areas mapped by the community members and the scientific maps developed by the suitability models, suggests that these values must be taken very seriously. Unfortunately, the community remains reluctant to map areas of cultural significance, due to past impingement on the land and culture of their ancestors. Therefore, a more complete cultural survey may show a stronger predictive ability of the community to identify environmentally sensitive sites. The more selective definition of areas using the cultural assessment may also relate to a better understanding of the local environment than is possible from a scientific method, based on air photo interpretation. Mapping outside of the boundary is also essential, since many areas of cultural significance exist outside the borders and are threatened by development.

Although more precise cultural and biophysical data are needed to make detailed decisions regarding residential suitability, it is clear that this community requires more land to properly locate and build two hundred homes. The plan is intended to describe areas where residential development should not occur. Detailed site planning is required before construction begins on areas suitable and acceptable for residential development. The community is able to use the suitability maps as a toolkit to make informed choices regarding future development.

5.2 Conclusions

The literature and the writers' experiences demonstrate that landscape suitability analysis is a powerful planning and management tool. The process can be used to describe and model the relationships between surface features that are of relevance to the full range of land use planning decisions. Geographers and regional scientists have used the technique to describe and model surfaces such a population, economic flows and even innovation waves.

This investigation takes the process one step further. It tests the correlation between culturally/spiritually based data and data derived from an ecological context. It tests for the correlation in a residential planning exercise where the coherence between the culturally/spiritually and ecologically based findings were subject to close community scrutiny. This study shows that sacred landscape spaces in this community are closely correlated with ecological diversity, richness, and sensitivity to disturbance. The consensus of the Big Cove community proves the veracity of the correlation. This, in turn, illustrates the close relationship between landscape ecology and aboriginal community values.

This finding is extremely important for many reasons. First, it suggests that we constructed a bridge to the Big Cove community that linked at least part of the community's value system to the more 'scientific' planning processes. Second, it showed ecologically based findings were complementary to the findings based on aboriginal cultural/spiritual beliefs. Different perspectives could lead to a common goal. Third, Big Cove saw, figuratively and literally, that the residential siting plan reflected their history and spirituality. Implementation was facilitated. Fourth and more generally, the investigation provides evidence that traditional planning methods, predictive landscape analysis via suitability modelling, quantitative assessment and community values and beliefs can be effectively treated in a integrated planning process. Beginning with a variety of perspectives reveals different scenery on the way to a common goal. The perspectives provide checks and balances and proof of a consensus result.

ACKNOWLEDGEMENTS

The authors wish to acknowledge the contributions of the community and Band Council of Big Cove, especially Eva Sock and Charles Levi, who managed this work on behalf of Big Cove and Marilyn Simons-Ingram for her assistance in mapping of culturally important and sacred sites. The authors also acknowledge the valuable contributions of the EDM team members, especially Jim Scott, who conducted interviews with members of the community and Trevor Hume and Dave Pitcher, who provided additional GIS analysis expertise and advice.

REFERENCES

Snell, E.A. (1984). Regional targeting of potential soil erosion and Non-point source sediment loading. *Journal of Soil and Water Conservation*, (40) 6.

Chapter 11

COMMUNICATING LANDSCAPE DEVELOPMENT PLANS THROUGH SCENARIO VISUALIZATION TECHNIQUES

Bärbel Tress & Gunther Tress
Alterra Green World Research, Landscape Centre, The Netherlands

1. INTRODUCTION

Agricultural land use has predominated European landscapes for centuries. It still accounts for large parts of European national territories, however, new functions are now demanding a place of their own and are ready to take over. As urbanization of the countryside takes place, European landscapes come under increasing pressure. The urban influence extends far beyond town limits and the boundaries of the urban and the rural become blurred. Urban lifestyles, transportation systems and architecture expand into the countryside that offers space for housing, business and recreation, but also functions and lifestyles that traditionally were found in rural areas enter urban areas (Brandt et al. 1996; Antrop 1998a, 2000a, 2000b). Although agriculture is still one of the main instruments to manage the countryside, it changes rapidly and questions arise on the shape of future landscapes.

Landscape researchers face the challenge of predicting the consequences of land use change and future landscapes. Policy-makers and the public in general, are interested in the outcomes of policy decisions. Besides agriculture, interests in recreation, housing and business, and also nature conservation are competing for space in the countryside. But as rural areas cannot be enlarged, more and more functions and interests have to be integrated simultaneously in a given landscape. The competition for space is putting high demands on individuals that are involved in the decision-making and research process. Planners and researchers must be able to

H. Palang and G. Fry (eds.), Landscape Interfaces, 185–209.

identify and coordinate functions while respecting and integrating the interests of many different groups. However, who can predict future development with any accuracy? And given the many variables affecting development, who should decide how future landscapes should be shaped? Of course, landscape researchers have considerable expertise that is immensely helpful in designing future landscapes, but they can hardly make reliable predictions because environmental change is not driven solely or even primarily by science, but rather by policies based on economy, public preferences, values, and politics.

Landscapes are an interface where natural and cultural values meet. It is not easy to define which values should be prioritized in future policy-decisions, but choices and decisions have to be made for planning and managing future landscapes, because any future landscape will be a product of individual choice (Nassauer 1997). Communication with people about changes in landscapes is therefore crucial for making sound and effective decisions for the future. Then, the different natural and cultural values attached to individual landscapes can be considered.

The aim of this chapter is to present visualization techniques of landscape scenarios as a tool to communicate alternative future development for planning and management. The proposed scenario approach combines stakeholder participation with visualization techniques of landscape scenarios in order to identify and illustrate the interests of different stakeholders in a landscape. The photorealistic visualization technique presented illustrates past and future landscape changes and the underlying cultural values of the different future alternatives.

2. LANDSCAPE SCENARIOS AND STAKEHOLDER PARTICIPATION

Working with a scenario approach has the advantage that one can deal with future landscape development. The term *scenario* is becoming widespread and used to describe very different kinds of future-oriented studies: trend analyses, prognoses, forecasts, variant analyses, sensitivity analyses, or snapshots of what might be (Hoss & Honton 1987; Schwartz 1992; Schoemaker 1993, Schoute et al. 1995; Emmelin 1996, 2000; May 1996; Wollenberg et al. 2000). By contrast, we propose to follow Van den Berg & Veeneklaas's (1995) definition of a scenario as a description of the current situation, of a possible or desirable future state as well as a series of events that could lead from the current state of affairs to this future state. Scenarios present stylized constructions of possible future developments: "sometimes quite deliberately in the form of stereotypes, archetypes, optimum or doomsday situations, or other extremes" (Van den Berg & Veeneklaas 1995: 12). Following this definition, scenarios do not necessarily present the most realistic future state; they are not prognoses, predictions, or forecasts, but a set of possible future developments for a landscape.

In contrast to prognoses, the scenario approach we propose allows the development of several alternative future landscapes while being aware of the uncertainties. Scenarios focus on 'what will happen if' and not on 'what will happen'. This shift in emphasis illustrates the move from the purely scientific-oriented approach to a participatory scenario approach for future landscapes. Scenario approaches are built on general sets of assumptions and rules covering distinct environmental, social, economic, political, and administrative developments.

With the term landscape scenarios, we refer to scenario studies that deal with the future shape of landscapes as a whole. In determining the chapter's aim to identify and integrate interests for planning and managing future landscapes, we acknowledged that all landscapes have a natural and a cultural dimension. Today's landscape would not exist without humans taking an active role in shaping and creating them (Tress & Tress 2001). Landscapes have evolved as a result of complex interactions with and between people. For the landscape scenario term we apply a holistic landscape concept that includes landscapes' multiple dimensions in a system. Following this landscape concept, landscape scenarios focus on the question 'what will happen if' including all aspects that are relevant for shaping future landscapes: natural and cultural assets as well as people's preferences and perceptions. In this way, designing landscape scenarios demands an interdisciplinary research effort, where several unrelated academic disciplines cooperate in a way that forces them to cross subject boundaries to solve a common research goal (Fry 2001; Fry et al. 2002; Tress & Tress 2002). Recent studies of comparable landscape scenarios have been conducted by FMST (1999), Barbanente et al. (2002), Höll et al. (2002) and Nassauer et al. (2002).

For landscape scenarios, we work with sets of assumptions and rules that define certain developments in the field of economy, society, policy and administration and environment. For each scenario we define assumptions, which represent different sets of choices and are based on different underlying values of nature and culture. Different ways to identify the necessary preferences and choices that determine scenario assumptions can be selected. The assumptions can follow policy-decisions, research recommendations or – as demonstrated in this chapter – stakeholders' preferences.

By stakeholder participation, we add a transdisciplinary perspective to designing future landscapes, in that we integrated both academic researchers from different disciplines and user-group participants to reach a common goal, the design of future landscapes (Fry et al. 2002; Tress & Tress 2002).

Stakeholders were identified as the people that either live in the area that is under study, people that are affected by changes in the area or people that have an interest in the future shape of this area. Participation can strengthen the commitment for planning and management activities, it can increase user satisfaction and can create realistic expectations of outcomes. Stakeholder participation can also contribute to building trust among the different actors in the planning process (Wates 1985; Whyte 1989; Sanoff 1990; Towers 1995; Al-Kodmany 1999; Khakee 1999; Luz 2000; Bryner 2001; Swearingen White 2001). For researchers and planners, stakeholder participation means getting access to local expertise and knowledge that can be of benefit for the research and decision-making process (Roe 2000).

3. COMMUNICATING THROUGH VISUALISATION TECHNIQUES

Choosing for an interdisciplinary scenario approach combined with stakeholder participation requests a good communication of research results and plans to the stakeholders. It is challenging for researchers to communicate about future landscapes with non-experts in the field. For coping with this challenge, visualization of landscape scenarios is a helpful and powerful tool that makes people easier to understand the changes that result from single choices or preferences.

Visualization of landscape scenarios can be done in numerous ways. Useful techniques include charts, tables, plans, maps, drawings, photos, videos, and GIS-based modeled landscape surfaces that all have advantages and limitations (Jansens 1990; Primdahl 1990; O'Riordan et al. 1993; Emmelin 1996, 1998; Swetnam et al. 1998; Herrmann & Osinski 1999; Kurz et al. 2000, Palang et al. 2000; Dolman et al. 2001; Ervin 2001; Höll et al. 2002; Lim & Honjo, in press). We have chosen photorealistic visualization technique based on aerial and land photos as we considered this as being most suitable for communicating landscape development plans to stakeholders, planners and decision-makers (see Figures 2-4). Photo design software was used to prepare the visualizations. The photographic basis for the visualizations was aerial photos taken from a slope angle. The landscape changes for the scenarios can be visualized by adding layers to the chosen pictures including details taken from other aerial photos or land photos taken in the area. Elements that cannot be found in the surrounding of the chosen scenario area can be taken from other sources or constructed. The software allows altering features like perspective, size, colors, contrast light, and shade for each single landscape element that is changed. Furthermore a landscape-scale was chosen for the visualizations, as this allowed being more detailed and to communicate even small changes (e.g. removal of single hedgerows) to the stakeholders.

For this chapter we use visualization technique not only for illustrating alternative future developments but include also a visualization of a historic state, which illustrates a past situation of the landscape for which we created the future scenarios.

Figure 11-1. Map of sample areas in Denmark

4. ALTERNATIVE FUTURE PLANS EXEMPLIFIED BY THE CASE AREA STAVNSHOLT OVERDREV

For illustrating the application of the landscape scenario approach combined with stakeholder participation and visualization techniques, we guided a study for future development in two rural areas in Denmark. One is the area around the small village of Kravlund in Southern Denmark with about 50 inhabitants, 5 km north of the Danish-German border (Tress & Tress 2003). The other area, presented here, is the area of Stavnsholt Overdrev, situated in the north-eastern part of the island Zealand at the northern fringe of the metropolitan area of Greater Copenhagen (see Figures 1 and 2). The research area is 4 km^2 large and Stavnsholt Overdrev situated between the municipalities of Birkerød in the east and Farum in the south. About 150 people live in the area. Figure 3 shows the Stavnsholt Overdrev area from a satellite image, Figure 4 from a slope angle photography, which was used for the visualization of the alternative landscape scenarios. The angle of vision is marked on Figure 2 and 3.

4.1 Set up of the study

The study was conducted in three phases (Tress et al. 1999). In phase 1, we created four alternative scenarios, four distinctively different developments for the area. The term extreme was used because each of the scenarios show a monofunctional and therefore extreme development. The extreme scenarios show how the landscape would appear in the year 2020 if each of four land uses were to dominate. Included were (1) industrial farming, (2) recreation and tourism, (3) nature conservation, and (4) residential expansion. The extreme scenarios describe how the sample area could look if a certain land use was to dominate future development. We decided to realize four scenarios because these four land uses already today place great demands on the countryside and compete with each other in Denmark as well as in other European countries (Coeterier 1994; Antrop 1998b, 2000b; Vos & Meekes 1999). In general, all possible interests in the countryside in Denmark can be subordinated to these four land uses, and all can be seen as processes spreading from urban to rural areas.

In phase 2, the scenarios were presented to, and discussed with, different groups of stakeholders in a meeting in the town Birkerød, close to the Stavnsholt Overdrev area. All stakeholders with interests in the area were invited to the meeting: local farmers, inhabitants, members of environmental organizations, hunting clubs, sporting clubs, or other leisure-time organizations; and representatives from administration and planning agencies. In the meeting, we explained the setting and purpose of the scenario study to the participants and then described each extreme scenario with the help of animated visualizations of the photorealistic landscape scenarios. After the scenario presentation, all participants were supplied with questionnaires asking for comments on each scenario. The questionnaire surveyed individual details and general evaluations of each scenario as well as data about the respondent. Respondents could assess eight landscape elements that changed in each scenario as positive, neutral, or negative. The corresponding elements were explained in our presentation of the scenarios. Additionally, they could add comments on each scenario. The survey was followed by an open discussion about the presentations and about participants' wishes, ideas, and fears for future development in the sample area. Aspects that could not be expressed in the questionnaires could be aired here.

In phase 3, we developed *integrated* scenarios, which illustrate different options for integration of land uses and landscape elements. The integrated scenarios are a direct outcome of stakeholders' preferences as aired in the stakeholder meetings and documented on questionnaires.

4.2 Sets of assumptions for extreme scenarios

The developed scenarios are not utopian but rather illustrate developments that could happen around the sample area. Therefore, the set of assumptions made within each scenario had to built on logic, coherence and consistency. The assumptions covered future development regarding political, juridical, economic, social, environmental and administrative aspects. The assumptions were made on a national level, whereas the visualizations show the local consequences. Each set of assumptions represents different underlying cultural and natural values. None of the scenarios was designed to be more realistic than others. We avoided a 'business-as-usual' scenario, that is, a further development of present conditions (O'Riordan et al. 1993). This scenario could attract people more than others, as it is closest to the present state. It was, however, our intention to identify the interests of stakeholders and therefore it was reasonable to design the scenarios equally attractive.

4.2.1 Industrial farming dominating

In the first scenario, industrial farming, the countryside in the Stavnsholt Overdrev area has become a resource for intensified industrialization in agriculture (see Figure 5). The four features of concentration, specialization, mechanization and intensification characterize industrial farming. Production is concentrated in fewer but larger farms in the agriculturally most favorable areas. The farms do not have a varied production, but specialize in cattle farming, pig farming, crop farming or market gardening. All parts of the agricultural production process are mechanized. The productivity per unit area increases as long as there is a difference between potential and real yields.

In the year 2020, fewer farms produce higher yields, so a few large farms dominate land use. A sharp specification and regionalization has taken place with solely cattle and dairy farming in the western parts of Jutland and crop farming and market gardening in the eastern parts of Jutland and on the Danish islands, also in the Stavnsholt Overdrev area. Accordingly, large areas of less favorable agricultural areas have been taken out of production and have become fallow land. Where production has been concentrated, the countryside has effectively been restructured towards an optimal production landscape. This means larger parcels of land and fewer biotopes. In those areas, which have been taking out of productive use, successive vegetation has been taking over so that sparse mixed woods cover large areas.

Large parts of society support technical, chemical and biological innovations and consequently these are promoted. Farmers sow genetically modified plants, which are resistant to pests. Therefore this highly intensive way of farming does not require a higher input of external aids such as chemical fertilizers and pesticides. The few substances that are still used are made in such a way that they do not affect the environment. This is possible due to developments in bio-chemical technology. Consequently, the environment is no longer a matter of public concern. Society focuses more on economical issues. An orientation towards profit optimization characterizes farmers' attitudes. Thus a widespread attitude is that highly industrialized production can present great advantage for the environment.

Farmers are oriented towards economic rationality. The turnover per farm is much higher than in the early 2000s. Preconditions for this are not only the higher productivity per area unit, but also increased specialization and mechanization. Production is smaller as well as diversified farming has been abandoned, because it has become unprofitable. Areas of the smaller farms with high productive value are rented out to larger farms. Less favorable areas have been left to fallow. Concentration based on increased productivity per man-hour is coupled with reduction of the agricultural work force. Only a few farmers remain in the agricultural sector, others seek work in different sectors. But the farmhouses are kept for residential purposes.

Legislation and regulations, which obstruct industrialization are loosened or abolished. Agricultural areas are at the full disposition of farmers enabling them to maximize profit from their farms. Restrictions are in place, which prevent a general application of the agricultural areas in the agricultural zone. These have the consequence that buildings on smaller farm properties can be used for no other purposes than those of habitation. However exceptions are made and permission is given to some activities.

4.2.2 Recreation and tourism dominating

In the second scenario, recreation and tourism, the area has a primarily recreational function (see Figure 6). Tourism is necessarily connected to movement; recreation in general is not. Whereas tourism covers the actions and visions of people traveling from one place to another, recreation also includes the actions and visions of people with a permanent residence in the locality in question. Tourism is based on a contrast to the everyday, recovery and recreation is part of the everyday.

The countryside is seen as a resource for recreational and leisure time purposes. It is used for all kinds of outdoor activities such as walking, outings, golf, riding, fishing, angling, hunting, and nature observation. People perceive the landscape as the kind of place they like to spend their leisure time. The landscape is both accessible and attractive. People want to use the landscape and come into contact with nature. But it is necessary to have additional infrastructure that makes it possible to experience the landscape. This implies all kinds of accommodation, tracks and roads, transport facilities and other tourist facilities such as golf courts, restaurants or sport facilities. Hotels, youth hostels, second homes and other tourist accommodation are established. Especially, second home areas are expanded and used by national and international tourists. Besides the above, the visible elements of landscapes become more important. Landscape architecture and design play a distinct role in shaping the landscape. Small villages and farm buildings are renovated. The appearance of houses and villages is improved.

In this scenario recreation and leisure time activities have a high standing in society and are important elements in individuals' lives. People increasingly demand for leisure activities and travel. This is due to the fact that leisure time increases. Leisure time and activities in leisure time have greater importance than they have had in 2000. People become more mobile and take more but shorter trips, suited to the individual's demands. Mobility, fun and (authentic) landscape experiences dominate recreation and tourism. Furthermore, Denmark also continues attracting tourists from abroad. Because of the large tourist pressure on the coast, tourist activities spread into the interior of the country. Tourists discover the countryside that is not characterized by the coast, but by agriculture and forestry as an attractive destination. This process is based on an idealization of life in the countryside. Cultural elements in the landscape are preserved for tourist purposes. People still live and work mainly in the cities and the urban areas but because of the increased mobility they use the countryside for recreational purposes.

Tourism is regarded as a driving force of the national economy. Therefore promoting tourism in Denmark has a large national interest. The aim is to support both domestic and international tourism. But tourism becomes important also on the regional and local level. On the local level tourism and recreation offer new possibilities for generating income. Farmers and local people provide accommodation and other services. Tourism offers jobs outside the traditional field of agriculture. The basic steering mechanism is economic optimization. International tourism is a bringer of foreign exchange and will be spread where there is a potential for it.

New recreational areas (e.g. for second home areas) are not established along the coast. Tourism development has continued in the interior of the country in order to ease the pressure on the coastal areas and to attract and accommodate more tourists in Denmark. The planning legislation is still in force, but the counties have the possibility of changing boundaries, i.e. to extend some areas. In this way, recreational areas arise in the countryside too. In order to promote tourism in the countryside the residence requirement for farm owners and the legislative obstacles for purchase of land by foreigners are removed. Public access to private land is extended. Regional and local planning authorities develop plans for locating recreational infrastructure.

4.2.3 Nature conservation dominating

In the third scenario, nature conservation, the countryside is regarded as a resource for the realization of nature conservation (see Figure 7). By conservation we understand all measures, which contribute to increase the diversity on the landscape, biotope and species level. Conservation takes place not only in segregated areas, but also in areas with intense agricultural production. Rather conservation is as far as possible integrated with other activities, e.g. farming. Either individuals (farmers, landowners) or the community can take measures in order to increase diversity. Concerning the individuals there are different ways of acting for respectively spare time and full time farmers.

Areas that have been managed intensively by conventional farming have decreased significantly. Correspondingly the areas, which are managed by organic farms, are increased. Organic farming in this case is built up, in such a way that the farmed areas are managed extensively. Areas with biotopes are also increased. Existing biotopes are protected and enlarged and additionally are established new ones. The geographical connectivity of small biotopes is improved; they are part of a biotope network, connecting biotopes of the same kind to each other. Part of the former agricultural areas is also used for afforestation with mixed forests or for renaturalization projects such as reestablishment of heaths, wetlands, bogs, and streams. But the aim was not only to develop existing natural elements, or those, which formerly existed. As the landscape is perceived of as a dynamic, continuously changing phenomenon, there was also room for the establishment of landscape elements (hedgerows, ponds) and for finding new ways of developing and arranging these elements.

An underlying assumption of the nature conservation scenario is that there is a public awareness that nature as the basis of livelihood is in danger. Thus environmental awareness increased and environmentally friendly ways of acting have been promoted and are generally accepted. This change in attitudes was also the cause of society no longer being willing to pay for the environmental damage caused by agriculture, industry or private persons. In spite of the land use changes in the agricultural areas, people's residential and working place remains at the same place as it has been before.

In this scenario, the economical strategy is to include the external costs, which result from environmental pollution. If a company uses techniques in its production process, which pollute the natural resources of water, air or soil, it has to pay for that. Fundamental to this policy is the opinion that nature is a valuable good and when natural goods are exploited, they have to be paid for. As a consequence environmental costs were increasingly taken into account in economic planning. Hereby actions that pollute or destroy became evermore expensive, whereas environmentally friendly techniques became cheaper and environmentally friendly behavior is supported. Following this, protection of the environment, e.g. in agriculture, is not only a precondition for receiving financial subsidies, but also has an economic importance in itself. Organic agriculture became the most profitable way of farming. Part of the agricultural area is no longer used for productive purposes, but for the protection of nature. The farmers, who allocate land to nature conservation, are paid by the public to manage these areas. Hereby the monofunctional orientation towards production only is relinquished in favor of an integration of agricultural and nature conservation schemes. Agriculture as it was perceived traditionally – in terms of production – has no high standing anymore. It did not disappear completely thus there is a certain number of organic farmers, who still produce food. But a large share of the area is used for the purposes of nature conservation. However, it is still the farmers, who manage these areas and receive payment for that service.

The scenario assumes that there exist national nature conservation strategies for the whole country of Denmark, which are parts of a larger network of natural areas all over Europe. The regional administrative area, the counties, is required to implement the overall aims of these strategies; they coordinate the plans at a local level with the national strategies. The local plans give detailed information about the land use in the whole community. They say, which areas or individual objects are used for what kind of purpose. Besides this plan, general protection of all biotopes is ensured. The distinctive types of biotopes are listed in nature protection legislation. High taxes are on pesticides as well as on chemical fertilizers and their application is prohibited in more sensitive areas, e.g. close to streams. The regulations for application of animal fertilizers are also tightened. In general, there is a minimum standard of biotope management and environmentally friendly production modes, which has to be met on each farm. The subsidy system is progressive, which means, the more farmers contribute to the local nature protection plan, the more subsidies they receive.

4.2.4 Residential expansion dominating

In the fourth scenario, residential expansion, the area around Stavnsholt Overdrev is regarded as a resource for housing and habitation (see Figure 8). Residential expansion is one sub-process of the urbanization process itself, the spreading of single houses and settlements. Whereas urbanization is a broad process, which includes population increase, increase in number of towns, extension of built-up areas and spreading of urban lifestyles and cultures, the aspect of residential expansion focuses only on the extension of built-up areas for residential purposes. Here, urbanization is not seen as a concentration of people and dwellings in towns, but as the opposite, a decentralized spreading over the countryside. The process that is described here could be regarded as part of the urbanization of the countryside. The process describes the movement of people from highly urbanized areas to the countryside. At the same time people in the countryside are confronted with urban functions and urban lifestyle.

The countryside in 2020 is primarily used for housing and residence. Areas, which formerly have been used for agricultural purposes, were subdivided and sold to private investors. The urban fence of existing settlements has been enlarged. Housing has expanded on the edges of those settlements. Agricultural areas close to settlements have been transferred to built-up areas. The built-up density in formerly existing settlements did not increase, but the consumption of area by housing has increased. Small villages and towns became 'urban centers' of the residential expansion. Many new one-family homes and some apartment buildings with more dwellings have been built as well as already existing buildings are used for housing and business purposes. Homes are surrounded by gardens. Areas close to the residences have been afforested. With spreading of houses, urban infrastructures such as roads and tracks, information networks, stores and shops, and cultural facilities are extended and improved in the countryside. Industrial buildings have been established in the countryside too.

The number of houses and population in the countryside has increased. People from urban areas with high population density have moved to the countryside. This is due to the fact that the former suburban residential areas have lost their attractiveness as quiet and exclusive residential areas. Social conflicts and ghetto development have surfaced in the urban areas as well as in the suburban areas, where high-income groups and middle-class people lived. These groups feel now attracted to the countryside. They want to have larger houses, larger gardens and space around their houses and the possibility of having pets or other animals. They settle in old farms, village houses or build new houses. People's mobility in relation to housing and working has increased. People live in the countryside but mainly work in the towns and urban areas.

Life in the countryside is less expensive than in towns. Ground prices, prices for renting houses and taxes are lower. Many people in the countryside still have their job in the towns and commute everyday between work and residence. Commuters are especially orientated towards Copenhagen and other bigger cities. Industries, business and services have partially moved to the countryside too. Especially, information technology industries and other footloose industries have left the urban areas. For them the countryside is attractive and cheaper too. Therefore job opportunities in the countryside have increased, especially for people within the information and technology sector, but also for people within services.

The designation of building land and the building in the countryside for housing and business purposes is allowed. It is people's private interest, which has a main influence on the appearance of their properties. In general exist more and better possibilities for business and industry to move to the countryside and into small villages or residential areas. Planning restrictions concerning the division between residential and industrial areas have been removed, as has the division between urban, recreational and agricultural zone.

4.3 Stakeholders' reactions and preferences

The photorealistic scenario visualizations were used for discussing the alternatives for future landscapes with the stakeholders in the area and also with the representatives from planning and administration authorities. The atmosphere at the stakeholder meeting was friendly, but not all participants felt immediately comfortable with the scenarios or future development of the sample area. The reason for this is the drastic changes evident in the photorealistic visualization of the scenarios to an area they knew intimately. Such drastic implications of daily changes had not been anticipated. People find it difficult to imagine a future that is different from the present (Nassauer et al. 2002).

Experiences from the stakeholder meetings and results from participant questionnaires demonstrated that the communication process among researchers, planners, decision-makers and stakeholders has shortcomings. The stakeholders were not used to be contacted by planners and decision-makers when discussing the future landscape development of the area.

Stakeholders expressed doubt that planning authorities would consider their interests. We identified a lack of confidence in academia and in planning authorities in general. Academics were perceived as overly distanced from the problems and focused on the single issues taken out of the context. Stakeholders suspected the landscape researchers of working in the interests of planning authorities only, not in the interests of the stakeholders. The future of their landscapes was an issue that stakeholders never had reflected upon and they did not believe or did not understand their own potentially influential role in future planning decisions.

To better understand stakeholder responses, we assessed group composition. Participants represented all invited groups. Three-quarter of the stakeholders lived in the area (local stakeholders), one-quarter of the participants were from nearby areas (nearby stakeholders). Eight of the participants were representatives from administration and planning authorities. In the questionnaire, participants were asked how they felt about the scenarios. Every respondent could rank each of the four scenarios on a scale between very good and very bad.

The industrial farming scenario as a future perspective was evaluated most positively by representatives from authorities and local stakeholders and got the lowest support from nearby stakeholders. The recreation and tourism scenario was evaluated most positively by nearby stakeholders, but got lowest support from representatives from authorities. The nature conservation scenario was most welcome among all stakeholders, but got the lowest support from representatives from authorities. The residential scenario got the highest ranking from nearby stakeholders, but lowest support from local stakeholders (see Table 1).

Table 11-1. Support to the four landscape scenarios by stakeholder group (n = 65)

	Industrial farming	*Recreation and tourism*	*Nature conservation*	*Residential expansion*
Local stakeholders	*35.0 %*	*30.0 %*	*92.5 %*	*7.5 %*
Nearby stakeholders	*6.3 %*	*56.3 %*	*87.5 %*	*12.5 %*
Representatives from authorities	*37.5 %*	*25.0 %*	*75.0 %*	*12.5 %*
All groups together	*27.7 %*	*36.9 %*	*87.7%*	*9.2 %*

The most apparent result of the evaluation of the scenarios of the Stavnsholt Overdrev area was that the measures that led to environmental improvement such as increase of forest and natural areas, were assessed positively by a wide majority of all analyzed groups.

In their comments many of the stakeholders stated their interest in the development illustrated by the nature conservation scenario, but expressed doubt that the authorities would give their support. The support for nature conservation measures increases the closer people live to the area. This is different from the results from the same study in the second sample area, Kravlund, in Southern Jutland. There, support for nature conservation diminishes the closer people live to the area (Tress & Tress 2003). However, this situation reflects the different preferences and attitudes of stakeholders in different rural environments. Whereas in Kravlund, an increase in nature conservation was less preferred by local stakeholders and considered as a obstacle for economic development, in Stavnsholt Overdrev locals appreciate the improvement of the rural setting with nature conservation measures. The same natural and cultural landscape elements have different values for the residents of the two sample areas. In both sample areas, the residential expansion scenario got the lowest support from all four scenarios. In Stavnsholt Overdrev, the sample area was for most residents an area that they have chosen for living. Most of the residents leave the area for working but highly value the area because of its quality of life. A future development predominated by nature conservation would therefore fit best to their choice. A recreation and tourism predominated development is not so much appreciated by the local stakeholders because they perceive the area as already characterized by a high recreational value. A different picture appears for the stakeholders living not close to this area; they would appreciate a development towards more recreation and tourism. Future development predominated by residential expansion is not appreciated at all for many of the residents in the area, who moved from more urbanized areas around Copenhagen to the sample area to escape increasing residential expansion.

There was a general appraisal of the status-quo situation instead of future landscape change of all respondents. In comments on the questionnaires and in the discussion, many stakeholders expressed their desire for no change at all. In the first reactions, change was perceived as a threat, but in reflecting on past development they realized that landscapes had changed and accepted that they would continue doing so.

4.4 Landscape scenario integrating different values and preferences

Whereas phase 1 and phase 2 of the study aimed at identifying different interests for future landscape development in the area of Stavnsholt Overdrev, phase 3 focused on an integration of the different interests for future development. Based on the results from the stakeholder responses to the four presented scenarios we designed a scenario that aimed at integrating as far as possible the different interests. In the questionnaire survey that was conducted at the stakeholder meetings, the participants had not only the chance to rank the four scenarios as a whole, but also to assess single elements that were changed in each scenario. The integrated scenario is based on these assessments. It illustrates only one among many possible ways to realize stakeholders' preferences in the landscape.

None of the participants preferred future development towards one of the presented extreme developments – industrial farming, recreation and tourism, nature conservation, and residential expansion – only. If the stakeholders could design their own future landscapes they would take some elements from each of the four scenarios. Integration of interests is therefore the challenge for future planning in the sample area.

When looking at the scenario evaluation as a whole, all participants in the stakeholder meeting ranked nature conservation highest. We decided therefore that the integrated future landscape scenario should consider mainly the development proposed in the nature conservation scenario, however, supplemented with the most appreciated elements of the recreation and tourism scenario and the industrial agriculture scenario, and a few elements of the residential expansion scenario.

Figure 9 presents again the situation in the Stavnsholt Overdrev area in the year 2000. Figure 10 shows one integrated scenario designed according to the above-mentioned aspects. It is one alternative of a future development, as all stakeholders together would appreciate it, but it is not the only possible one.

4.5 A historic visualization of Stavnsholt Overdrev

In the stakeholder meetings it proofed to be difficult, to make participants reflect about landscapes changes, which are evident from a researcher's point of view, but are not perceived by stakeholders in their daily contact with the landscape. As one possibility to better and more convincingly communicate landscape changes, we therefore tried to design a historic visualization. This historic visualization was not presented in the stakeholder meetings, as the discussion there should only focus on future perspectives. But we found out, that in fact, a historic photorealistic 'picture' of a landscape may be a suitable tool, to communicate to stakeholders that landscape changes over time lead to a significant different landscape appearance. The development of a historic visualization links to a study of Caspersen & Fritzbøger (2002) that applies a retro-perspective approach to the investigation of landscape change in Denmark. When comparing historic visualization with a present photograph, stakeholders could at one glimpse get an impression of the changes that had happened and would probably be more open to anticipate future changes.

We designed a picture that illustrates how the landscape of the sample area today (Figure 11) could have looked liked about 120 years ago (Figure 12). The visualization is based on historic maps from the second half of the 19th century and on descriptions of historic landscape appearance and structure as provided by Fritzbøger (1998).

Around 1850, a major reallocation of land (*udskiftning* in Norwegean), which has been conducted in entire Denmark, had come to its end. The major reallocation of land had a tremendous effect on the landscape, because the earlier villages and their common-fields had been dissolved. Farmhouses were moved from the villages to the former outfields and property was allocated around them. The historic visualization of Stavnsholt Overdrev shows this new landscape structure. The term *overdrev* meaning 'common grazing area' still indicates the former function. With the end of the land reallocation a rapid increase in production took place. It was the spirit of the time that the unordered nature should finally be convicted by human rationality and harmony (Fritzbøger 1998). Geometric and rectangular forms of a productive agricultural landscape substituted all accidental or chaotic mosaic landscape. A fundamental agro-technical innovation was the use of marl, which improved soil fertility, leading to much higher production rates. Marl was also dug out in Stavnsholt Overdrev. In the background of the visualization a huge marl pit is visible.

In the second half of the 19th century, there were not more natural areas in Stavnsholt Overdrev than today. The former outfields (woods, heath lands or grazing areas) had been converted to single farmers' properties. And the new landowners, aimed at filling, draining or otherwise removing the many waterholes. This process had however not come to an end in Stavnsholt. The visualization shows that there were still distinctively more waterholes left than today. For the land reallocation, the old beech woods were removed and transferred to fields. Around 1850, Denmark had only 5% of its territory covered with woods (today 10%), whereas 73% was agricultural area (today 63%) (Fritzbøger 1998). Conifers, which were newly imported to Denmark, were planted on soils too poor to be used for agriculture.

With the emergence of private land ownership, an immense demand for marking property boundaries occurred. Whilst before reallocation hedgerows or stone dikes may have marked boundaries between two villages, they became now common for the single property. The visualization also shows a higher density of roads. Although they may have been used much less than today, there were more small roads connecting single farms with each other, thus accessibility of the countryside was higher then today.

Photorealistic visualization technique can illustrate also here the long-term dynamics in landscape change. The power of the visualization of historic landscapes can convince stakeholders that landscapes are changing continuously and thus open for the debate about future landscapes.

5. DISCUSSION

5.1 The landscape scenario concept

Landscape scenarios are presented in this paper as an interdisciplinary tool to discuss future landscape development in a holistic way, encompassing a broad set of assumptions and aspects that have influence on future development. Scenarios have become a broad and frequently used term for all kinds of future-directed studies, but the expression landscape scenario is introduced here as a specific concept to deal with alternative spatial developments on a landscape scale. The strength of the landscape scenarios is their potential to present different alternatives instead of one alternative, which in most cases would be a forecasted future development from the perspective of one particular specialization. In contrast, the multiplicity of alternatives bears the potential to calculate with future situations that one has not though about before. From policy and society, interest is large in getting an answer from landscape research on what will happen in a certain time in a certain area, but future landscapes cannot be predicted. The study at hand demonstrates that future development can be influenced if different options are known. It is easier to design a set of possible future landscapes, based on different underlying assumptions or influences, than to predict one future landscape. Landscape scenarios – when combined with participatory landscape planning – open for a discussion on these options.

The combination of extreme and integrated scenarios as well as visualizations of future and historic situations is well suited to stimulate the discussions on future options. Extreme scenarios present several options that would otherwise not be considered. Extreme scenarios can be used as a kind of brainstorming for possible options. Integrated scenarios give one option to present the preferences of stakeholders, however, in a real planning situation lots of integrated options would be possible that still take stakeholders' preferences into consideration. The usage of visualizations for historic and future situations allows tracking landscape change in general but also to focus on single cultural and natural aspects. Here again, the implicit preferences of stakeholders can be made explicit.

5.2 Degree of stakeholder participation

The study presented here was not a real planning situation but a study to investigate the methods and approaches. However, in a real planning situation the question would arise what the optimal degree of stakeholder participation in the decision-process would be. Different levels of participation can be distinguished, reaching from 'giving information to stakeholders but do not include them in decision-making', the level with lowest influence, to 'delegating full power for decision-making and actions to the public', the level with the greatest influence (Warburton 1998; Asselt Marjolein & Rijkens-Klomp 2002). On the one hand, to be successful in practice, stakeholder participation must transfer some power from planners and decision-makers to stakeholders. On the other hand, we have to acknowledge that successful stakeholder participation does, however, not mean that all power is delegated to stakeholders. The ideal degree of stakeholder involvement has to be determined for each context in which it is used.

The landscape scenario approach combined with stakeholder participation was a good instrument to identify the different preferences and interests in the landscape of the sample area. It was helpful that we avoided using a 'business-as-usual' scenario, that is, a further development of present conditions (O'Riordan et al. 1993). This scenario could have attracted the stakeholders more than the other options, as it is closest to the present state. Keeping future development as close as possible to the current state was the first reaction of most stakeholders. The historic visualization of the sample area can be used to convince stakeholders that landscape change and that future development can be steered and their preferences can be included. The unexpected preferences of stakeholders are a strong argument for applying a participatory approach and taking the stakeholders' opinions into account. But stakeholder involvement does not release planners and decision-makers from making decisions and prioritizing one choice before others. Also in a participatory approach, elected representatives have responsibility to decide.

5.3 Communicative abilities of photorealistic visualization

Photorealistic visualization technique can avoid disadvantages of other techniques such as charts and tables, maps, plans, or GIS-based visualizations, which may accurately portray the extent of change that should be realized with the scenario but are not easily interpretable by stakeholders. Artists' renderings or sketches could be an alternative as they may be equally good interpretable by stakeholders, but lack concreteness (Orland 1994; Lange 2001). Since photographs are a very close representation of reality, little interpretation is needed to convey the message to stakeholders (Al-Kodmany 1990; Orland et al. 2001). A higher degree of abstraction leaves an empty, unknown area that is filled by people's own imaginations, which could easily lead to misinterpretations and consequently complicate communication. In the process of planning future landscapes, concrete tools that visualize the effects of each change are very helpful. People can react very easily on visuals because they demand less specialist knowledge to understand than maps and plans. Additionally, visualization helps people to grasp very quickly the possible consequences of future landscape development.

Aerial photos taken at a slope angle communicate better to stakeholders than vertical satellite or aerial pictures. This perspective allows for both easy identification of the area by the stakeholders and the necessary manipulation of the picture.

In a landscape scenario designing process that involves non-experts, some of whom will rely on visualization, small landscape-scale visualizations are more appropriate than large-scale visualizations because they have a very realistic appearance and are very detailed (see Orland 1992, 1994), thus facilitating communication about the scenarios.

The visualization technique has also limits. It is descriptive only and does not provide an evaluation of the visualized development. Non-visual aspects of landscape change have to be communicated in different ways, for example in written scenario descriptions or oral explanations in stakeholder meetings. Furthermore, photorealistic visualizations could be suspected for influencing and manipulating stakeholders by the realistic outcome (Sheppard 2001). Researchers, planners and designers have to be aware of the responsibilities of using these tools. But one must not overestimate the power of photo design in manipulating stakeholders' opinion. Stakeholders have a good knowledge of the area under discussion and will therefore discover any inconsistency or intended manipulation in the design process.

5.4 Stakeholders' difficulties to acknowledge landscape changes

Barbanente et al. (2002) support thee findings of this study that stakeholders are not used to reflect on the issue of landscape change or future landscape development because of missed or lacking opportunities. The use of a historic visualization that shows the degree of past change is a good opportunity to communicate how fundamentally landscapes may change in a certain period of time and thus be very convincing to stress the need for reflection on future changes. Once stakeholders realize their potential influence on future landscape development they can contribute to protecting landscape continuity and cultural characters of certain landscapes or help detecting which cultural elements should be managed in which way.

6. CONCLUSIONS

Scenario visualization can be used to identify stakeholders' interest in and preferences for future landscape development. It can be used as a tool to communicate alternative futures to experts as well as to non-experts. By this, successful communication on future development plans can take place among all involved stakeholders. Photorealistic visualization has the advantage of being easier understandable than other visualization tools. It is of particular value in a stakeholder participation process where people have to reflect about landscape changes. The photorealistic visualization cannot only be applied to present different alternative future developments but also to demonstrate the landscape changes that have happened in the past. This helps stakeholders to understand that changes will come also in the upcoming decades. Having reached this awareness, stakeholders can be actively involved in the process of defining choices for future development. The study has shown several extreme developments but also a future alternative that could express stakeholders' preferences and non-preferences. Having the visual image by hand makes the discussion and definition process easier, but does not releases from the fact that decisions have to be made.

Scenario visualization contributes to communicating the consequences of several policy decisions, cultural values, or illustrates the possible visual impact of policy scenarios as, for instance, conducted by Höll & Andersen (2002). The technique can be of high value for policy-makers as they can test the consequences of several future alternatives. In the presented case study, it was surprising that stakeholders preferred a future development that has clear components of a nature conservation scenario. The power of visualization made people reflecting about the alternative consequences and led to a result that policy-makers and researchers would not have expected. This again stresses the importance of stakeholder involvement for future landscape planning. Future landscapes are a question of choices. The possible options and consequences have to be made clear and this is what the presented approach is best suitable for.

ACKNOWLEDGEMENTS

The case study reported here was a part of the interdisciplinary research initiative *Value, landscape and biodiversity* under the programme *Man, landscape and biodiversity* financed by four Danish Research Councils. We would like to thank Jesper Brandt and our colleagues at the Centre for Landscape Research and the Department of Geography and International Development Studies at the University of Roskilde, Denmark, for their support with the stakeholder meeting. In particular, we would like to thank Anders Rademacher Svendsen, Birgitte Norby Jensen, and Morten Lehgaard Madsen for helping to prepare for the meeting. Topographical map and satellite image copyrighted by Kort- og Matrikelstyrelsen, per agreement G23-99. Finally, we are most grateful to Gary Fry and the anonymous referees for their comments.

REFERENCES

Al-Kodmany, K. (1999). Using visualization techniques for enhancing public participation in planning and design: process, implementation, and evaluation. *Landscape and Urban Planning, 45,* 37-45.

Antrop, M. (1998a). Impact of urbanisation upon the countryside. In P. Agger, R. Bjerregaard, & J. Brandt, (Eds.), *The farmer as a natural resource-manage.* Landskabsøkologiske skrifter nr. 10, (pp. 24-41). Roskilde: Centre for Landscape Research.

Antrop, M. (1998b) Landscape ecology and planning. In: P. Agger, R. Bjerregaard & J. Brandt (Eds.). *PhD course in landscape ecology.* Roskilde, 29. May–2. June 1995. Landskabsøkologiske skrifter nr. 9, Roskilde, Centre for Landscape Research, 17-29.

Antrop, M. (2000a). Background concepts for integrated landscape analysis. *Agriculture Ecosystems Environment, 77,* 17-28.

Antrop, M. (2000b). Changing patterns in the urbanized countryside of Western Europe. *Landscape Ecology, 15,* 257-270.

Asselt Marjolein, B.A.v. & Rijkens-Klomp, N. (2002). A look in the mirror: reflection on participation in Integrated Assessment from a methodological perspective, *Global Environmental Change, 12,* 167-184.

Barbanente, A., Khakee, A., & Puglisi, M. (2002). Scenario building for Metropolitan Tunis. *Futures, 34,* 583-596.

Brandt, J., Primdahl, J., & Reenberg, A. (1996). *Rural land use and landscape dynamics – analysis of driving factors in space and time.* Landskabsøkologiske skrifter nr. 2, Centre for Landscape Research, Roskilde.

Bryner, G. (2001). Cooperative instruments and policy making: assessing public participation in US Environmental regulation. *European Environment, 11,* 49-60.

Caspersen, O.H. & Fritzbøger, B. (2002). Long-term landscape dynamics – a 300-years case study from Denmark. *Danish Journal of Geography, Special Issue 3,* 13-27.

Coeterier, J.F. (1994). Non-agricultural use of the countryside in the Netherlands: what is local opinion? *Landscape and Urban Planning, 29,* 55-69.

Dolman, P. M., Lovett, A., O'Riordan, T. & Cobb, D. (2001). Designing whole landscapes. *Landscape Research, 26* (4), 305-335.

Emmelin, L. (1996). Landscape impact analysis: a systematic approach to landscape impacts of policy. *Landscape Research, 21* (1), 13-35.

Emmelin, L. (1998). Strategic environmental assessment for tourism methodological lessons from landscape impact analysis. *Environmental Education and Information 27* (3), 227-242.

Emmelin, L. (2000). The confusing multiplicity of futures and the beguiling logic of history. In P.G. Møller, P. Holm, & L. Rasmussen (Eds.) *Aktører i landskabet* (pp. 27-55). Odense: Odense Universitetsforlag.

Ervin, S.M. (2001). Digital landscape modeling and visualization: a research agenda. *Landscape and Urban Planning, 54,* 49-62.

Federal Ministry of Science and Transport (FMST) (1999). *Szenarien der Kulturlandschaft.* Research Initiative Cultural Landscapes, report no. 5, Vienna.

Fritzbøger, B. (1998). *Det åbne lands kulturhistorie.* Frederiksberg, DSR forlag.

Fry, G. (2001). Multifunctional landscapes – towards transdisciplinary research. *Landscape and Urban Planning, 57* (3-4), 159-168.

Fry, G., Tress, B., & Tress, G. (2002). *Interdisciplinary landscape studies: matching policy with research expectations.* Presentation at 20th Session of the Permanent European Conference for the Study of the Rural Landscape, Tartu-Otepää, Estonia.

Herrmann, S., & Osinski, E. (1999). Planning sustainable land-use in rural areas at different spatial levels using GIS and modelling tools. *Landscape and Urban Planning, 46,* 93-101.

Höll, A. & Andersen, E. (2002). Landscape impact of three agricultural policy scenarios. *Danish Journal of Geography, Special Issue 3,* 59-75.

Höll, A., Andersen, E., Rygnestad, H., Dalgaard, T., Johansson, T.D., Sørensen, E.M. & Nilsson, K. (2002). Scenario analyses for cultural landscape development – a Danish, interdisciplinary research project. *Danish Journal of Geography, Special Issue 3,* 1-12.

Hoss, I.R., & Honton, E.J. (1987). Scenario planning: what style should you use? *Long Range Planning, 20 (4),* 21-29.

Jansens, J. W. (1990). Landscape development scenarios for planning and implementing agroforestry: A case study in the semi-arid lands of eastern Kenya. In W.W. Budd (Ed.), *Planning for agroforestry, Developments in landscape management and urban planning 6C* (pp. 267-292). Amsterdam: Elsevier.

Khakee, A. (1999). Participatory scenarios for sustainable development. *Foresight, 1* (3), 229-240.

Kurz, W.A., Beukema, S.J., Klenner, W., Greenough, J.A., Robinson, D.C.E., Sharpe, A.D. et al. (2000). TELSA: the Tool for Exploratory Landscape Scenario Analyses. *Computers and Electronics in Agriculture 27,* 227-242.

Lange, E. (2001). The limits of realism: perceptions of virtual landscapes. *Landscape and Urban Planning, 54,* 163-182.

Lim, E.-M., & Honjo, T. (2003). Three-dimensional visualization forest of landscapes by VRML. *Landscape and Urban Planning* (in press).

Luz, F. (2000). Participatory landscape ecology – a basis for acceptance and implementation. *Landscape and Urban Planning, 50,* 157-166.

May, G.H. (1996). *The future is ours. Foreseeing, managing and creating the future.* Westport, Conneticut: Praeger.

Nassauer, J.I. (1997). *Placing nature. Culture and landscape ecology.* Covelo, Washington D.C.: Island Press.

Nassauer, J.I., Corry, R.C., & Cruse, R.M. (2002). Alternative future landscape scenarios: A means to consider agricultural policy. *Journal of Soil and Water Conservation 57* (2), 45-53.

O'Riordan, T., Wood, C. & Shadrake, A. (1993). Landscapes for tomorrow. *Journal of Environmental Planning and Management, 36* (2), 123-147.

Orland, B. (1992). Evaluating regional changes on the basis of local expectations: a visualization dilemma. *Landscape and Urban Planning, 21* (4), 257-259.

Orland, B. (1994). Visualization techniques for incorporation in forest planning geographic information systems. *Landscape and Urban Planning, 30,* 83-97.

Orland, B., Budthimedhee, K., & Uusitalo, J. (2001). Considering virtual worlds as representations of landscape realities and as tools for landscape planning. *Landscape and Urban Planning, 54,* 139-148.

Palang, H., Alumäe, H., & Mander, Ü. (2000). Holistic aspects in landscape development: a scenario approach. *Landscape and Urban Planning, 50,* 85-94.

Primdahl, J. (1990). Heterogeneity in agriculture and landscape: from segregation to integration. *Landscape and Urban Planning, 18,* 221-228.

Roe, M. (2000). Landscape Planning for Sustainability: community participation in Estuary Management Plans. *Landscape Research, 25* (2), 157-181.

Sanoff, H. (1990). *Participatory Design: Theory and techniques.* Raleigh, NC, Bookmasters.

Schoemaker, P.J.H. (1993). Multiple scenario development: its conceptual and behavioural foundation. *Strategic Management Journal, 14,* 193-213.

Schoute, J.F.T., Finke, P.A., Veeneklaas, F.R., & Wolfert, H.P. (Eds.) (1995). *Scenario studies for the rural environment.* Boston: Kluwer.

Schwartz, P. (1992). *The art of the long view: scenario planning – protecting your company against an uncertain world.* London: Century Business.

Sheppard, S. R. J. (2001). Guidance for crystal ball gazers: developing a code of ethics for landscape visualization. *Landscape and Urban Planning, 54,* 183-199.

Swearingen White, S. (2001). Public participation and organizational change in Wisconsin land-use management. *Land Use Policy 18,* 341-350.

Swetnam, R. D., Ragou, P., Firbank, L. G., Hinsley, S. A., & Bellamy, P. E. (1998). Applying ecological models to altered landscapes: scenario-testing with GIS. *Landscape and Urban Planning, 41,* 3-18.

Towers, G. (1995). *Building democracy: community architecture in the inner city.* London: University College London Press.

Tress, B., & Tress, G. (2001). Capitalising on multiplicity: a transdisciplinary systems approach to landscape research. *Landscape and Urban Planning, 57* (3-4), 143-157.

Tress, B., & Tress, G. (2002). Disciplinary and meta-disciplinary approaches in landscape ecology. In O.Bastian & U. Steinhardt (Eds.), *Development and perspectives in landscape ecology* (pp. 25-37). Dordrecht: Kluwer.

Tress, B., & Tress, G. (2003). Scenario visualisation for participatory landscape planning – a study from Denmark. *Landscape and Urban Planning* (in press).

Tress, B., Tress, G., Brandt, J., & Agger, P. (1999). *Scenarier for landskabsudvikling og flersidig brug af det åbne land. – Forslag til scenariestruktur indenfor projektet "Værdi, landskab og biodiversitet".* Centre for Landscape Research, Roskilde.

Van den Berg, L.M. & Veeneklaas, F.R. (1995). Scenario building: art, craft or just a fashionable whim? In J.F.T. Schoute, P.A. Finke, F.R. & H.P Veeneklaas (Eds.) *Scenario Studies for the Rural Environment* (pp. 11-13). Boston: Kluwer.

Vos, W., & Meekes, H. (1999). Trends in European cultural landscape development: perspectives for a sustainable future. *Landscape and Urban Planning, 46,* 3-14.

Warburton, D. (1998). Participation in conservation: grasping the nettle. *Ecos, 19* (2), 2-11.

Wates, N. (1985). Co-op consolidation. *The Architectural Review, 177* (1058), 57-64.

Whyte, W.F. (1989). Advancing scientific knowledge through participatory action research. *Sociological Forum, 4* (3), 367-385.

Wollenberg, E., Edmunds, D. & Buck, L. (2000). Using scenarios to make decisions about the future: anticipatory learning from the adaptive co-management of community forests. *Landscape and Urban Planning, 47,* 65-77.

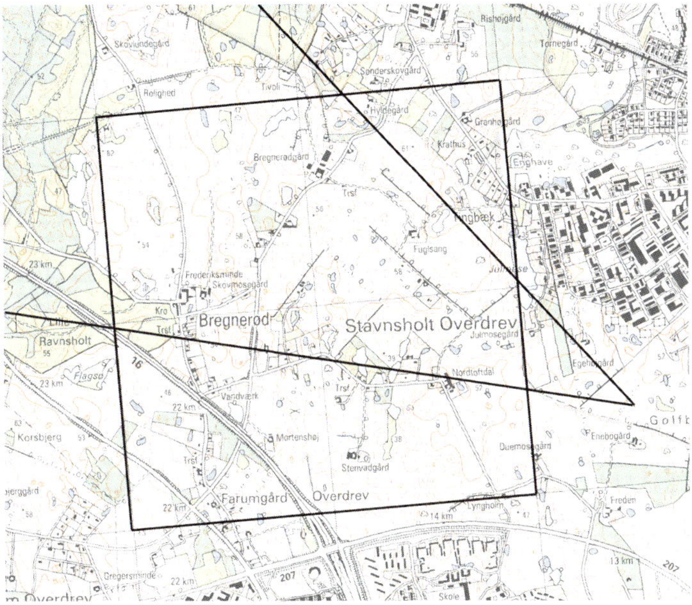

Figure 11-2. Map of Stavnsholt Overdrev

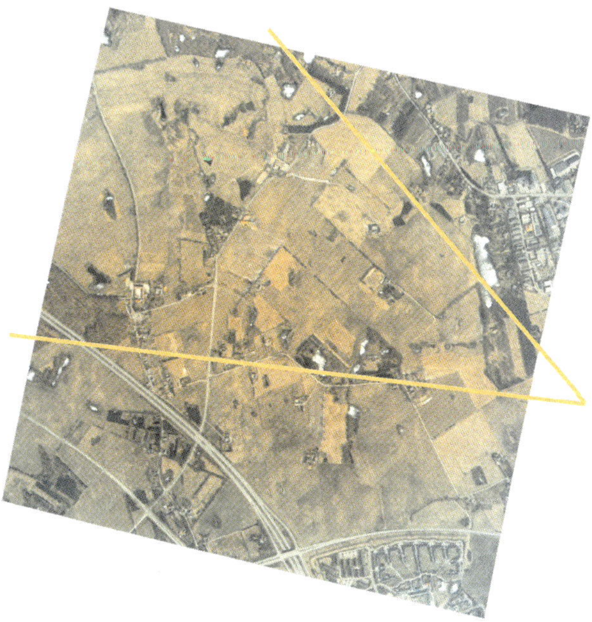

Figure 11-3. Stavnsholt Overdrev (1996), satellite image

Figure 11-4. Stavnsholt Overdrev (1999), slope angle aerial photography

Figure 11-5. Industrial farming dominating Stavnsholt Overdrev in 2020

Figure 11-6. Recreation and tourism dominating Stavnsholt Overdrev in 2020

Figure 11-7. Nature conservation dominating Stavnsholt Overdrev in 2020

Figure 11-8. Residential expansion dominating Stavnsholt Overdrev in 2020

Figure 11-9. Stavnsholt Overdrev in 2000

Figure 11-10. Integrated scenario for Stavnsholt Overdrev 2020

Figure 11-11. Stavnsholt Overdrev in 2000

Figure 11-12. Photorealistic simulation of the Stavnsholt Overdrev area in the second half of the 19th century

Figure 12-1. The 1874 cadastral map of the farm Grinde in Leikanger municipality including the holding investigated, Grinde-Eineberg

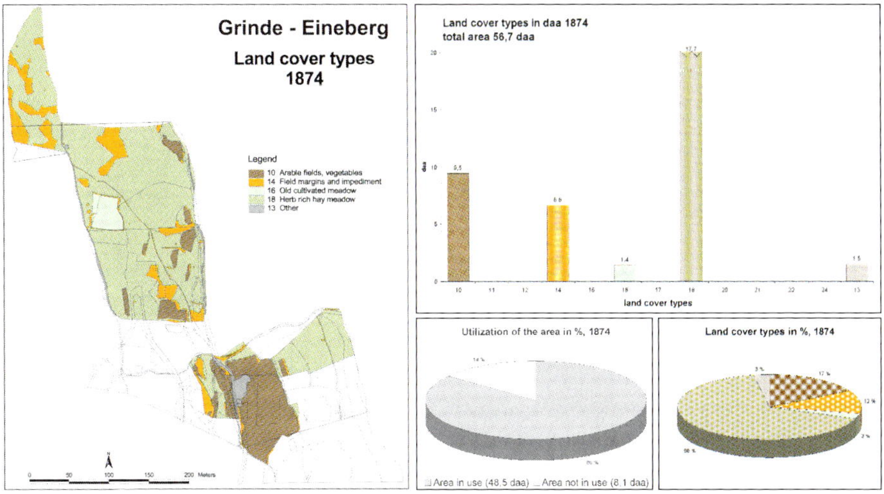

Figure 12-2. Digitized cadastral map of Grinde-Eineberg showing tilled fields, meadows and impediments, size and percentage distribution of land cover types in 1874. Area in use consists of the categories 10, 16 and 18

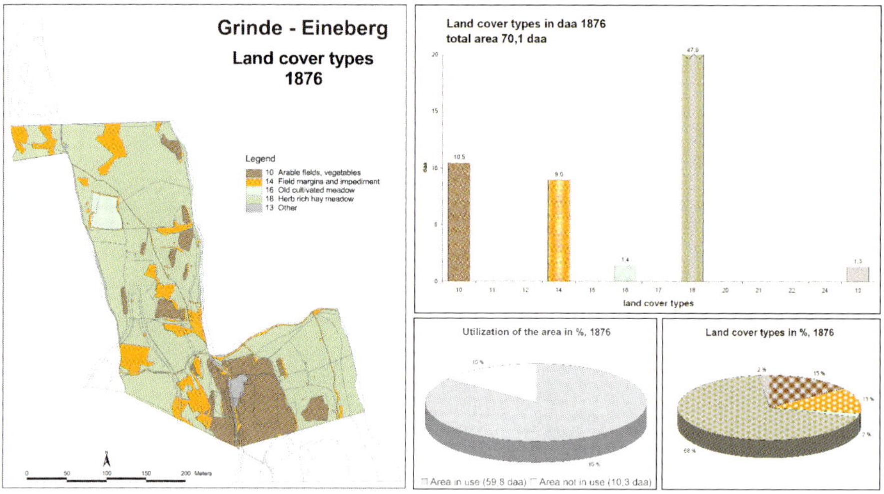

Figure 12-3. Digitalized cadastral map of Grinde-Eineberg showing tilled fields, meadows and impediments, size and percentage distribution of land cover types in 1876. Area in use consists of the categories 10, 16 and 18

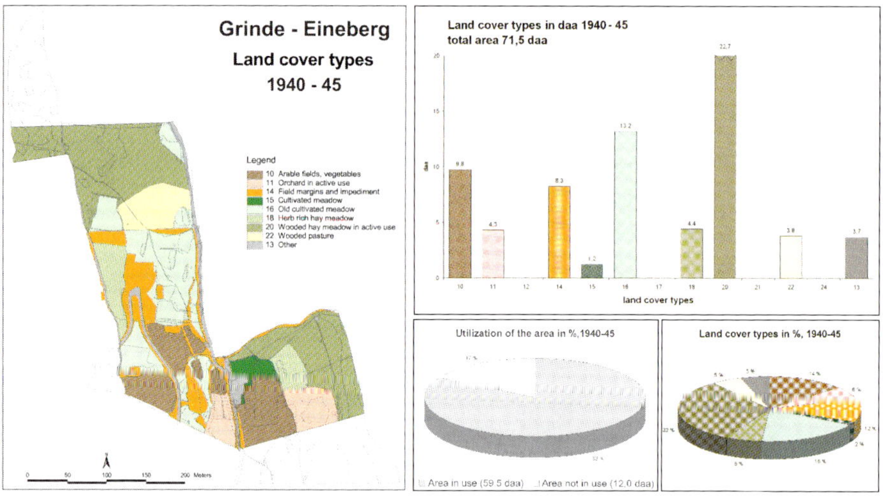

Figure 12-9. Distribution and size of areas and land cover types at Grinde-Eineberg 1940-1945. Area in use consists of the categories 10, 11, 15, 16, 18, 20 and 22

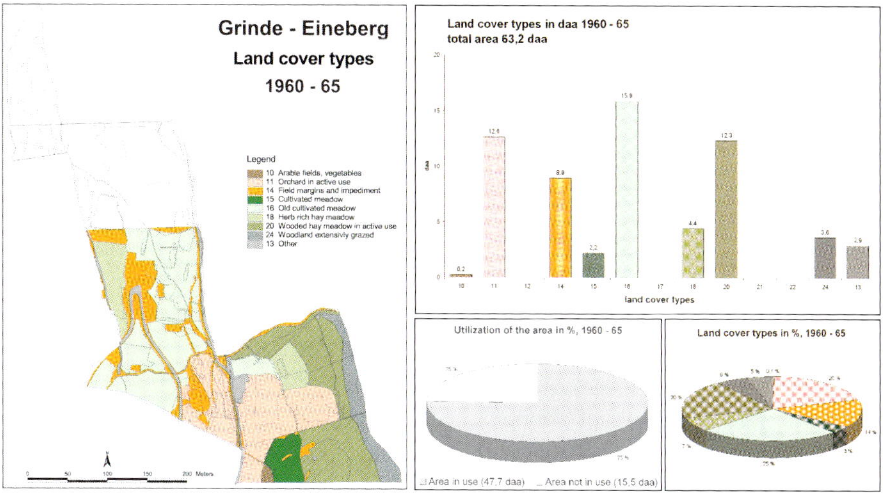

Figure 12-10. Distribution and size of areas and land cover types at Grinde-Eineberg 1960-1965. Area in use consists of the categories 10, 11, 15, 16, 18 and 20

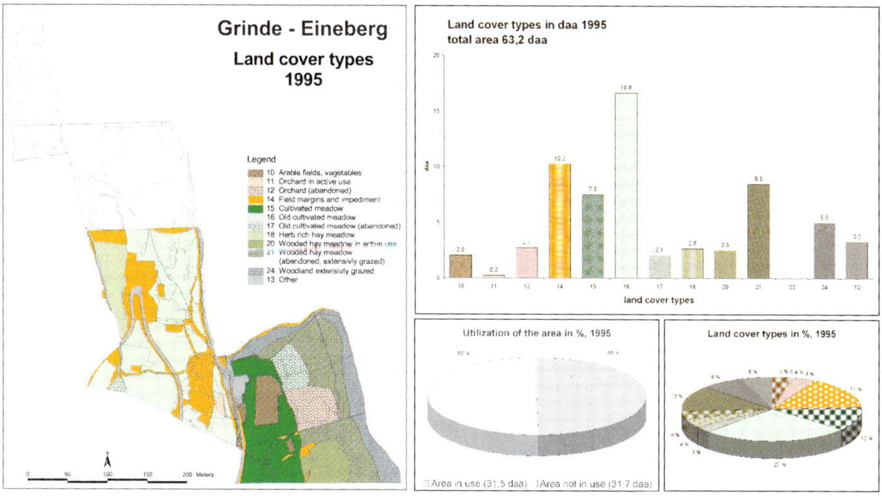

Figure 12-11. Distribution and size of areas and land cover types at Grinde-Eineberg 1995. Area in use consists of the categories 10, 11, 15, 16, 18 and 20

Chapter 12

HISTORICAL CADASTRAL MAPS AS A TOOL FOR VALUATION OF TODAY'S LANDSCAPE ELEMENTS

Stein Tage Domaas, Ingvild Austad, Jan Anders Timberlid & Ann Norderhaug
Sogn og Fjordane University College, Norway

1. INTRODUCTION

Physical and biological traces of human activity in the landscape such as various stone structures, buildings, irrigation channels, semi-natural vegetation types and land use patterns can give us information about the past and particular historical periods. However, it can often be difficult to fix the exact age of such traces, and to evaluate their continuity, because fragments dating from different parts of a long history may be recovered alongside each other in today's landscape. Historical cadastral maps may be very useful tool for evaluating such landscape elements. They can be used to distinguish between older and younger man made structures, and may thus tell us, which structures survived land consolidation, which structures were built later on, and which have been lost. Knowledge of the types and locations of previously existing structures may also help us to understand other traces in today's landscape. An understanding of the history of the landscape gives us information on areas with a long continuity of land use and management. Historical materials and historical maps have therefore been used in a number of ecological and geographical studies (Foster 1992; Kienast 1993; Skånes 1996; Ihse & Blom 2000; Cousins 2001). In Sweden, Riddersporre (1995) described the old agricultural landscape from before the 17th century and Borgegård (1994) demonstrated how historical maps could be used in the reconstruction of an old agricultural landscape. He also

H. Palang and G. Fry (eds.), Landscape Interfaces, 221–236.

discussed the possibility of dating various structures in the cultural landscape. In Norway, Jerpåsen, Sollund and Widgren (1997) did a pilot study producing overlays on the basis of late 18th and early 19th century cadastral maps to make simple comparisons between the landscape then and the present-day landscape to locate archaeological sites and to study the effect of the land consolidation.

The aim of the present study was to develop a more advanced methodology connecting the information from historical cadastral maps to the landscape situation of today (1995) using new technology (GIS) both to obtain a better understanding of the history, dynamics and development of the farm and to identify the oldest traces and most important areas for protection and management. One aim of the study was to determine whether cultural landscape structures that are generally perceived as old are in fact the result of recent developments or long-lasting processes. It was necessary to carry out a detailed land survey in order to get the needed information on today's landscape situation. Due to the time/cost factor this was done on only one holding at the farm for this first investigation.

2. MATERIAL AND METHODS

2.1 The study area

The farm under investigation, Grinde, has a long history of human activity and settlement, traced back to 2505-2415 BC (collection of twigs and leaves) and 1875-1680 BC respectively (Austad et al. 2001). The farm is located on the northern side of the Sognefjord in Leikanger municipality, Western Norway, and is situated on a terrace 100-125 m above sea level. The farm is a typical fjord-hill farm with a vertical distribution of farmland areas and a small-scale landscape structure. According to the survey of valuable cultural landscapes carried out by the Directorate for Nature Management (*Direktoratet for naturforvaltning* 1994), the cultural landscape of the farm is one of the most valuable and authentic in the County of Sogn og Fjordane. This study comprises one of the holdings on the farm Grinde, named Eineberg, which is thought to be one of the most valuable and authentic on the farm.

The infield area of the farm has been fairly stable covering an area of 630-650 da (63-65 ha)[1]. On the other hand the infield area of the holding used for this study (Eineberg) has changed from 56-70 and from 70-63 da due to purchasing of land and land consolidation.

[1] Due to the small size of the research area all measures are given in da. 1 da = 0.1 ha.

2.2 The cadastral map of 1874

In 1857, a new Land Consolidation Act was passed in Norway, and brought about extensive reorganization of agricultural areas, partly because land consolidation procedures were required even if only one of the farmers in an area requested this. Before this, strip-farming had been practiced on many farms, especially in western Norway, where it was a result of long-lasting agricultural practices. This began at quite an early date to ensure that land, particularly areas suitable for grain production, was fairly partitioned between all those entitled to a share of the farmland (both tenants and owners) (Frimanslund Holmsen 1976; Sevatdal 1991). Rules for allocating strips of land may be found in the old Norwegian regional laws that were later collected in King Magnus the Lawmender's national code of 1274, and after that in the 1687 Norwegian Law Code (*Norsk Lov* in Norwegean).

The 1857 Land Consolidation Act required land surveys and the production of maps and written protocols. The cadastral maps that resulted provide information on land use and different types of management. They usually describe not only the distribution of agricultural fields, meadows and pastures, wetlands, heathlands and impediments, but also indicate point elements such as buildings, heaps of stones and linear elements such as footpaths, roads, fences, streams and rivers. The written protocols make such maps even more informative.

In this investigation we used a historical cadastral map from the cluster farm Grinde dating from 1874. The written protocol gives information about the infield areas of the farm, including ownership, land use and production capacity/soil quality. Production capacity/soil quality (*bonitet* in Norwegian) was used to classify parcels of land according to how much they could produce, to ensure that when a farmer lost a particular area during land consolidation, the land received in compensation would give an equivalent amount of agricultural production. At Grinde production capacity/soil quality was divided into 42 classes on a scale from 1 to 200. Class 1 was the most valuable, and 2.75 or higher correspond to low production capacity. An area of land in class 2 would produce half as much as an equal-sized area in class 1. The most valuable areas (tilled fields and high quality hay meadows) were very finely divided, so that 12 classes were used for the quality classes between 1 and 2.

The cadastral map covers an area of 630 da, where there were eight holdings and three cotters' farms in 1874. One more farmer and two cotters farmed land in the area investigated, but lived outside it. The buildings belonging to most of the holdings (seven) were situated in a small hamlet in the centre of the area (Figure 1). One of the holdings, Grinde-Eineberg, and the three cotters' farms were located outside and to the north of the hamlet in 1874.

In this investigation GIS was used to construct a series of detailed maps for the holding Grinde-Eineberg based on the historical cadastral map and the protocol.

2.3 Land Survey of 1995

To be able to describe and compare the landscape picture at different points in time after the land consolidation, it was necessary to carry out a detailed land survey. For the holding Grinde-Eineberg such a survey was completed in 1995. The coordinate system NGO1948 used for Norwegian economic maps was chosen and the accuracy was better than 10 cm (Domaas & Ihse 2003).

2.4 GIS

Modern computer technology and geographical information systems, GIS, have been used to handle the huge volumes of information provided by the historical land consolidation map and the written protocol. The varying geometry and local coordinate systems of historical cadastral maps make it necessary to transform them onto a modern coordinate system. For this investigation the coordinate system NGO1948 used for Norwegian economic maps was chosen. A simple GIS-based method for transformation was developed and tested out in this area. The transformation accuracy proved to be better than 2 m, the same or as good as the standard set by the Norwegian Mapping Authority for modern economic maps (Domaas & Ihse in print). The use of GIS to link historical maps with registrations of elements and structures in today's landscape has thus been found to be a suitable method for evaluating the age and continuity of various traces of human activity in an area.

The use of GIS made it possible to study the holdings in detail, for instance the distribution and patterns of different types of farmland and parcels of land, the size of tilled fields compared to hay meadows, and how poor soils and high quality soils were distributed on the different properties. The landscape picture could then be compared with other important historical information to give a more precise understanding of the situation for each individual holding. In this study, we focused on the years 1874 and 1876, just before and after land consolidation, the situation during WW II (1940-1945), the period 1960-1965, and the current situation (1995), identifying changes in land use and man made structures on the holding.

3. RESULTS

3.1 Land consolidation period 1874-1876

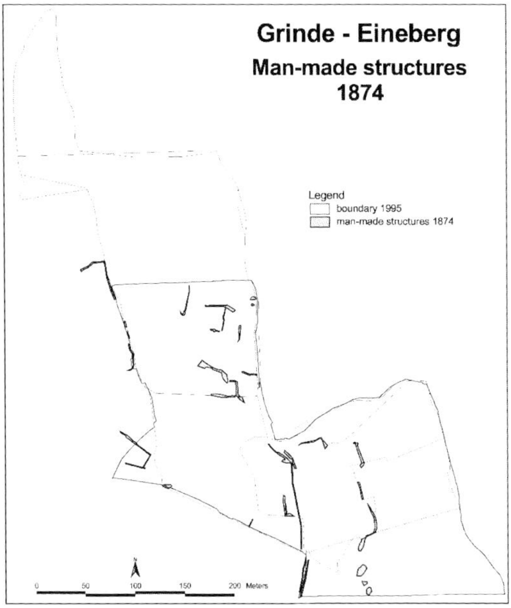

Figure 12-4. Man made structures on the holding of 1995 in 1874

Figures 2 and 3 present the infield areas on the holding Grinde-Eineberg in detail before and after land consolidation, in 1874 and 1876 respectively. The size and percentage distribution of the various types of land cover are also shown. The ratio between tilled fields and meadows is 1 to 4. Tilled fields made up 10 da in both 1874 and 1876, and were surrounded by hay meadows, which accounted for most of the fenced land. Impediments (bare rocks, cliffs, scree, roads and field margins) were included in the survey. The total infield area was 56.7 da in 1874 and increased to 70.1 da in 1876, since extra land was purchased during the consolidation process. At the same time, the length of boundaries dropped from 1891 m in 1874 to 1549 m in 1876, demonstrating that the fenced infield areas became more compact and more coherent in structure, even though most of the infields were still situated to the north of the buildings, as they were in 1874. The main road to the outfields passed through the Grinde-Eineberg farmyard, up through the Grindsdalen valley. The road was also used as a droving track, and was partly bounded by dry stonewalls. Another droving track ran along the western boundary of the property. Many of the tilled fields had terraces in the lower part, and both terraces and dry-stone walls can be identified on the 1874 map (Figure 4).

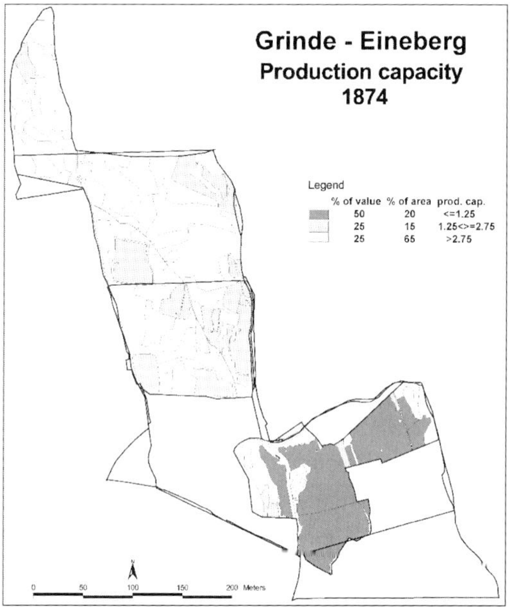

Figure 12-5. Distribution, areas and percentages of soil quality/production capacity classes on the holding Grinde-Eineberg in 1874

Figures 5 and 6 show the distribution of productivity/soil quality on the holding before and after land consolidation, based on information from the written protocol. In 1874 the holding had a relatively small area (10.7 da) of high quality soil of 1.25 or better. Soil quality/production capacity was poorer than 2.75 on most of the infield area (37.3 da). Infield areas of low quality classes were found in the upper hay meadows to the west and northwest. The map indicates that they have a surface structure characterized by a high content of stones, and this is confirmed by their description as 'stony' in the protocol (Figure 7). This may be why they are classified as being of low quality (see Figures 1 and 7).

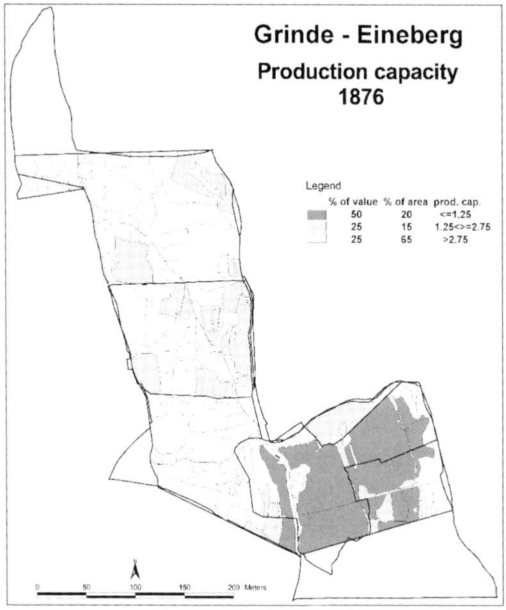

Figure 12-6. Distribution, areas and percentages of soil quality/production capacity classes on the holding Grinde-Eineberg in 1876

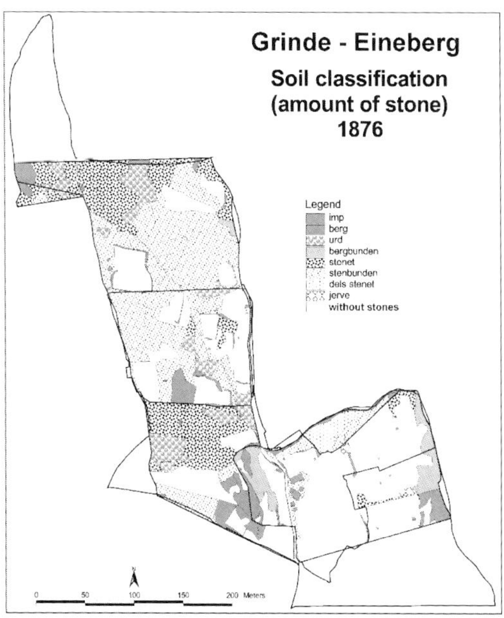

Figure 12-7. Soil classification map (amount of stone) for Grinde-Eineberg in 1876

The 20% (10.7 da) of high quality soil (tilled fields and hay meadows) accounted for 50% of the production value (production capacity) of the holding and the 65% (37.3 da) of poor quality soil accounted for only 25% of the production value (Figure 8). The situation was the same after the land consolidation, confirming that the basic principle was followed: that is, the value of a farm should be more or less the same before and after a land consolidation transaction (Figure 8).

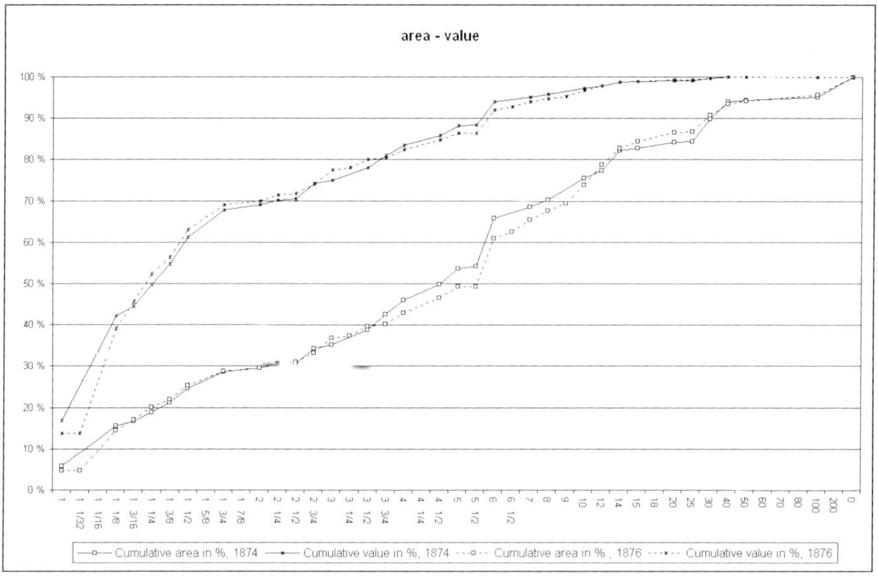

Figure 12-8. Area and corresponding value of the holding in percentage for the years 1874 and 1876. Based on the production capacity

In 1875, eight people lived in the holding. The livestock was varied, and included one horse, nine cows, six calves (0-3 years), 21 sheep and two pigs. Barley and potatoes were grown. The holding was reported to consist of much stony ground and scree in the infields. The latter half of the 1800s was a period when the resources in infields, outfields and mountains were exploited to capacity or even overexploited (Austad et al. 2001).

3.2 Changes in Grinde-Eineberg 1940-1945

Figure 9 shows the situation during WW II. The boundaries of the holding were the same as in 1876. However, there had been great changes in man made structures, surface structure (clearing of stones) and land cover types. Some areas to the north still remained as unfertilized wooded hay meadows or wooded pastures with an intact surface structure. However, stones were cleared from the surface of several of the upper, stony hay meadows in the 1930s (Mjaatvedt 1993). Several dry-stone walls that earlier bounded the smaller, scattered tilled fields were also removed, while new heaps of stones were piled up in the infields. Most of the dry-stone walls and terraces found in the landscape in 1995 were built in 1930s.

The management of the meadows in 1940-1945 was more varied, with seminatural unfertilized (herb-rich) hay meadows and wooded hay meadows, fertilized meadows, and old tilled fields used as meadows. Pollarded ash, elm, birch and willow trees were characteristic features of the infield areas. Some of these trees were planted in the 1930s. Leaves and twigs were important as winter fodder for domestic animals during the 19th century and were still used during WW II.

Two horses, 8-10 cows with calves and some 20 sheep were kept through the winter in the holding. Outfields and mountain areas were used intensively during this period as well. Fruit cultivation had begun on the holding, and according to the 1939 statistics, there were 190 fruit trees on Grinde-Eineberg. During WW II, a new road to the neighboring farm Engeseter was built, and the old road through Grindsdalen was improved and relocated. There were also changes in the farm buildings. The farmhouse was enlarged towards the north and west and a second floor was added in 1925, most likely all around the original small log cabin. The combined barn/cowshed was altered several times, and in 1930 an extension to the south was built. Other buildings recorded in 1940-1945 were a storehouse on pillars and an old stable (Austad et al. 2001).

3.3 Changes in Grinde-Eincberg 1960-1965

A second land consolidation transaction was carried out in 1958-1959. The infield areas were reduced to 63.6 da in 1960, and the boundaries reduced to 1,453 m. Property to the north was exchanged for two parcels of land to the south, and the infields became even more concentrated around the farm buildings (Figure 10). Most areas were under some form of management, but the wooded meadows along the river Grindselva were abandoned and scything and pollarding almost ceased by the end of the 1950s. The areas of tilled fields were small, grain production came to an end, and fruit plantations and hay production became dominant. In the 1959 agricultural survey, as much as 26 da of fertilized meadow was recorded, and 33 da was reported to be semi-natural (herb-rich) hay meadows. In this period, no man made structures were removed, nor were any new heaps of stone added. The farmyard was still the same, but some buildings were altered and new ones built, including a sawmill in 1958. After 1963, the only livestock kept were sheep.

3.4 The situation of Grinde-Eineberg in 1995

Boundaries were the same in 1995 as in 1963 (see Figures 10 and 11), but management regimes and land use were still changing. The arable fields were now dominated by raspberry cultivation. Large areas of hay meadows were no longer in active use, only sporadically grazed by sheep in spring and autumn. This means that only half of the production areas were under active management. In 1970 the storehouse was removed, and later on also the stable, but no other man made structures had been added or removed since 1940-1945. The barn/cowshed was given a new roof of stone slabs in the early 1990s, and the farmhouse windows were replaced at the same time. Dry-stone walls and heaps of stones were still characteristic features of the cultural landscape in 1995 (Figure 12). In 1995, only two people lived in the holding Grinde-Eineberg. The livestock were sheep.

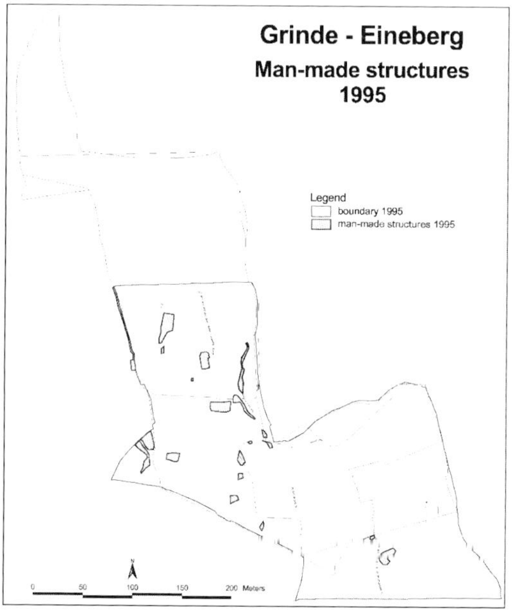

Figure 12-12. Man made structures in the holding in 1995

4. DISCUSSION

4.1 Continuity and age – farm area, property boundaries and types of land cover

The infields belonging to the holding Grinde-Eineberg were already more or less contiguous blocks in 1874. We therefore presumed that land consolidation would have led to few changes here. We expected to find continuity of use and ownership for a relatively large percentage of areas and boundaries on this holding. However, the results show that there have been considerable changes here, too. Only about a half of the infield area (28.6 da) in 1995 corresponded to infields in 1874 (Figure 13). The results also show that only a small proportion, 8% or 117 m, of the property boundaries in 1995 can be traced back to 1874. This is unexpectedly low. Half of this consists of natural boundaries that follow the terrain, for example the river Grindselva, while the rest consists of boundaries along droving tracks, which also mostly follow terrain lines. Other boundaries, especially those that cross terrain lines, have changed to a greater extent. The extensive changes in property boundaries were unexpected as such changes are generally associated with legal claims, and property boundaries are considered to be linear elements that show great continuity in the cultural landscape (Sky 1995).

The degree of scattering of infield areas as shown by old cadastral maps may prove to be useful in identifying older and newer properties. The holding Eineberg was established in 1854, when one of the other holdings that formed part of Grinde was sold and divided between different owners. This may explain why its infield areas were so concentrated even before the 1874 land consolidation.

Continuity of land cover can be identified in the wooded hay meadows along the river Grindselva and in the less fertile meadows where the surface structure is intact. Although these areas have been less actively managed since 1965 (Austad & Losvik 1998), they have kept much of their structure and semi-natural vegetation. Pollarded trees, many of which may be dated back to 1874 and earlier, add to the impression of an old, historical landscape. On the other hand, the tilled fields have been through many changes, from grain to potatoes and vegetables, to orchards then raspberry plantations, and most recently to cultivated and fertilized meadows.

The Grinde-Eineberg farmyard and buildings were established in 1854, some years before the farmyards and buildings belonging to most of the other holdings at Grinde, which were moved out from the cluster-type farmyard after the first land consolidation in 1874. However, surveying and registering the buildings existing today has revealed that although the farmyard itself is unchanged, the buildings have gone through comprehensive changes in size, localization, number and architectural design.

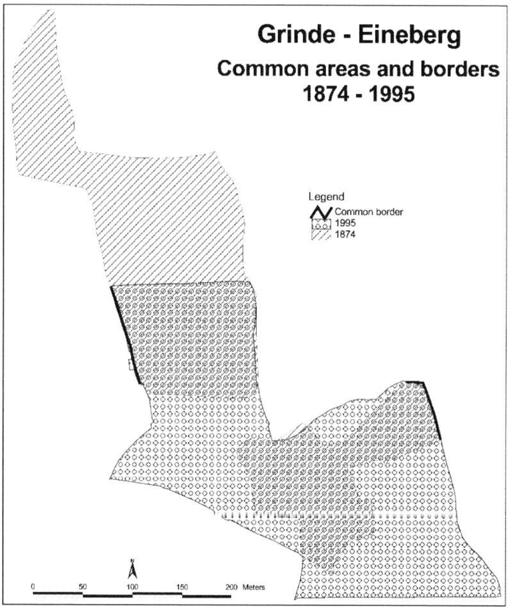

Figure 12-13. Continuity of property and boundaries at Grinde-Eineberg

An exchange of areas and new boundaries do not necessarily result in major landscape changes as long as the areas are managed in the same way on neighboring holdings. For example, the wooded hay meadows situated along the river Grindselva and the upper hay meadows were used in the same way up to the early 1960s even though property boundaries were moved during land consolidation. Later on however, management of the meadows, pastures and woodlots changed in areas no longer belonging to Eineberg, with intensified cultivation and adaptation to modern activities, planting of spruce, or abandonment. The result is that areas that used to be similar in quality class, structure and land use now appear to be totally different.

Changes in land use have been documented also on the holding Grinde-Eineberg. More than half of the areas were no longer in use in 1995, especially the steep areas and semi-natural vegetation types such as the wooded hay meadows. Other meadows on areas that were formerly ploughed near the farmyard are heavily fertilized and intensively used. However, the traditional cultural landscape is better preserved here than on many Western Norwegian farms, and quite a lot of the small-scale structure is still intact.

Some semi-natural (herb-rich) hay meadows are also in use and quite a number of monumental old pollarded trees are still being harvested in the traditional way.

4.2 Continuity and age: man made structures

There are large numbers of dry-stone walls, heaps of stones, droving tracks and terraces on the modern infields, but few of them are really old. Many of them are a result of the intensification of use and clearing of marginal areas encouraged by the government, especially in the 1930s (Mjaatvedt 1993). This was probably related to the introduction of 'modern' (horse-drawn) machinery. Most clearing activity can be identified in the upper meadows, where large heaps of stones, walls and terraces were built in addition to the already existing ones, or replaced the old ones. Figure 14 shows which of the man made structures found in the property in 1995 can be traced back to 1874, and where they are located.

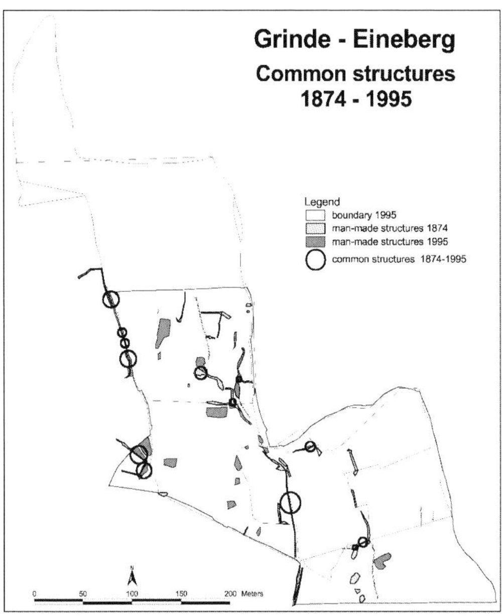

Figure 12-14. Man made structures from 1874 identifiable in the landscape in 1995

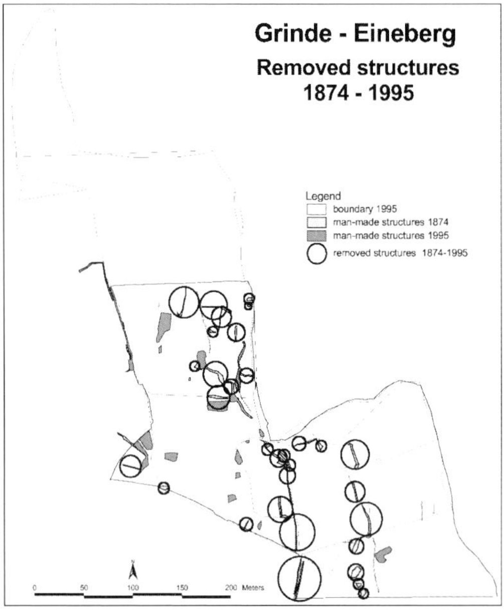

Figure 12-15. Man made structures removed 1874-1995

In 1874 there were 39 man made structures on this part of the infields. In 1995 there are 28. But only 13 of these can be traced back to 1874 and of those five contain only small parts of the structure of 1874. This means that 26 man made structures of 1874 have been removed, in addition five structures have been partly removed (Figure 15). On the other hand some of the structures have been enlarged so that what were several structures in 1874 appears as one structure in 1995.

4.3 Use of GIS: degree of detail

Based on traditional landscape interpretation methods, the cultural landscape at this farm could be assessed as one of the most authentic in the County of Sogn og Fjordane (*Direktoratet for naturforvaltning* 1994). The results of this study show that very few landscape elements can be traced back to 1874.

This indicates that geographic time studies based on so detailed materials open up for a whole new understanding of continuity and age for all parts of the landscape picture. Since traditional methods for interpretation of the cultural landscape are fairly international, these results may very well be valid other places as well.

5. CONCLUSIONS

This investigation reveals that even if a landscape appears to be ancient and to contain old man made structures, this may not be the whole truth. In this study we could identify continuity in landscape elements and land use as well as a number of changes since 1874. The extensive changes in property areas and boundaries were a result of the land consolidation transactions. The greatest changes in land use and management occurred after World War II, while the most extensive changes in man made structures were made 1876-1940. Many structures have been removed, but have often been replaced by new ones similar to the old ones. This underlines the importance of identifying the oldest landscape elements. The study also reveals that property areas and boundaries may change relatively swiftly, while land cover types and agricultural utilization of specific areas may be more long-lasting.

The study further confirms that digitised and transformed historical cadastral maps combined with detailed land surveys are very useful for evaluating today's landscape elements. They can help us to identify the most important areas for protection and further investigation, such as registration of archaeological remains, excavation and sampling for carbon-dating. Such investigations can focus on particular structures or parts of them, thus saving time and money. The knowledge gained from studies of this type is also valuable for the protection and management of valuable cultural landscapes, and enables us to avoid the destruction of important historical traces and structures. This type of detailed knowledge is also of interest for teaching and tourism, as well as for the local people as it may improve their sense of identity.

ACKNOWLEDGEMENTS

The results presented in this paper are from an interdisciplinary research project on the history and development of four fjord and hill farms (Norderhaug et al. 1996; Austad et al. 2001). The project has received financial support from the Research Council of Norway.

We would like to thank the landowner, Lars Grinde, for his enthusiastic support and for giving us a free hand to carry out the investigation. We are grateful to Kari Kvamme and Alison Coulthard who have revised the English text.

REFERENCES

Austad, I. & Losvik, M.H. (1998). Changes in species composition following field and tree layer restoration and management in a wooded hay meadow. *Nordic Journal of Botany, 18,* 641-662.

Austad, I., Øye, I., Andersen, R., Auestad, I., Domaas, S.T., Gjerland, B., et al. (2001). Den tradisjonelle vestlandsgården som kulturbiologisk system. In B. Skar (Ed.), *Kulturminner og miljø. Forskning i grenseland mellom natur og kultur* (pp. 135-205). Norsk institutt for kulturminneforskning.

Borgegård, S-O. (1994). Rekonstruktion av odlingslandskap med hjälp av historiska källor – exemplet Biskops-Arnö, Uppland. *Svensk Botanisk Tidsskrift, 88,* 341-352.

Cousins, S.A.O. (2001). Analysis of land-cover transitions based on 17th and 18th century cadastral maps and aerial photographs. *Landscape Ecology 16*, 41-54.

Direktoratet for naturforvaltning. (1994). *Verdifulle kulturlandskap I Norge. Mer enn bare landskap! Del 4.* Sluttrapport fra det sentrale utvalget for Nasjonal registrering av verdifulle kulturlandskap – DN. Trondheim.

Domaas, S.T. & Ihse, M. (2003). A simple GIS-method for transformation and evaluation of accuracy of historical cadastral maps onto a modern coordinate system. In I. Austad, E. Ådland & R. Andersen (Eds.) *Vestlandsgården og småskala landbruket. Trusler, verdier og utfordringer*. Bergen Museums skrifter. In print.

Foster, D.R. (1992). Land-use history (1730-1990) and vegetation dynamics in central New England, USA. *Journal of Ecology 80,* 753-772.

Frimanslund Holmsen, R. (1976). Steinsland. En vestnorsk mangbølt gård. *Dugnad 1976, 1,* 40–56.

Ihse, M. & Blom, G. (2000). A Swedish countryside survey for monitoring of landscape features, biodiversity and cultural heritage in Sweden. In Ü. Mander & R.H.G. Jongman (Eds.), Consequences of Land Use Changes (pp. 39-74). WIT Press.

Jerpåsen, G., Sollund M-L. B. & Widgren, M. (1997). Historiske kart og kulturminnevern. En metode for landskapsanalyse. *NIKU Fagrapport 003.*

Kienast, F. (1993). Analysis of historic landscape patterns with a Geographical Information System – a methodological outline. *Landscape Ecology 8* (2), 103-118.

Mjaatvedt, S. (1993). Kulturhistoriske undersøkelser på Grinde 1992. *NLVF/NAVF-prosjektet. Endring i biologisk mangfold i tradisjonelle kulturmarkstyper på Vestlandet ved gjengroing, tilplanting og skjøtselstiltak.* Høgskulen i Sogn og Fjordane, Avd. for naturfag. (Manuscript).

Norderhaug, A., Austad, I. & Hauge, L. (1996). Holistic studies of land use changes: a focus on the traditional Western Norwegian farm as a biological and cultural system. In R.H.G. Jongman (Ed.) Ecological and landscape consequences of land use change in Europe. *Proceedings of the first ECNC seminar on land use change and its ecological consequences.* Tilburg.

Riddersporre, M. (1995). Bymarker i backspegel. Odlingslandskapet före kartornas tid. *Meddelande från Lunds Universitets Geografiska Institutioner, avhandlingar 124.*

Sevatdal, H. (1991). Jordfellesskap og klyngetun. *Jord og gjerning* 1991. Årbok for Norsk landbruksmuseum (pp. 54-65).

Sky, P.K. (1995). *Eiendomsutforming, teigstruktur og jordbruksdrift. Geografisk informasjonssystem som analyseverktøy.* Thesis. Institute for Landscape Planning. Norwegian Agricultural University.

Skånes, H. (1996). *Landscape change and grassland dynamics. Retrospective studies based on aerial photographs and old cadastral maps during 200 years in south Sweden.* Stockholm: University dissertation series, 8.

Chapter 13

FROM OBJECTS TO LANDSCAPES IN NATURAL AND CULTURAL HERITAGE MANAGEMENT: A ROLE FOR LANDSCAPE INTERFACES

Gary Fry
Institute for Land Use and Landscape Planning, Agricultural University of Norway, Norway

1. INTRODUCTION

This essay sets out to explore some of the debates concerning the interface between two landscape research disciplines; landscape archaeology and landscape ecology. The former is a long established field within archaeology whereas the latter is a relative newcomer that amalgamates spatial and temporal landscape interests. Landscape ecology has grown from inputs from a wide range of disciplines including ecology, geography and landscape architecture. Landscape archaeology, although more established, has had its ups and downs, reaching a high in the 1960s and 70s when rapidly developing technologies and numerical methods provided new tools in the form of pattern analysis techniques. However, the pursuit of pattern as an end in itself with little interpretation or links to social processes gave rise to criticism and internal conflicts. Pattern analysis was often associated with the extreme end of environmental determinism. During the last decade there has been a clear revival of interest in landscape studies in general that has resulted in major landscape research programmes often incorporating both archaeology and ecology. I believe that this popularity of landscape approaches owes much to two movements that are probably not independent. Firstly, the practical relevance of landscape issues to areal planning and, secondly, the advances in landscape perspectives including theoretical advances in landscape research. Nevertheless, it is in this area of theory

H. Palang and G. Fry (eds.), Landscape Interfaces, 237–253.

development that we also find a great void. Advances in landscape theory have been slow. Even though there have been significant development in single subjects, there has been little progress in theory development related to holistic perspectives. I would argue that to advance integrative theory remains one of our greatest research challenges, especially theory that crosses epistemological paradigms and may bring new insights to old questions.

Although this chapter concentrates on the interfaces between the cultural and natural heritage aspects of landscape, I wish to emphasise that these are but two of the many interests competing for territory in future European landscapes. They are, however, two aspects of landscapes that, in one way or another, are closely related to most of the other interests - whether production systems (farming and forestry), recreation, aesthetics etc. The focus on natural and cultural aspects of landscape also reflects that they are very closely interwoven in all European landscapes. In addition, they are currently two of the major policy arenas. Europe possesses an enormously rich variety of cultural landscapes, with many-layered small-scale mosaics reflecting local social or tenure patterns. These intricately pattered landscapes also contribute to a sense of place and local identity for the people who live there; aspects that are increasing in importance across Europe. This may explain why there is widespread public support for the maintenance of such landscapes as an important component of cultural and natural heritage. Nature management and cultural heritage management have another perspective in common, they have both moved their focus from sites to landscapes. Many countries are developing spatial standards for strategic planning. Incorporation of a landscape perspective in sectoral action plans, especially biodiversity action plans, as well as action plans specifically in response to the Landscape convention (see Chapters 6, 7 and 17). Likewise, we see a growing movement within landscape archaeology to include the context of important sites whether Stone Age rock art or sites of modern industrial archaeology. As a response to these policy changes for the wider countryside, we also observe a significant increase in interdisciplinary landscape planning projects in Europe where nature and culture are central themes (Tress et al. 2003).

2. INTERFACING LANDSCAPE ECOLOGICAL CONCEPTS AND THEORIES

Subject disciplines tend to develop quite different jargon and technical terms for similar concepts and processes. This is true of landscape ecology and landscape archaeology, but with one difference, an important difference.

The remit of landscape ecology has always included the historical development of landscapes and hence the cultural perspective has been an important aspect of landscape ecology during its evolution as a research theme. In many ways landscape ecology is an overarching subject as defined by the International Association for Landscape Ecology:

> "Landscape ecology is the study of the interactions between the temporal and spatial aspects of a landscape and its flora, fauna & cultural components" (Dover and Bunce 1998).

This has naturally caused some problems in defining the core of the subject (Hobbs 1997) and for internal communication between scientists from widely different knowledge cultures. But this breadth has also been one of the great strengths of landscape ecology. The subject has evolved as interdisciplinary and taken on working methods and concepts from a wide range of academic subjects spanning both qualitative and quantitative approaches. This has opened the door for theoretical development across subject boundaries and, even more importantly, fostered transepistemic approaches. One such area of cross-over conceptual development is between landscape archaeology and landscape ecology. Here we find much common ground even though the terminology may be different. The role of spatial analysis techniques in both disciplines has been important, perhaps even more so after the realisation that patterns have little intrinsic meaning or value in themselves only in their ability to either stimulate the generation of hypotheses or as explanatory variables.

3. MAPPING THE THEORETICAL COMMON GROUND

Natural and cultural heritage studies share many common aspects but have also kept their own research worlds with natural heritage relying mainly on the natural science traditions, while cultural heritage has held closely to its humanistic research traditions. This remains one of the great divides in academia. Geographers have long studied spatial aspects of historical landscapes and the societies that populated them. They also have contributed much to the typology and understanding of morphological features such as ancient field systems and been responsible for the development of most of our understanding of the diffusion of technology, farming and ideas. They have also been at the forefront of interpreting spatial landscape patterns. Yet, to maintain this position in landscape research in the future the challenge is in how to maintain this integrated approach when so many geography departments throughout Europe are already or threatened by internal separation of physical and social/human geography. Maintaining the integration and working across academic interfaces is especially important for the training of the coming generations of landscape planners.

Nevertheless, recent research has provided several exciting interchanges between disciplines and even different knowledge cultures within disciplines. We have seen cross-over studies examining the importance of landscape ecological processes for people, e.g. the debate on authenticity in vegetation restoration, and the meaning of hedgerow networks for farmers studied from a humanistic perspective (Gustavsson & Peterson Ch 17 in this book; Oreszczyn & Lane 2000). Similarly, many of the approaches of landscape archaeology, especially those used to describe spatial aspects of human behaviour, have been mirrored by parallel conceptual developments of space use and resource exploitation in landscape ecology, e.g. the concepts of habitat complementation and supplementation (Taylor et al. 1993). Phenomenological perspectives contributed by archaeology offer insights into the cognition of the view and the interactions between cultures and landscape that have moved pattern analysis beyond environmentally deterministic landscape theory (Gaffney et al. 1996). This provides another parallel in the increasing interest within landscape ecology of the way animals perceive landscapes and how their behaviour responds to different landscape structures (Dover & Fry 2001; Hehl-Lange 2001). Research on animals, ranging from mammals to insects, has found that perceptual cues as often as physical barriers affect animal space use.

The way landscapes are perceived and interpreted by different cultures has defined the structure of landscapes and changing land use patterns (Aldenderfer & Maschner 1999; Ashmore & Knapp 1999). Cultural processes often based on the perception and interpretation of the surrounding landscape produce spatial patterns, and these patterns can be identified and mapped as historical landscapes. At a regional level, natural features such as land form are important in determining landscape patterns, although even at this scale the patterns of nature are often integrated into local perceptions and belief systems (Ashmore & Knapp 1999). At the site level, the archaeological focus has been more on the properties of material culture and social systems in describing space use (Tilley 1994). These processes also have parallel concepts in landscape ecology where environmental parameters decide regional scale animal distribution patters but at the landscape scale and site scales we see a greater influence of behavioural interactions between individuals and populations with landscapes. In Table 1 is a list of potential cross-over concepts which have direct parallels between ecological and archaeological landscape studies.

Table 13-1. Examples of landscape ecological themes that seem to have wide application to both nature and culture aspects of landscapes

connectivity
corridors
nodes
habitat supplementation & complementation
heterogeneity
continuity
size and shape of habitat patches
scale issues
viewsheds

Cross-over landscape ecological themes with wide application to both nature and culture aspects of landscapes. This is just a selection of landscape research themes to provide a discussion of the interfaces between cultural and natural sciences approaches. These and other landscape concepts are important for human space use, settlement and survival as well as species distribution and survival, and offer the basis for theory development across subject boundaries.

3.1 Connectivity

Connectivity is a fundamental process in landscape ecology. It relates to the functional linkages in a landscape and differs from connectedness, which refers to the mere physical connections between landscape elements. The concept of connectivity includes the resistance to movement caused by barriers or by land use types in a landscape. Connectivity is also important in cultural studies where the perception of time and space relate to the mode and characteristics of transport networks and how these have altered over time. Connectivity in landscapes is important for the ways animals or man moves around in the landscape and will have significance for resource availability and the frequency of cultural interaction and contact with other social groups. The concept involves the flows of materials, nutrients, energy, people and wildlife and includes both mental and physical barriers to movement, as well as the physical resistance to movement of different land uses/cover. Incorporating mental barriers to movement is one of the research challenges in developing movement models of man and wildlife.

3.2 Corridors

Corridors are linear landscape features that increase the flow of individuals, materials or energy between resource patches or suitable habitat/settlements and are important in defining the movement infrastructure of both animals and people. Corridors are one of the most important ways of increasing landscape connectivity. They have been found to function for a wide range of animals and plants (via wind and water spreading of propagules or through seed vectors). The human parallel to landscape corridors is, of course, transport infrastructure including roads, railways, waterways, paths and trails. Transport infrastructure is the most significant human corridor system, increasing access to and availability of physical and mental resources (Dover 2000). An important research theme is the way transport infrastructure and modes of transport affect our concepts of place and space.

3.3 Nodes

Nodes are intersections in movement corridors or barriers that result in important meeting places. They are especially of significance for determining the number of alternative ways (routes) individuals can take to move around and orientate themselves in a landscape. Nodes are therefore important in determining flows of species, nutrients and energy around landscapes (Forman & Godron 1986). If we examine a typical hedgerow network, we find nodes where hedges meet or cross and form T or X or more complex types of junction. It is interesting to note that for a wide range of animal and plant taxa, the number of species that accumulate at nodes is richer than the surrounding landscape (Fry 1991; Sarlöv-Herlin & Fry 2000). The ecological interpretation of this phenomenon is that the diversity of species is a result of the favourable (food, shelter and less disturbance) habitat at nodes and the significance of nodes for the landscape scale dispersal of organisms. Movement models predict greater visitation rates to nodes than other sections of hedgerow networks. The importance of nodes to geographers and archaeologists is much older than the new-found relations in ecology. The role of nodes in transport infrastructure has long been a major explanatory variable in the locating of defences, settlements and industry. Multiple infrastructure nodes such as waterways and roads have tended to be especially important at different historical periods.

3.4 Supplementation and complementation

Supplementation and *complementation* are key landscape ecological concepts that reflect different strategies by which individuals and populations sustain themselves with essential resources in fragmented landscapes (Taylor et al. 1993). The process of *supplementation* relates to obtaining necessary resources from several small sources within accessible distance. We can think of an animal such as the wolf that has a large home range; it may be able to use several forest areas to meet all its needs as long as the forests are all accessible and within travelling distance.

Complementation is a similar, but subtly different, concept that describes the acquisition of several quite different but essential resources, which must be 'available' i.e. close enough to utilize without expending too much energy or being exposed to risks such as predation. 'Availability' of resources in human terms is related to both environmental and social processes. It is thus possible to discuss the theory of complementation in archaeological terms as the basis for many locational models (Fry et al. 2003).

3.5 Heterogeneity

Heterogeneity is a complex concept and much used in landscape analysis for a wide range of purposes. Heterogeneity sums up two aspects of landscapes their grain size (the size of fields, forest patches etc.) and the variety that exists. Perhaps because of this dual influence on heterogeneity, it is not so easy to capture by numerical indices (Cale & Hobbs 1994; Fjellstad et al. 2002). Despite difficulties in capturing heterogeneity, it is an important concept that ensures the availability of a wide range of resources in a spatially restricted area and is correlated to a wide range of wildlife population and community characteristics. Because of this, landscape heterogeneity has been a central theme in the development of landscape indicators for biodiversity (Olff & Richie 2001; Leitão & Ahern 2002; Schneider & Fry 2002). It is also a major feature in human landscape preference studies and used in landscape characterisation. In a paper based on the results of a multiple-interest landscape monitoring project, Dramstad et al. (2001) found that bird and plant diversity, the density of pre-historic grave mounds and human visual preferences all responded positively to increasing heterogeneity of landscapes. In other words, the finer the scale of the landscape the higher the biodiversity, cultural heritage and visual value of landscapes. The concept of heterogeneity is thus likely to provide an interesting starting point for theory generation across disciplinary boundaries in an attempt to explain its wide influence in man and nature at the landscape level.

3.6 Continuity

Continuity is an aspect of the time depth and state of cultural heritage interests in landscapes. Although many landscapes have a great time depth, for landscape experience we are most concerned with the visual time depth. The depth that is readable by the public or the trained landscape historian is what is important. Yet, the hidden time depth may be more important. The historical depth of a landscape may be hidden in the sense that it is under the surface, but it may also be hidden because it is not a visible aspect of the landscape: For example, the sites of myths, battles or the home of artists or authors. The visual time depth is clearly appreciated by the general public and is an important visual quality (Coeterier 1996; Ode & Fry 2002). Time depth is an important landscape ecological theme. Old forests tend to be more important than new forests. They have more species and more important species than younger forests (Hermy et al. 1999). But it is not only the age of some habitats that make them important for biodiversity. There may be other links with the past. Historical cultures may have altered the soil conditions or drainage characteristics of an area, ploughed fields, grew crops etc. Many of these earlier activities leave their imprint in the vegetation of today. In addition some of the structures erected by earlier societies may now be important substrates or habitats for rare species such as the lichens that grow on gravestones.

3.7 Size, shape and pattern of habitat patches

The size and shape of habitat patches are important patch variables affecting many ecological and social processes. They are especially important for population viability and immigration and emigration dynamics in the study of populations of man and wildlife. Since the early theories on island biogeography we have been aware of the role that patch size has on biodiversity and later on population stability. The extensions of these theories to include metapopulation and source-sink dynamics have made conservation biologists aware of the importance of the spatial pattern of habitat patches. Especially important are the movement parameters of immigration and emigration rates leading to inter-patch dispersal. Both large and small scale demographics of human populations have similar traits with access to resources (physical or culturally determined) also being a major force driving movement patterns from the scale of daily commuting to the scale of inter-continental emigration.

3.8 Scale issues

Scale issues are important to both natural and cultural heritage management debates through the links to hierarchies of spatial and temporal processes and considerations of the scale of management units. For both natural and cultural heritage there has been a search for more meaningful management boundaries than current administrative boundaries and for a move upwards in scale from site to landscape and regional perspectives. There are many reasons why administrative boundaries do not function well as management units but the main one is simply that they do not match well the boundaries of natural and cultural resources. Distributions of rare plants and animals or special habitats often do not coincide with administrative boundaries, thus confounding resource management efforts. In nature conservation, the area of search used in the evaluation of site and species status tends to be administrative rather than natural and thus may have little relevance to the needs of the resource. Sometimes historical landscapes reflect administrative landscapes but not always. When there is a mismatch it can be difficult to develop integrated management strategies for whole landscapes. Scale issues are also related to spatial hierarchies that are important for many landscape processes (King 1997).

3.9 Viewsheds

Viewsheds are perceptual units and are important for interpreting visual relationships between archaeological sites, determining the visual character of an area and determining the space use of animals. They are also important in identifying the bounds of cultural environments. The central concept of many forms of visual landscape analysis is the landscape room (Keller 1993). A landscape room is a perceptual unit bounded by a floor, walls and a ceiling. Flat areas such as fields constitute the floor, while elements such as hills or forest edges form the walls. The sky is the ceiling. Landscape rooms are important for human orientation in landscapes (Lynch 1960; Bell 1999). The concept is central to research in landscape architecture as it defines the way people structure the landscape and through this structure provide us with insights into the aesthetic structure in the landscape. An aim of landscape archaeology is to reveal and understand the cultural structure of landscapes, which is often reflected in how different cultures chose to furnish a landscape (Gansum et al. 1997). A landscape ecological perspective might include theories related to landscape grain size and heterogeneity. The grain size of landscapes is significant for the ways different species can use the landscape and how they can move through it. Viewshed analysis is a technique that identifies the area of a map visible from any given point. The calculation of viewsheds (the area visible from a given monument) have become well integrated into landscape archaeological research (Kvamme 1990, 1991; Wheatley 1995; Llobera 1996). Sometimes, we wish to explore more advanced types of viewsheds such as the accumulated viewshed for al points along a ridgeline in the landscape or from a road. More often in archaeology the emphasis is on inter-site visibility to establish visual relationships between archaeological sites. Viewsheds can be considered as visual territories and exist as nested sets in a hierarchy of scales where landscape rooms increase in size with increasing height. The boundaries of visual units match well historical and current social boundaries especially those marked by physical structures in the landscape of today such as streams, field systems, and transport infrastructure (Fry et al. 2003). Landscape room boundaries can sometimes be related to social boundaries such as pre-historic farm-ownership boundaries. Perceptual units clearly play an important role in the way we use space and have used it in historical times. The concept is likely to be linked to that of defensible space in human ecology and territorial behaviour in wildlife ecology.

4. LANDSCAPE MANAGEMENT: DEFINING FUNCTIONAL UNITS

The scientific and political acceptance of the importance of landscape scale management has called for new planning tools, ones related to the wider countryside rather than being restricted to just the special places. The trend also calls for new methods for identifying the bounds of cultural environments. Defining what is an archaeological/historic landscape is not easy. In addition, management is often faced with the question of whether the boundary of cultural landscapes coincides with the boundary of an ecologically defined landscape when attempting to develop trans-sectoral policy. Both ecological and cultural boundaries will be related to landscape functions and often reflect these functions. This is generally true for the visible functions in landscapes but not for invisible cultural aspects. The invisible cultural heritage provides functions that are less easy to map as lines or polygons. Culturally rich landscapes may gain their importance from their symbolic value, for example as the site of a famous battle, work of art, or the home of a famous person. In such cases, the role of nature in defining the landscape may be very weak unless, for example, the view or topography had clear links to the cultural importance of the area.

Defining the bounds of historical environments or landscapes is very difficult. The older the site and the less complete the material evidence, the less likely we are to be able to interpret or delimit the cultural environment. This is not to say that delimiting landscapes based on their ecological functions is any easier. Ecologists have long accepted that it is the process they are studying that defines the landscape, rather than physical boundaries. It is likely, therefore, that nested sets of historical and ecological landscapes may exist in any single cultural landscape.

One approach to the integration of values in countryside management has been the development of landscape character mapping schemes (Usher 1999). At the national to regional scales, these methods tend to be based on physical and biogeographical variables such as topography, climate and vegetation types. Whereas, at the local landscape scale, cultural aspects including infrastructure, buildings, land use and settlement patterns are important. The value of developing landscape typologies is that they provide a context for management decisions that are often more meaningful than administrative boundaries (Fry et al. 1999). Landscape character is formed by assessing many different aspects of landscapes and therefore represents an integrative approach to resource assessment.

Cultural heritage aspects are incorporated into most landscape character assessment schemes to a greater or lesser extent. In addition, a variety of methods specifically addressing the needs of historical landscape characterisation are being developing for use in landscape planning. Fairclough & Rippon (2002) provide a review of European practice in cultural heritage management at the landscape level. The authors provide important inputs to the debate on how we cope with moving from sites to landscapes, as well as the relationship of historical mapping to *the European Landscape Convention* and wider issues of preserving a sense of place, local identity and quality of life in changing landscapes. The approach argues that the archaeological perspective brings to landscape management the awareness of long-term change and a knowledge of historical processes that has human agency at the forefront of explanation. In this way, archaeology explains the human and historic reasons for the current appearance of the landscape (Fairclough & Rippon 2002). An appearance that has undergone long-term changes and will continue to change in the future. Linking these approaches with visual, ecological and other perspectives will help us develop integrated landscape character assessment methodologies.

These methodologies are still under development and face a number of challenges including coping with the very wide diversity of landscapes throughout Europe, the poor state of inventory in many countries and the dominance of the visual aspects of landscape character assessment. Nevertheless, landscape characterisation is likely to continue to develop as an important tool in landscape planning and management and be of great benefit in changing the focus of landscape policies from objects to landscapes.

5. NATURE AND CULTURE: WHEN INTERFACES COLLIDE

At both academic and policy levels, we find a broad support for linking nature and cultural values and for increasing the management units from points to landscapes. Yet there will remain management conflicts. In this section I want to describe very briefly the type of conflicts that can arise between nature and cultural interests in landscape management. Fortunately, most of the conflicts appear at detailed levels of landscape management where the interface between nature and culture may be dominated by competition for space. In other cases, the management needs of cultural heritage operate at a different scale or with different priorities than as those of natural heritage. The following Norwegian examples expand on these points and you may recognise parallels in other landscapes.

5.1 Vegetation and grave mounds

Norway has a rich heritage of grave mounds from the Bronze and Iron Ages. These can be quite large and often found in clusters in grave fields. Many of the larger mounds and fields have line-of-sight relationships with other grave mounds or topographic features. Over the centuries these visual inter-sight relationships are lost through the development of visual barriers such as urban development or by forest planting. Sometimes, the forests involved are now old enough to have developed important nature qualities. In most cases the forest remains, and the goal of restoring spatial relationships between grave mounds cannot be achieved.

A more direct threat to the graves is the planting of trees (commonly birch) on the top of large grave mounds. Both the roots of the trees and the risk of uprooting during storms are very serious threats, yet the trees may be important habitat patches or found attractive. In such cases the different values must be weighed against each other and priorities set. In making such decisions, we must always be aware that different values have different levels of support at different times. Values are very dynamic and we need to reflect on the consequences of decisions that permanently reduce the quality of landscapes for one value or another.

5.2 Walls

Problems often occur during the restoration of ancient walls, banks and cairns to reveal old field systems or important landscape boundaries. Restoration aims can conflict with the use of walls as a substrate for rare species of moss and lichen. Some walls are so overgrown that restoration management requires the removal of shrubs and trees to prevent further deterioration of the wall. The vegetation may be important or have important nature conservation values as habitat or a movement corridor for animals. In many cases it will be difficult to find a suitable landscape level compromise. In such cases, one interest must take priority over the other or, alternatively, a spatial zoning proposed that would leave sections of the wall overgrown for its biological interests while others are restored.

5.3 Access, visitors, interpretation, education

Part of the aim of modern cultural heritage management is to increase public awareness of cultural heritage by the provision of interpretation and the use of sites for education by schools and the public. Creative management may be needed to avoid serious conflict with nature interests in some sites, for example, to avoid vegetation damage, soil erosion, disturbance to breeding birds etc.

In most of these potentially conflicting areas, the challenge is to identify how nature and culture interests are related and how they can complement each other. On the other hand, we must also face up to the fact that it will not be possible in every case to optimise management for both nature and cultural interests at the same place and time. Sometimes, and more often than management wishes to believe, real conflicts exist that call for priorities to be made based on clear criteria. Fortunately, the majority of conflicts do relate to detailed site management questions rather than the integration of values at the landscape level. Perhaps in moving from sites to landscapes we will be more likely to view these management conflicts at a larger scale where greater opportunity for spatial zoning, selective restoration and other measures are available to provide potential solutions.

6. BROADENING LANDSCAPE INTERFACES IN MOVING FROM POINTS TO LANDSCAPES

One of the very active arenas in landscape management is in the field of transdisciplinary research where researchers, management agencies and stakeholders work together to find solutions. Some of the important landscape scale topics currently under investigation are:

- identifying the cultural and natural heritage resource in the wider countryside. What is it we should be valuing in everyday landscapes and why?
- landscape conservation is very fixated on looking at the past, but the challenge is how do we develop a more forward looking perspective and how can we identify the values that will be important in the future and understand how they will interact with other values?
- how do we develop of the concept of environmental capital to include a wider range of countryside interests?
- how do we identify the contribution that different landscape values offer to the quality of life for local residents, visitors and other user groups
- how do we cope with the bias for the visible in landscape characterisation?

- how do we co-ordinate efforts in reviewing the concept of authenticity? It is needed in many aspects of the landscape management from forest certification to the development of tourist strategies.
- how do we develop robust and communicative scenario development methods, identifying which landscape policies will affect which landscape interests?
- how do we identify the circumstances when different landscape values conflict and when they are harmonious or synergistic?

In facing these challenges in moving the focus from sites to landscapes we are forced to increase the number of interactions and degree of interfacing with other landscape disciplines and knowledge cultures.

7. DISCUSSION

There is much to gain through moving policy and management to the landscape level and the further integration of culture and nature in landscape management. This change of perspectives reflects the wishes of policy makers and the public and is built on diverse research and management projects across Europe. Research is increasingly moving towards interdisciplinary landscape studies where cross-over theories are likely to arise. The prospect for greater fusion of landscape interests in the future is very positive. Challenges relate to how we cope with the dynamic aspects of landscape in spatial planning, especially how we incorporate the cultural aspects of landscape, which are so connected to the past with the evolution of current landscapes. We also need to ask how well our landscape research and policies incorporate the potential for new forms of sustainable land use. Cultural features of different ages can visually dominate and often replace cultural or natural features that were important to earlier societies (Renfrew 1992). Many pre-historic monuments had a more striking impact on the landscape of the past than they do today where they compete with remains from historic and recent times. Which brings us back to questions about what is the authentic landscape, and which time period should gain precedence and why.

The interface between archaeological and ecological landscape studies is long and complex with some areas of study hardly touching while other parts of the interface have wide and interacting fronts. In between these two categories exists a wide academic terrain of shared and borrowed concepts, analytical approaches, ideas and methods crossing from one or both directions. At best, this process has formed new alliances, constellations and networks of researchers, although at worst, borrowing at the interface can lead to uncritical application in an unsuitable context. Yet, overall the nature-culture interface seems one of the most productive and active of the landscape interfaces and one that will undoubtedly develop in size and depth.

ACKNOWLEDGEMENTS

I thank my colleagues in NIKU for discussions in Oslo and in the field regarding linking landscape ecology and landscape archaeology.

REFERENCES

Aldenderfer, M., & Maschner, D.G. (Eds.) (1999). *Anthropology, Space and Geographic Information Systems*. Oxford: Oxford University Press.

Ashmore, W., & Knapp, A.B. (Eds.) (1999). *Archaeologies of Landscape: contemporary perspectives*. Oxford: Blackwell.

Bell, S. (1999). *Landscape: Pattern, Perception and Process*. London: E&FN Spon.

Cale, P.G., & Hobbs, R.J. (1994). Landscape heterogeneity indices: problems of scale and applicability, with particular reference to animal habitat description. *Pacific Conservation Biology 1*,183-193.

Coeterier, J.F. (1996). Dominant attributes in the perception and evaluation of the Dutch landscape. *Landscape and Urban Planning 34*, 27-44.

Dover, J.W., & Bunce, R.G.H. (Eds.) (1998). Key Concepts in Landscape Ecology. *Proceedings of the 1998 European Congress of the International Association of Landscape Ecology*. IALE (UK).

Dover, J. (2000). Human, Environmental and Wildlife Aspects of Corridors with Specific Reference to UK Planning practice. *Landscape Research 25*, 333-344.

Dover, J.W., & Fry, G. (2001). Experimental simulation of some visual and physical components of a hedge and the effects on butterfly behaviour in an agricultural landscape. *Entomologia Experimentalis et Applicata 100*, 221-233.

Dramstad, W.E., Fry, G., Fjellstad, W.J., Skar, B., Helliksen, W., Sollund, M.-L.B., Tveit, M.S., Geelmuyden, A.K., & Framstad, E. (2001). Integrating landscape-based values – Norwegian monitoring of agricultural landscapes. *Landscape and Urban Planning 57*, 257-268.

Fairclough, G. & Rippon, S. (Eds.) (2002). *Europe's Cultural Landscape: archaeologists and the management of change*. Brussels: EAC.

Fjellstad, W.J., Dramstad, W.E., Strand, G.-H., & Fry, G.L.A. (2001). Heterogeneity as a measure of spatial pattern for monitoring agricultural landscapes. *Norwegian Journal of Geography 55*, 71-76.

Forman, R.T.T., & Godron, M. (1986). *Landscape Ecology*. New York: John Wiley & Sons.

Fry, G.L.A. (1991). Conservation in agricultural ecosystems. In. I. F. Spellerberg, F.B. Goldsmith, & M.G. Morris (Eds.) *The Scientific Management of Temperate Communities for Nature Conservation. British Ecological Society Symposium Volume 31* (pp. 415-443). London: Blackwell.

Fry, G., Puschmann, O., & Dramstad, W.E. (1999). Geographical Information for Research and Policy: A Norwegian Landscape Perspective. In M. Usher (Ed.) *Landscape Character: Perspectives on Management and Change*. Edinburgh: SNH/HMSO.

Fry, G., Skar, B. G., Jerpåsen, B., & Bakkestuen, V. (2003). Predicting archaeological sites: a method based on landscape indicators. *Landscape and Urban Planning,* in press.

Gaffney, V., & van Leusen, M. (1995). GIS, environmental determinism and archaeology: a parallel text. In G. Lock, & Z. Stančič (Eds.) *Archaeology and GIS: a European perspective*. London: Taylor & Francis.

Gaffney, V., Stančič, Z., & Watson, H. (1996). Moving from Catchments to cognition: Tentative Steps Toward a Larger Archaeological Context for GIS. In M. Aldenderfer & H.D.G. Maschner (Eds.) Anthropology, Space, and Geographic information Systems (pp. 132-154). Oxford: Oxford University Press.

Gansum, T., Jerpåsen G., & Keller, C. (1997). *Arkeologisk landskapsanalyse med visuelle metoder*. Arkeologisk Museum Stavanger.

Hermy, M., Honnay, O., Firbank, L., Grashof-Bokdam, C., & Lawesson, J.E. (1999). An ecological comparison between ancient and other forest plant species of Europe, and the implications for forest conservation, *Biological Conservation 91*, 9-22.

Hobbs, R. (1997). Future landscapes and the future of landscape ecology. *Landscape and Urban Planning, 37*, 1-9.

Hehl-Lange, S. (2001). Structural elements of the visual landscape and their ecological functions. *Landscape Urban Planning 54*, 107-115.

Keller, C. (1993). Visuelle landskapsanalyser i arkeologien. *UO Årbok 1991/92*, 59-68.

King, A.W. (1997). Hierarchy Theory: A Guide to System Structure for Wildlife Biologists. In J.A. Bissonette (Ed.) *Wildlife and Landscape ecology: Effects of Pattern and Scale* (pp. 185-212). New York: Springer.

Kvamme, K.L. (1990). The fundamental principles and practice of predictive archaeological modeling. In A. Voorrips (Ed.) *Mathematics and Information Science in Archaeology: A Flexible Framework* (pp. 297-305). Bonn: Holos-Verlag.

Kvamme, K.L. (1991). Terrain form analysis of archaeological location through geographic information systems. In J.G.R.M. Lock (Ed.) *Computer Applications and Quantitative methods in Archaeology* (pp 127-136). Oxford: Tempus Reparatum.

Leitão, A.B., & Ahern, J. (2002). Applying landscape ecological concepts and metrics in sustainable landscape planning. *Landscape and Urban Planning 59*, 65-93.

Llobera, M. (1996). Exploring the topography of mind: GIS, social space and archaeology. *Antiquity 70*, 612-622.

Lock, G.S., & Stančič, Z. (Eds.) (1995). *Archaeology and Geographical Information Systems: A European perspective*. London:Taylor & Francis.

Lynch, K.F. (1960). *The Image of the City*. Cambridge, Mass: MIT Press & Harvard University Press.

Ode, A.K., & Fry, G.L.A. (2002). Visual aspects in urban woodland management. *Urban Forestry & Urban Greening 1*, 15-24.

Olff, H., & Richie, M.E. (2002). Fragmented nature: consequences for biodiversity. *Landscape and Urban Planning 58*, 83-92.

Oreszczyn, S., & Lane, A. (2000). The meaning of hedgerows in the English landscape: Different stakeholder perspectives and the implications for future hedge management. *Journal of Environmental Management 60*, 101-118.

Renfrew, C. (1992). Explanation Revisited. In C. Renfrew, M. Rowlands & B. Deagraves (Eds.) *Theory and Explanation in Archaeology* (pp. 5-24). New York: Academic Press.

Sarlöv-Herlin, I., & Fry, G. (2000). Dispersal of woody plants in forest edges and hedgerows in a Southern Swedish agricultural area: the role of site and landscape structure. *Journal of Landscape Ecology* 15, 229-242.

Schneider, C., Dover, J., & Fry, G.L.A. (2003). Movement of two grassland butterflies in the same habitat network: the role of adult resources and size of the study area. *Ecological Entomology 28*, 219-227.

Taylor, P.D., Fahrig, L., Henein, K., & Merriam, G. (1993). Connectivity is a vital element of landscape structure. *Oikos 68*, 571-573.

Tilley, C. (1994). *A Phenomenology of Landscape. Places, Paths and Monuments*. Oxford: Berg.

Tress, B., Tress, G., van der Valk, A., & Fry, G. (Eds.) (2003). *Interdisciplinary and transdisciplinary landscapes studies: potential and limitations*. DELTA Series 2, Wageningen.

Usher, M.B. (Ed.) (1999). *Landscape Character: Perspectives on Management and Change*. London: The Stationery Office Ltd.

Wheatley, D. (1995). Cumulative viewshed analysis: a GIS-based method for investigating intervisibility, and its archaeological application. In G.S. Lock & Z. Stančič (Eds.) *Archaeology and Geographical Information Systems: a European perspective* (pp. 171-185). London: Taylor & Francis.

Chapter 14

LANDSCAPE ARCHAEOLOGY AND MANAGEMENT OF ANCIENT CULTURAL HERITAGE SITES

Some Notes Based on Finnish Experiences

Päivi Maaranen
National Board of Antiquities, Finland

1. INTRODUCTION

Cultural heritage is more than sites, landscapes, monuments, pictures, paintings or other material things. It also consists of non-material cultural phenomena like memories, customs, traditions and values. Even presentations of the past are part of the cultural heritage. So there is not one or unique cultural heritage, but rather several cultural heritages of different origins. According to Molyneux (1994) we are not just preserving our past but creating, explaining, structuring and reinforcing it all the time. This tangled preservation process with all interpretations and representations is important because it seems to be necessary to define man's place in daily life and even in whole world.

Heritage is more than an inheritance that individuals or groups receive from the past generations. It is a view from the present, either backward to a past or forward to a future (Graham et al. 2000). This view makes differences, which are important when classifying different phenomena and defining their significance. The classifying process can be called for 'a circuit of heritage', too, where different factors like heritage as means of representations, cultural identity and economic modification, production, consumption and regulation are all linked together by ever rolling circle or wheel (Graham et al. 2000). The wheel presents the way in which things get their meanings and how those meanings are modified and replaced when

H. Palang and G. Fry (eds.), Landscape Interfaces, 255–271.

time goes by. It is thus human beings and societies that define and remold their idea of cultural heritage and its contents.

Heritage sites and objects provide cultural and physical links with the past, with the history of human habitation and with the evolution of physical landscape as well as plants and animals (Lennon et al. 2001). Archaeological cultural heritage including ancient artifacts, ruins and monuments is part of that heritage providing links between past and present. Archaeological material remains are also among sources to recreate and reconstruct past landscapes and human impact on them. However, ancient remains do not include any kind of active data, which could be in use for landscape studies as such but need to be conceived, sorted and understood to make interpretations. There are always options for different and even unreliable presentations resulted from interpretation process because the relics of past time help people not just to know the past but bend it to their own uses (Lowenthal 1998).

In this chapter I discuss the archaeological cultural heritage sites and monuments from a point of view of landscape studies. In general there is a belief that knowing the history will help to maintain the present landscape and its elements for the future too in some extent. Knowledge of landscape history is acquired by landscape studies, which should include studies of natural and cultural phenomena from prehistory to historical times. In this chapter, the study of landscapes is mainly considered from the point of view of landscape archaeology, which is briefly outlined and connected to environmental archaeology, too. Mainly, the chapter deals with questions of heritage and its qualities as source material for landscape studies and as a target of different management activities. We also discuss how to advance the site management planning process because the traditional planning is not usable enough. The chapter ends with some observations concerning mitigation of the impact of development. Conversation of the subjects in article is mainly based on author's experiences from the Finnish site management activities and studies concerning the development of Finnish landscape from past to present (Maaranen 2000b; Maaranen 2002).

2. ARCHAEOLOGICAL HERITAGE IN LANDSCAPE

Archaeological sites and monuments are situated everywhere in cultural and natural landscapes of Finland (Figure 1). Some are thousands of years old, some from the 20th century, but all of them are results of human activity and modified by natural processes. Ancient sites and historical archaeological remains sign the human impact on environment, which is directly connected to landscape history. Study the past of landscapes as a whole is sometimes neglected, which results that many details of information are not recognized, but lost. Especially, when planning land use or assessing significance of areas for special care or protection, the knowledge of past is preliminary. It is not only the present landscape, but also the past incidents, which should be in focus of the studies. This fact is especially important because many landscapes are results of long and ongoing human activity. In addition, if the focus is to preserve that landscape as a unique visual scenery or interesting combination of different elements, the awareness and knowledge of different elements and their origins as well as other past incidents is needed. There are valuable cultural landscapes in Finland, which carry on patterns of medieval land use and traffic routes as basic characters, for instance. These patterns characterize the landscape and give it a special air. To maintain those patterns and structures of the landscape it is important to clear out and link the medieval past to the new uses of land.

Figure 14-1. The landscape photographed from the Iron Age hillfort of Hakoinen in Janakkala includes several invisible archaeological sites. There are settlement sites and cemeteries from the Iron Age as well as Medieval sites and other historical remains. Photo: Päivi Maaranen

Landscapes can be seen as summaries of past and present processes, elements and those ideas and opinions, which are connected to it (see Tuhkanen 1994). Sometimes different elements and other traces are visible and easy to find by observing visual landscapes, but in many cases archaeological remains are invisible and hidden partly or totally. That is why landscapes are sometimes described as layered constructions caused by landscape change through centuries (Helmfrid 1994). From time to time there emerge new processes, which produce new 'layers' on the top of older ones and remold the land and space causing disappearance and destruction of previous 'layers'. This layered structure can be compared to layers or deposits of archaeological sites, which lie one above the other with natural and human disturbance since they were originally deposited. This kind of idea of landscape is very near the Sauer's tradition, where the landscape is seen as a whole consisting of different elements and processes of culture and nature (Rikkinen 1977; Coones 1992; Raivo 1997).

Landscapes can be also understood as magical mirrors that maintain and reflect different impressions of the past. When looking at the mirror, every scholar sees different images, which mean that every interpretation is different, and depends on scholar's way to understand human activity and changes in landscape. This approach is near to cognitive and phenomenological interpretations as well as hermeneutics (e.g. Bender 1992; Tilley 1994; Ashmore & Knapp 1999). In the context of phenomenological studies the landscape has gradually come to be regarded as a complex idea or category with which to analyze past human societies (Johnston 1998). In addition, the landscape can be seen as storage of past history, which is not just possible to see but read and rewrite too by studying different phenomena. The landscape is thus "a palimpsest on to which each generation inscribes its own impressions and removes some of the marks of earlier generations" (Aston & Rowley 1974: 14). According to Simmons (1996) landscape can be envisaged as a text, the meaning of which is layered and where the deeper layers are to be explained with experts help.

Landscape as a context connects together people, society and nature, which means that landscapes are not separate entities of present-day but multidimensional spaces of visible and invisible remains and material and non-material dimensions (Keisteri 1990). From a point of view of landscape archaeology there cannot be cultural landscapes without past incidents and history of human activity. This aspect is important for management and assessment of landscapes as well as planning of the land use. It also leads to the question whether there is not just natural and cultural heritage but also landscape heritage too worth preserving and protecting. The management of landscape heritage and the assessment of significance based on landscape studies can help to mitigate impacts of development and to balance competing economic, social and cultural interests. It can be combined to the cultural resource management with objectives concerning value judgment, maintenance and use of the area (Banning 2000).

3. ARCHAEOLOGICAL HERITAGE AS SOURCE MATERIAL

Archaeological heritage includes material remains like sites and artifacts, which are resulted from past human activity. In addition, there are interpretations and representations produced by scholars, which affect our common knowledge concerning the past people and society and interactions between man and nature. For landscape studies the most important and interesting are sites and monuments as well as environmental remains from different time periods. When using archaeological heritage as source material the problems to take into consideration deal mainly with number, distinction, character and age of the archaeological evidence (Maaranen 2000a).

Figure 14-2. The Medieval fortress of Sibbesborg in Sipoo is partly destroyed because of agriculture. However, a major part of the rampart is still in good condition. Photo: Päivi Maaranen

The number and distinction of archaeological evidence are connected to the different formation processes, which have remolded sites during decades (Renfrew & Bahn 2000). The formation processes vary many ways depending on land use practices and natural processes characteristic for different geographical areas. The formation processes can change the site or hide it by covering surfaces, or even destroy sites partially or totally. The most effective processes are often land use practices connected to agriculture and development (Figure 2) (Kristiansen 1985; Sporrong 1985; Darwill & Fulton 1998). The distinction of data is also connected to fact that not all sites and monuments are discovered by archaeologists. Therefore often a sample of archaeological sites represents not only the past activity but also the intensity of modern land use and research activity (Kirkinen 1999).

Definitions of character and function of ancient sites can also cause trouble when studying landscape change. When interpretations of studies are based on those properties, accurate definitions are extremely important. Archaeological excavations usually detect the character of an archaeological site, but sometimes they may not give accurate knowledge (Figure 3). Especially when the excavation does not cover the whole site, there are always options to loose some important source material. Estimating functions of non-excavated sites produce always guesses with varying reliability. Monument types that have not been in focus of archaeological interest or are newly discovered types are just more difficult to define. Most difficult to define are usually non-excavated sites, which have only few or none visible features on the ground level.

Figure 14-3. Archaeological excavations cover the whole area of the Iron Age site in Virala, Janakkala. Today the highway runs over the place. Photo: Päivi Maaranen

Also, the age of sites can be crucial for landscape studies, but it might be difficult to clear out because of lack of evidence suitable for dating. Dating is clarified by relative or absolute dating methods, which give different dating ranges (Renfrew & Bahn 2000). Absolute dating methods are usually connected to a need to know absolute ages in calendar years. Radiocarbon dating is thus in common use to clear up ages, but concerning medieval or younger historical sites there are options to date them by historical written sources. Relative dating is connected to typological features of artifacts found from the site. Sometimes the types of the archaeological sites help dating, if there is common knowledge concerning relative or absolute dating for other similar sites. Goals for all dating procedures are to create chronological contexts to understand changes in time and space and put past incidents in chronological order.

In conclusion, when using archaeological material for landscape studies there must be some basic knowledge concerning archaeology to derive the best benefit from evidence. Some understanding concerning the distinction, age and function of archaeological sites help to estimate reliability of source material. Common knowledge of past social structures, technological skills and cultures help to understand the interaction between man and environment which is important when tracing marks of human impact on landscapes. In general, present interaction between man and nature is very different compared to past because improved technology allows us to dominate environment many ways. In the past people were living in nearer relations to environment and carried different worldview and ideas of nature and its functions. This different point of view is important to remember too when analyzing archaeological remains and their meaning in landscape studies.

4. LANDSCAPE ARCHAEOLOGY

A fragmentary tale of human life and societies in a landscape context is traced especially by landscape archaeology in which scholars are working for production of visible past from invisible or limited traces. A common task of landscape archaeology is to recover the story of an area and present the landscape utilizing every trace of human activity and natural processes. The major challenge of traditional landscape archaeology, or the so-called total archaeology, is to study the landscape as a whole, where individual sites have significance only as components of the area in which they lie (Bray & Trump 1980) Thus the total archaeology deals with questions of change and continuity, and it uses all suitable methods from archaeology, history, geology and geography.

Traditional landscape archaeology focuses on structures, elements and anomalies of landscapes. Fieldwork involves basically observing and recording of relict features with aims on explaining an evolution of patterns and shapes in landscapes (Aston & Rowley 1974). Every single element and landscape as a whole can be seen as important and equal when tracing landscape history. Thus the practical landscape archaeology includes ever-going searching, documentation and interpretations activities, which can give both interesting general knowledge concerning landscapes and help to explain the origins of patterns and functions. Amongst important aspects there are also studies of causes and consequences of different processes and how they are connected to each other.

Modern landscape archaeology does not give up tracing the landscape history but adds more systematic and dynamic perspective to studies. There are dynamic landscapes and anthropogenic environments in the focus, which are reached by several approaches (Fisher & Thurston 1999). Regarding the modern approach, it is important to notice personal interpretations, where archaeologists set out to experience the humanly shaped landscape as it has been modified and formed by human activities (Renfrew & Bahn 2000). Landscape has thus a cognitive dimension with social and spiritual meaning. Interaction between people and landscapes is also important, so goals of the studies are often connected both to landscapes as dynamic entities and to people as social and ideological actors. The modern landscape archaeology tries to take into consideration many social, cognitive and political processes, which affect to different activities of society and can be visualized in landscape scenery too. The landscape is merely the context where different interaction processes are observed and where space, time and society are in network-like discourse with each other. Explanations can be based on materialistic ideas that landscape is a resource complex or there can be more phenomenological idea that stresses subjective experiences of human beings (Ersgård & Hållans 1996). It is important to study landscapes not just by counting, measuring and sorting different elements but by feeling and understanding.

Landscape archaeological studies are often connected to or completed with environmental archaeological studies, which are focusing on both the use of natural resources and how past societies adapt to the ever-changing environment (Evans & O'Connor 1999). It is not just physical adaptation to environment but more cognitive perspective, which is included in environmental archaeology. Places, people and their activities are in the focus of its attention and that is how environmental archaeology seeks to understand the changing dialogue between people and their environments over time. By studying material environments of man it is also possible to understand the options of human action in different environments.

In conclusion, by combining approaches of modern landscape archaeology and environmental archaeology, it is possible to place people into right milieu and understand them as dynamic actors. The aims of landscape archaeological studies can vary widely from creating simple environmental reconstruction to producing visualized past landscapes. The modern landscape archaeological approach is well suited for problems that elucidate man's critical, shared connection to the physical and cognitive environment. Connections between archaeology, landscape studies and environmental aspects form an interesting approach that gives a broad, inclusive and holistic view to include humans, their anthropogenic ecosystem and manners with which humans conceptualize, experience and symbolize the landscapes (Fisher & Thurston 1999).

5. MANAGEMENT OF ARCHAEOLOGICAL SITES AND MONUMENTS IN FINLAND

The archaeological cultural heritage in Finland covers all prehistoric and historical archaeological sites and monuments from the Stone Age dwelling places to defensive works of WW I (Figures 4 and 5). The decree on *the Protection of Ancient Monuments* was given in 1883 and the current *Antiquities Act* of 1963 is largely based on the decree of 1883. *The Antiquities Act* defines types of ancient monuments and without permission stipulated in that act it is forbidden to excavate, cover, change, damage or remove ancient monuments or disturb them in any other way (*Antiquities Act* 1963, § 1-2). Site management and maintenance is based on *the Antiquities Act*, according to which the board can restore an ancient monument and carry out clearance and other measures required by protection and care of the monument in its location (*Antiquities Act* 1963, § 10). Site management is mainly duties of the Section for Site Management, which was organized in the end of the 1980's by the board when new *Employment Act* provided funding for hiring site managers to plan, develop and organize site management (Koivisto & Tiitinen 1995).

The main goal of the Finnish site management is to preserve and protect the archaeological cultural heritage as intact and original condition as possible. The main method is to prevent damages by carefully planning the maintenance and use of the site before any activities are carried through. These tasks and methods are very near to international ones, although financial resources and therefore options for multidisciplinary cooperation are limited (Feilden & Jokilehto 1993; Moisanen 1999). Finnish site management tries to establish management that covers all planning, programming and budgeting concerning the maintenance and use of the sites, although not all this has yet been achieved. The need to improve the site management and its planning is all the time more important because of growing use of heritage places. Thus the general idea of the site management and improvement of management planning are followingly discussed.

Figure 14-4. Many sites of the Stone Age are situated in the woods. There are no visible traces of the past human activity, only some low depressions, which indicate ancient hut places. Photo: P. Maaranen

The successful site management requires both broad multidisciplinary studies and basic knowledge concerning the preventive conservation (*E.C.C.O. Professional Guidelines*). According to Feilden & Jokilehto (1993) the planning process should be a multidisciplinary and continuing activity with various experts and it should be resolved by examining the implications of all possible alternatives. Cooperation and discussion between experts are essential when mapping matters relating to the significance of the site. Because the ancient monuments are fragile and non-renewable source, the aim of sites is to maintain their authenticity and values. The management and use plan is an essential tool for the site managers to preserve, manage and study archaeological sites and monuments. The plan should contain several essential parts to give thorough and reliable knowledge of the site and matters related to it (e.g. Feilden & Jokilehto 1993; *Ten steps to Protecting Heritage Places*).

Traditionally the management and use plans contain following information of the site and matters relating to it (e.g. Feilden & Jokilehto 1993; Maaranen 2000b; *Ten steps to Protecting Heritage Places*):

Figure 14-5. The Antiquities Act does not preserve defensive works of WW II. However, there is growing interest both to preserve them and use them as tourist attractions. This construction is part of the bigger defensive system in Sommarö, Tammisaari, and needs site management activities. Photo: National Board of Antiquities/P. Maaranen

- *a description of the site* presents an age, type and history of the site and it usually includes drawings or photographs of special constructions and elements of the site;
- *a description of location and size of the site* includes explanations concerning local topography, location, roads and routes. A map presentation is important too to show a precisely spot and surroundings of the site;
- *a report of vegetation and geological factors* presents mostly knowledge of soils, geomorphology, possible erosion processes, valuable animal or plant species and so on. The last mentioned is very important because of many rare and valuable plant species, which can indicate a past human impact;
- *a report of known and predicted land use and development projects* is important too because it gives information concerning different development processes and helps to take long-term planning in consideration. The development can cause serious damages for the sites or environment around them, so it is must be monitored and regulated (Darwill & Fulton 1998);
- *a detailed conservation plan* which is needed from a point of view of conservation and restoration of the site. It is reporting of condition of the

site and takes into consideration all international recommendations concerning the conservation of monuments and sites. The conservation procedures must not change the lay out of the monuments and no new construction, demolition or modification, which would alter the site, is allowed (*The International Charter for Conservation and Restoration of Monuments and Sites*);

– *a detailed maintenance program* is important too to give instructions for every measure, forest planning operations and logging of the site. A good maintenance strategy can prevent a great deal of damaging and decay so it is essential for preventive conservation (Feilden & Jokilehto 1993);
– *a detailed use plan* supplements other measures by describing infrastructure like paths, signs, guides and so on. Mapping of risks caused by the use of the site is too included (Stovel 1999). More general but as important parts of the management and use plan are different reports of financial and other resources;
– *a detailed budgeting plan* should contain both short-term and long-term planning to guarantee the funding of the site management procedures;
– *a list of experts and organizations that are related to the management project* contains all information of the specialists, which are connected to specific sites. Important part of the planning process is also planning of documentation and following effects of the maintenance and use, which can be separate or included in the management and use plan.

Every part mentioned before are important in management and use plans but the problem is that they may be difficult to bound together. That is why there is usually a summary of the whole management and use plan too, which helps to present the site management process in general. Despite of the summary can traditional management and use plans be easily entireties that describe different parts of the process but do not connect them to each other.

In Finland, the Section for Site Management stores the management and use plans in a database based on Microsoft Access, which enables to handle large amounts of data. However, the database is from year 1996 and requires reformation because the system does not allow storing all information. Almost all information of the plan is gained by observing and describing the situation in archaeological heritage places. There is no place for analytical studies of landscape or matters relating to it but descriptions of natural and cultural phenomena. On the side of the database reformation in basic, there is need to develop GIS as a tool for the site management too because the old database does not contain options to operate with digital maps and store them to the database.

The reformation of the Finnish site management planning system is aimed to create better and more effective planning tool for the site management. There are also needs to weigh a point of view of landscape archaeological studies when creating the new database to get more logical tool to study sites and plan management and use of them. Landscape studies as well as many other studies are seen very important when preparing and updating the management and use plan. Combining both landscape archaeological viewpoints and landscape historical approach to planning process could solve problems of the traditional management and use plans in Finland. GIS and options to use digital sources and store data in digital form are also appreciated. Needs for combining the site management process and land use planning are also topical because of the growing land use activities with threats against archaeological cultural heritage.

6. SOME NOTES CONCERNING MITIGATION OF THE IMPACT OF DEVELOPMENT

In Finland there is a long tradition of citizens' participation in the development of their own living environment, which is connected to an idea that a human settlement is in key position in a creation of more sustainable future (Making future now). A development of urban living conditions on a socially, economically and ecologically sustainable basis is a worldwide challenge, however. Good practices are always bound up with the social conditions from which they have arisen and from this point of view have landscape studies and monitoring of environment an important role providing more knowledge for decision making of citizens and administrators (compare to *Guidelines for Programme Design* 1997). Many present settlement centers have been important through decades, but an ongoing growth of cities cause disappearance of the remains of past human activity. Amongst others are archaeological cultural heritage sites under these threats, which are taking away humans history forever. Pressures of development and natural processes have raised questions concerning the preservation and management of cultural and natural heritage (Stovel 1995). In general, research and management of the environment, heritage and biodiversity require wide-ranging monitoring information, which is providing a basis for more sustainable management. Information at the landscape level is particularly important giving a wider context for objective monitoring of environment's change and its causes (*TemaNord* 2001). Traditional division between natural and cultural may not be reasonable in present-day, when there is a need for more holistic and wider approach. In addition to natural and cultural heritage, there is also landscape heritage, which should be evaluated, understood and preserved by assessing the significance of a landscape and mitigating the impact of development by careful and many-sided land use planning.

The landscape heritage and its history are gaining more interest from common people and planners. This interest is connected to helping people to adjust to new circumstances in growing cities, for instance. Surroundings of suburbs are changing and there is a need to plan more comfortable and pleasant atmosphere for inhabitants. People are seeking safety and continuity. Many development procedures are usually very rapid and demanding in respect of the landscape heritage, which can lead to the destruction of historical continuity and feelings of familiarity from a citizen's point of view. A good planning based on landscape studies and discussion between different actors, like citizens, planners, researchers and administrators leads to more monitored and considered development processes. When preserving a general landscape structure with elements of past and present, there are more options to inhabitants to engage to their surroundings and milieu. This raises common satisfaction and helps to value both landscape heritage and its preservation.

Sustainable management of the landscape heritage including natural environment and cultural remains requires enormous amount of data, therefore planning and land use activities must be continuously supervised on relevant spatial levels and in a landscape context to get better results. Proactive approach towards landscape heritage monitoring is connected to identification, recording and reporting of the changes over time. It is also important to influence the decision making of preservation and land use planning with up-to-date knowledge based on landscape studies. Imitating Ward (1995), heritage monitoring creates a decision-making feedback tool, which ensures that the impact of plans, policies and measures are evaluated and continued all the time. Subjectivity is minimized and decision makers must concentrate on the long-term status of the landscapes. In general, the landscape studies linked to monitoring and management of the cultural heritage promote a greater awareness of long-term change and provide valuable data for future generations too. There are real options to assess the significance of landscape heritage and many impacts of development, which help both to mitigate the impact and salvage the information before the landscape change. As conclusion, there are both opportunities and needs for the landscape resource management, which requires more open attitudes towards protection and preservation of the valuable landscapes. An extensive co-operation between citizens as well as planners and scholars from different branches must be also encouraged.

REFERENCES

The Antiquities Act. (1963). *Archaeological Heritage Management in Finland* (pp. 25-30). The National Board of Antiquities, Department of Archaeology. Helsinki: Helsinki University Press.

Ashmore, W., & Knapp, A.B. (Eds.). (1999). *Archaeologies of landscape. Contemporary perspectives*. Oxford: Blackwell Publishers Ltd.

Aston, M., & Rowley, T. (1974). *Landscape archaeology: an introduction to fieldwork techniques on post-roman landscapes*. Newton Abbot London Vancouver: David and Charles.

Banning, E.B. (2000). *ANT 409F – Cultural Resource Management and Site Significance*. Regional Analysis in Archaeology – Cultural Resource Management (CRM). Computing in the Humanities and Social Sciences. http://www.chass.utoronto.ca/~banning/409files/409Fch11.htm

Bender, B. (1992). Theorizing landscape and the prehistoric landscapes of Stonehenge. *Man, 27* (4), 735–755.

Bray, W., & Trump, D. (1982). *The Penguin Dictionary of Archaeology*, 2nd edition, Aylesbury: Penguin.

Coones, P. (1992). Landscape Geography. In A. Rogers, H. Viles, & Goudie, A. (Eds.), *The Student's Companion to Geography* (pp. 70-76). Oxford: Blackwell.

Darwill, T., & Fulton, A. (1998). *Mars: The Monuments at Risk Survey of England 1995*. Main report. School of Conservation Sciences, Bournemouth University and English Heritage. Bournemouth: Bournemouth University.

E.C.C.O. European Confederation of Conservator-Restorers' Organisations *Professional Guidelines (I): The Profession*. Promoted by the European Confederation of Conservator-Restorers' Organizations E.C.C.O. and adopted by its General Assembly (Brussels, 11 June 1993). http://palimpsest.stanford.edu/byorg/ecco/library/guidel.html

Graham, B., Ashworth, G.J., & Tunbridge, J.E. (2000). *A geography of heritage. Power, culture and economy*. London: Arnold.

Ersgård, L., & Hållans, A.M. (1996). Medeltida landsbygd. *Riksantikvarieämbetet, arkeologiska undersökningar. Skrifter nr* 15. Stockholm: Riksantikvarieämbetet.

Evans, J., & O'Connor, T. (1999). *Environmental archaeology: principles and methods*. Thrupp: Sutton.

Feilden, M.B., & Jokilehto, J. (1993). *Management Guidelines for Cultural Heritage Sites*. UNESCO, ICOMOS. Rome: ICCROM

Fisher, C.T., & Thurston, T.L. (Eds.). (1999). *Dynamic landscapes and socio-political process: the topography of anthropogenic environments in global perspective: introduction*. Antiquity, Special section. http://intarch.ac.uk/antiquity/landscapeintro.html

Guidelines for Programme Design, Monitoring and Evaluation. (1997). Ministry for Foreign Affairs, 7.11.2001. http://global.finland.fi/julkaisut/yleis/pdme/design.htm

Helmfrid, S. (Ed.). (1994). Kulturlandskapet och bebyggelsen. *Sveriges Nationalatlas*. Kulturgeografiska institutionen, Stockholms Universitet. Höganäs: Bokförlaget Bra Böcker.

The International Charter for Conservation and Restoration of Monuments and Sites. II International Congress of Architects and Technicians of Historic Monuments. Venice, 1964. Adopted by ICOMOS, 1965, http://www.ovpm.org/ovpm/english/charter/venice.html

Hall, S. (Ed.). (1997). *Representation: cultural representations and signifying practices*. London: Sage/Open University.

Johnston, R. (1998). Approaches to the perception of landscape. Philosophy, theory, methodology. *Archeological Dialogues* (pp. 54-68). Dutch perspectives on current issues in archaeology. Leiden: Archaeological Dialogues Foundation.

Keisteri, T. (1990). The study of changes in cultural landscapes. *Fennia 168,* (1), 31-115.

Kirkinen, T. (1999). The influence of modern land-use on the formation of archaeological Record. Source critical problems of Regional Studies. In M. Huurre (Ed.), *Dig it all. Papers dedicated to Ari Siiriäinen* (pp. 21-28). Helsinki: The Finnish Antiquarian Society.

Kristiansen, K. (1985). Economic development in Denmark since Agrarian Reform. *Archaeological Formation Processes. The Representativity of Archaeological Remains from Danish Prehistory.* København: Nationalmuseet.

Koivisto, L., & Tiitinen, T. (1995). The Management of Archaeological Sites and Historical Monuments in Finland. *Karhunhammas* 16, (pp. 99-106). Department of Cultural Studies, Archaeology. Turku: University of Turku.

Lennon, J., & Associates Pty Ltd. (2001). State of the Environment Australia 2001. Natural and Cultural Heritage Theme Report. *Australia State of the Environment Report 2001.* Commonwealth of Australia, CSIRO Publishing on behalf of the Department of the Environment and Heritage. http://www.ea.gow.au/soe/2001/heritage

Lowenthal, D. (1998). *The heritage crusade and the spoils of history.* Cambridge: Cambridge University Press.

Maaranen, P. (2000a). Arkeologinen kulttuuriperintö osana suomalaista maisemaa. (pp. 25-34). In J. Saarinen, & P.J. Raivo (Eds.) Näkökulmia suomalaiseen maisematutkimukseen ja -suunnitteluun. *Metsäntutkimuslaitoksen tiedonantoja* 776 Metsäntutkimusalaitos, Rovaniemen tutkimusasema. Saarijärvi: Gummerus Kirjapaino Oy.

Maaranen, P. (2000b). Questions of management planning and preventive conservation concerning the site management of the archaeological cultural heritage of Finland. In R. Koskivirta (Ed.), *Conservation Without Limits, IInd Nordic Group, XV Congress 23-26.8.2000 Helsinki Finland* (pp. 137-146). Loimaa: Loimaan kirjapaino.

Maaranen, P. (2002). Human touch, natural processes. The development of the rural cultural landscape in Southern Finland from past to present. *Fennia, 120* (1-2), 99-109.

Moisanen, J. (1999). *Monuments and sites, Finland.* ICOMOS, 12th General Assembly, Mexico. National Committee, Finland. Helsinki.

Making future now: Good practises in Finland. (1996). Associaton of Finnish Local Authorities. Vammala: Suomen kuntaliitto.

Molyneaux, B. (1994). Introduction: the represented past. In P.G. Stone, & B.L. Molyneaux (Eds.) *The Presented past. Heritage, museums and education* (pp. 11-13). One world Archaeology 25. Reprint 2002. London and New York: Routledge.

Raivo, P. (1997). Kulttuurimaisema: alue, näkymä vai tapa nähdä? (pp. 193-210). In T. Haarni, M. Karvinen, H. Koskela, & S. Tani (Eds.), *Tila, paikka ja maisema. Näkökulmia uuteen maantieteeseen.* Tampere: Vastapaino.

Renfrew, C. & Bahn, P. (2000). *Archaeology: theories methods and practice* (3ed.). London: Thames and Hudson.

Rikkinen, K. (1977). Maantieteen historia. In K. Rikkinen, & M. Eskola (Eds.), *Maat ja kansat. Otavan maantieteellinen tietosanakirja* (pp. 1076-1078). Helsinki: Otava.

Sporrong, U. (1985). *Mälarbygd. Agrar bebyggelse och odling ur ett historisk-geografisk perspektiv.* Stockholm: Kulturgeografiska Institutionen, Stockholms Universitet.

Simmons, I.G. (1996). *The Environmental Impact of Later Mesolithic Cultures. The Creation of Moorland Landscape in England and Wales.* Edingburgh: Edingburgh University Press.

Stovel, H. (1999). *Risk Preparedness: a Management Manual for World Cultural Heritage.* UNESCO, ICOMOS, WHC. Rome: ICCROM.

Stovel, H. (1995). *Monitoring World Cultural Heritage Sites*. ICOMOS Canada, Bulletins, *4* (3). http://www.icomos.org/canada/bulletin/vol4_no3_stovel_e.html

TemaNord. (2001). Strategic landscape monitoring for the Nordic countries. *TemaNord 2001, 523*. Nordic Council of Ministers, Denmark.

Ten steps to Protecting Heritage Places. Australian Heritage Commission, Canberra. The website is based on a publication *Protecting Local Heritage Places – A guide for communities* (1999) by Nicholas H. with the assistance of M. Parsons, M. Switzer, S. Blair, D. Young, M. Pearson, L. Thompson, A. Jelinek, K. Heffernan, M. Pantehis & L. http://www.heritage.gov.au/protecting.html.

Chapter 15

PRESSURE ON THE FRINGE OF THE CITIES

Grete Swensen
Norwegian Institute for Cultural Heritage Research, Norway

1. APPROACHES TO THE STUDY OF CULTURAL HERITAGE MANAGEMENT

One of the major challenges within today's social development is to find useful methods within physical planning. At the same time as planning is meant to append to an overall picture and be shaped within the form and functional ideals adjusted to both present and prospective demands, one has to assume that the cultural heritage is taken care of as a potential resource. These three demands are today strongly pulling in different directions and often seem to exclude each other. For a better understanding of the forces that commonly come forward in local planning, it is important to understand the role of cultural heritage.

1.1 Aiming towards an interdisciplinary model

We will start by giving a short account of the methods chosen in the project. They are categorized into two groups: methods creating a general view, and definite tools. To create a better understanding of how the potential conflicts related to questions concerning landscape preservation and cultural heritage values are handled on a local and regional level, we find an interdisciplinary approach useful. Interdisciplinarity can be recognized as insights generated through the process of co-operation. Practicing interdisciplinary research raises several challenges, of theoretical as well as practical and methodological kind. Besides discussing different theoretical frameworks that correspond with the four different professions represented in the study (archaeology, architecture, sociology and

H. Palang and G. Fry (eds.), Landscape Interfaces, 273–293.

ethnology), we also have to look for superstructural theories usable in an interdisciplinary approach (examples are the landscape concept, the discursive analytical approach etc.).

A useful, practical tool within preservation work is based in the concept of cultural milieu, and includes a special documentation method. By selecting cultural milieu as a useful concept, landscape areas as well as connected groups of buildings, roads etc. are considered units related to each other. The concept functions as a practical tool, where information registers as well as digital maps are being applied. The tool is helpful to promote a way of differentiating and analyzing areas apart from each other.

Among the various professional methods used, we will stress the importance of the qualitative use of interview. The method has been applied within branches of sociology as well as within ethnology. An interview is started with a special aim in order to get insight into the other person's knowledge, viewpoints and evaluations. The focus on the individual perspective and the range of variations that this method presents, gives an opportunity to look closer into a problem that otherwise is often closed for insight. The interesting questions often arise from interpreting the meaning and intentions individuals assign to their actions. Through the individual approach interviewing is suitable for explaining why phenomena, like planning process, take place in a special way. It is less reliable to use this method in order to explain how it actually is conducted.

1.2 Cultural heritage management as a sector based interest

Over the last 20 years a change in planning the cultural heritage work has taken place. The focus has altered and expanded from being based chiefly on preservation of single farms and buildings. Larger areas have come into focus, where the structural connections between the historical and physical elements, the importance of time-space connection as well as historical consciousness are being stressed. Introducing cultural milieu as a concept within the Cultural Heritage Act (1992) underlined the importance of emphasizing totality and unity in an area, and it encouraged cooperation between nature conservation, local planning and cultural heritage work.

We know too little how the cultural heritage interests are managed on the local level and how it is taken care of as a sector based interests within the local communal planning system. While some municipalities are used as magnificent examples of planning and obtain praise for good course of action, we have to admit that we know too little what the basic conditions for a successful recipe are – is success mainly depending on the political constellations within the local party politics, of particular persons ability to cooperate, or is the explanation based on the system level, in a particular ability to utilize and exploit today's regulations and planning tools?

There are many viewpoints on good planning available in forms of instruction guides, handbooks and so on. Despite this, we actually know very little how actual planning is taking place, and what is prioritized. This research project became even more actual after starting up, because political signals point to a wish to strengthen local cultural heritage administration.

Our main approach to face these problems is to raise the following question: what happens to the cultural and historical values in an area during a process of area development?

Attention is directed towards the position the cultural heritage interests hold, and what consequences this practice has for the landscape and the cultural heritage. To obtain insight into this question, it is relevant to look into the basis of knowledge, values and professional background the different actors have, since this affects the praxis. By contextualizing the actors the main approach in the project is presented.

1.3 Landscape components and the process of fragmentation

Another main question concerns the content of the landscape. Changes within land use influence the possibility to read the landscape history. Since landscapes contain components of different time depths, they have to be related to the specific historic connection concerning their appearance, continuity and change. Since there is often large disagreement about what is valuable in landscape, it is important to create a basis of knowledge concerning how landscapes are identified, limited, understood and valued.

Understanding the numerous cultural aspects of landscapes draws the attention to the time and space components in landscape. Within landscape research, focus has been directed along two paths: on the one hand towards the abstract concept of space, and on the other hand towards the concrete notion of place, as the locality people relate to on a personal level. A sense of place can be identified as the emotional attachments to localities developed by individuals and communities in the course of living and growing within the setting of the home. Places matter because they are the focus of personal sentiments, with the feeling for place permeating day-to-day life and experience (Muir 1999). These two aspects of landscapes have been discussed by a series of researchers (among others Relph 1976; Giddens 1984; Harvey 1989; Gregory 1994; Urry 1995). Human relations 'inscribe' themselves into landscapes as concrete signs. In this way the daily tasks that take place within a landscape appear like biographical encounters between people of today and past times (Tilley 1994). Landscape in this sense can be understood as an expression of the links between people and their environment and of the operation of social life within particular milieu (Coones 1985, referred in Muir 1999). Areas and landscapes are in this way found in social as well as individual understanding of time and space.

An aspect stressed by certain planners, is the coordination that takes place between the physical planning carried out by architects and planners, and the conversion and modifying actions that turn space into livable areas. De Certau (1984) has emphasized this process and defines it as 'space as practiced place'. The focus is put on the transformation that occurs when areas get redefined and are given new meaning through praxis. An important point is made of the interconnection between the physical structures and the operations taking place. Thereby the narratives and tales contribute to animate the area into living quarters. The stories visualize the actions that make it possible to mix different elements, and thereby they tell how space can be used, 'they are treatments of space', as De Certau (1984) puts it. Stories and tales hereby give a rich and varied basis for analyzing central aspects in the physical surroundings, like dimensions, affinity and orientation. Cultural milieus are not abstract and independent phenomena but areas people have a relationship towards and add meaning to. The meaning is incorporated to the stories attached to the place and functions as narrative space.

Our landscape analysis is based on the process of fragmentation that often hits the landscapes in area with modernization and population growth. Changes deform the traditional structures in the existing landscape and affect the function. Because of this, only parts or remnants of the traditional landscape formations can be recognized in today's landscape, where they form isolated and fragmented elements in otherwise uniform room. They may be difficult to spot, and the knowledge about them can simply have vanished. In some areas the traditional landscape is completely blurred because it has been replaced by new, rational land use techniques (Antrop 1997).

Based on easy attainable data (registers, maps, local historical literature and so forth), we are in the process of carrying out a landscape analysis that 'draws a picture' of the landscape in the way we presume appeared at different time periods. The results of this analysis are being linked to the planning procedure. By stating what elements are lost and what elements are preserved in different time periods, we intend to gain a tool to help us read the landscape. However, this arises a series of questions: is it still possible to gain information and knowledge about the cultural structure imbedded in the landscape? Are there recently added factors that interfere with our perception, or do we actually think that the recent changes have increased the landscapes' cultural qualities? What relationships and active mental and physical actions take place that contribute towards creating a sense of being an insider?

2. A STRONGER GRIP IN SUBURBAN AREAS

To analyze the pressure that municipalities on the fringe of bigger cities experience, two communities have been selected as case study areas. Madla in Rogaland County is a part of Stavanger municipality, and represents the outskirts of an old city located and strongly affected by the nearness to the sea. Madla has been dominated by agricultural activities for hundreds of years. During the last twenty years it has gradually been influenced the city, and the need for more building areas as well as areas for industrial purposes has been coming more evident.

The other case study area situates in Nannestad in Akershus County. This is the case we at the moment have most information from, and therefore a closer presentation of what we consider the major factors of pressure in this municipality will be put forward.

2.1 *A green community in rapid growth*

The slogan *A green community in rapid growth* is used by the community administration to characterize Nannestad. Nannestad is situated approximately 60 km north of Oslo. It is part of the best agricultural areas in Norway, with prosperous farms that have a long history. It still has an evident character of being agricultural area with large green belts and extensive forests. By January 1, 2001 the number of inhabitants in Nannestad reached 9,200. For hundreds of years it has maintained its character of being a 'sleepy countryside', without strong pressure from busy railways and elaborate rodents. Since 1992 this situation has changed. Norway's new international airport was planned in the approachable neighborhood and that brought along great changes. The airport was opened in 1998 and situated in the eastern part of the municipality.

In accordance with the situation for other flat Norwegian south-eastern agricultural areas, the farmlands in Nannestad in the last 20-30 years have gone through big changes concerning effectivity, specialization and mechanization. These changes have lead to big consequences for the landscape, both visually and ecologically. In these clayey soil areas the leveling out and closing of streams have changed the hilly and divided landscape into an open and gentle landscape suitable for mechanized grain cultivation. The farms emphasize the landscape with sparsely situated clusters of farmyards, placed on the ridges in the landscape.

Figure 15-1. When you approach Nannestad center from the south, the church is the first to catch your eye. The oldest part of it is built in 1193, and it is known as an important Medieval church. On its eastern side we find a cluster of older buildings: a farm building called 'The Sheriffs House', the old civic town hall and the old bank. Placed as it is in a smooth hillside, it represents a characteristic porch to the center, a short distance more north. Photo: NIKU 2001

2.2 In the wake of the newly established airport

Nannestad (Figure 1), together with the neighboring municipalities Ullensaker and Eidsvoll, has been strongly affected by the establishment of the new airport. Both the building of a new railway and extension of the road into a six-lane motorway have created good conditions for people previously working in the capital now moving to Nannestad. Three new housing estates have been established since the location of airport was decided. The local plan (*kommuneplan*) is set to recruit 300 new inhabitants on a yearly basis. The municipality has changed into a growing commune, where the expansion in inhabitants has to be followed up by establishing new schools and better health and social services. Another direct consequence of the airport (Figure 2) is the central order of building a new regional road connecting the eastern regions to the airport. This road is to be opened in 2003. The airport has lead to a situation where Nannestad is gradually turning into a suburban fringe, where the effects of the neighborhood to Oslo are getting evident. The area has changed into an area under pressure, which is a situation that can directly influence how the cultural heritage management is being tackled in the municipality.

Figure 15-2. This old Nannestad map from 1859 shows the closeness to Gardermoen. Today's international airport is situated where the old military drill grounds from the beginning of 1700 were stationed, marked in the eastern part of the map as Gardermoen Exercerplads. The first airport in the area was a military airport established in 1912, and a lively village gradually grew up near by. When the new airport was established, the old village was erased. Photo: Statens Kartverk/NIKU 2001

3. LEVELS IN DECISION MAKING

Within the cultural heritage work in Norway, two different sets of legislation acts exist, which influence the decision making and the planning procedure in different ways.

Starting with the heritage administration, it is organized on three levels. There is one central institution, the Director General of Historic Monuments (*Riksantikvaren*), and a regional administration based in each county. This regional cultural heritage administration is meant to function as a channel for information, a control apparatus and a forum for planning, related to the local planning activities. On local level, however, cultural heritage work is partly related to the general cultural work, partly to the planning and building activities. Very seldom special resources are specified for this field of activities.

The Cultural Heritage Act is administered from the central level, and it gives directives concerning how automatically preserved cultural heritage is taken care of, and how preservation of more modern cultural heritage can be brought forward. It represents a strong legislation act where the archaeological remnants are concerned, because it assumes that necessary examinations are done on developers' behalf before building is accepted. At the same time the state (in this case university museums) are responsible for the actual preservation of artifacts. When cultural heritage from earlier than 1537 is involved, special dispensation is necessary to enable the developer to build. In this case, the regional cultural heritage administration plays an important role in such procedures.

It is harder to base the preservation of cultural remnants younger than 1537 (for buildings the year is 1650) on *the Cultural Heritage Act*. Preservation orders have in such cases to be prepared either by the regional preservation administration or by the Director General of Historic Monuments (*Riksantikvaren*). This process is time consuming and intricate.

The way *the Cultural Heritage Act* is being administered, makes it less relevant for archaeologists within the administration to attend in the discussions about evaluations and priorities in the preservation work. Today's regulations concerning automatic preservation make the evaluation discussion unnecessary and thereby less relevant and interesting. People working with preservation of younger cultural heritage, however, have to refine their argumentation and continuously experience the difficulties involved in getting a breakthrough within the field.

The Cultural Heritage Act is still administered with a pronounced difference between the handling and control of archaeological remnants and newer heritage. This has contributed towards erasing the unfortunate division within the preservation field. The themes and perspectives attached to cultural history activities are widened and today include landscapes, nature conservation, places, processes, intentions and actions. Unfortunately the administration and implementation on the local and regional level are still tied up in a settled and fixed pattern, and parts of the cultural heritage work practiced in the local administration are still dominated by an object-orientated and museum-based profile.

The Plan and Building Act is the principal tool in local planning. As the responsibility for the technological planning in Norway is based in the local political authorities, the preservation of cultural heritage in such cases is considered and treated just as one of several sector based interests. One of the advantages in today's legislation is a large degree of flexibility. The fact however that many municipalities and counties have inadequate capacity to run their own planning policy, represents one of the most important defects in the act. The result is that they handle plans promoted from private developers in an unsatisfactory way. It gives a wide range of possibilities to use the planning tool in the direction one wishes. This is an advantageous position for municipalities with plenty of resources, but is a bad position for those with a limited economy. As such the planning process can be looked upon as a discursive arena where different interests are played out against each other.

4. PARTICIPANTS, BASIS OF KNOWLEDGE AND LEGITIMACY

We base our observations on the fact that the cultural heritage management's own decisions, interpretations and evaluations will influence their legitimacy (Nesheim 1999). Furthermore, if there is too much of a gap between the opinion held by 'the man on the street' about what can be considered valuable, and the opinions held by the cultural heritage professional, it is very likely to weaken their legitimacy. If, on the other hand, the chosen preservation procedures seem reasonable and adequate to the local understanding, this will strengthen their legitimacy.

In this project we studied what form of knowledge and values the preservation management based their decisions upon, since we consider it likely that the academic background and ideological basis held by officials within the cultural heritage management will influence the practice in this field. We wished to examine whether professional and discursive, struggles exist between different participants and different professions on the question of what should be preserved and why. Is there certain parties that seem to get more breakthrough for their arguments than others? The question of legitimacy will be interpreted and discussed from our experience, primarily focused on this empirical study by examining to what extent the cultural heritage interests are given priority before other sector-based interests (like agriculture, communication and health care).

Now the attention will be turned towards the different participants that work inside the cultural heritage field, and the focus is drawn to the fact that description and valuation within cultural heritage work is directly related to the persons involved in the story-telling. The analysis is based on interviews and conversations with people who have been sought out on behalf of the positions they hold. Their opinions reflect personal values, knowledge and factual expertise and are interesting because these people are centrally placed within the planning process. The local idealists are chosen for the study because they consider cultural heritage work and studying local history as important activities in their spare time. Different professional positions as well as various influences on decision making affect both how the stories are formulated and the importance each individual attach to the cultural heritage work. Through the focus on the individual perspective as well as through the range of variations this method presents, interviews give admission into a problematic field otherwise often closed for insight.

We know that cultural heritage work has several participants (Figure 3), and the degree of integration among these participants varies greatly. In areas where the interaction between the different fields of preservation work are well integrated and are regarded as a matter of course, these are the areas where we also find the criteria for success and a flourishing cultural heritage work. The situation we have found in our first case study, Nannestad, however, is the opposite – it is a minimum-standard that prevails. The discussion about the various fields will be based along a line from experience-near to experience-distant positions, or otherwise also known as personal space versus abstract expert space.

Figure 15-3. In Lauvåslia we find a small, but peaceful area with several old buildings. Despite its positive elements, it illustrates some of the insufficiencies of such private plans: lack of an overall directing perspective for the total area and too much responsibility placed on the local owners alone. Photo: NIKU 2001

Within the local preservation work, there are in large terms room for a series of participants with different positions, and I will, as follows, present the six different forms of participation. The degree of integration between the different levels or fields varies.

4.1 'Local idealists'

In Nannestad it is primarily the History Group that can be labeled as 'local idealists'. They have taken responsibility for arranging the yearly gathering in one of Nannestad's oldest farms, Sø-Kringler, that has become an important gathering place in summertime. Another project of the History Group has been to turn one of the classrooms in the old school into a museum (Figure 4). The craft group also plays a role within practical heritage work, since they rent the old school and are partly responsible for its maintenance.

Figure 15-4. Preservation work in schools and local public life are important to ensure the cultural heritage work a strong fundament. This building is one of seven schools that were built in the beginning of 1900. Today it is the only old school building left in Nannestad, and it contains an old classroom used as a museum as well as localities for craft courses. Photo: NIKU 2001

In some communities where preservation campaigns have taken place, such idealists have made it possible to save local memorials. The discussion in Nannestad at the moment concerns the preservation of the fundament of an old barn in the very center of Nannestad. Some politicians are opposed to keeping it and are prepared to tear it down in order to make room for shops and other new buildings. It has become a bit of 'a hot potato' locally at the moment.

There are several people who have come forward as examples of people with serious interests in preservation work and old things in general. An example is the owner of a large farm that received the county restoration prize a while ago. The farm was offered to the county council for sale to start up a local museum. But the price the county politicians offered was considered too low and the offer was turned down. Another example is a big forestry owner who has restored an old building and turned it into an impressing Hunting Lodge.

4.2 Preservation work in schools and local public life

The preservation work in schools and local public life is primarily tied up with the History Group. Today there is no local museum in Nannestad, but there are discussions along that line. An anonymous person has promised to support financially the establishment of a local museum. The precondition, however, is that it is to be localized in the very center of Nannestad, in a building erected for this purpose.

4.3 Preservation work integrated in local county council planning

It is necessary to differentiate between cultural heritage work as a sector with people employed specifically to look after heritage-projects alone, and others that have it as a sideline. In Nannestad it is definitely a sideline. This field of work is organized as one of the many tasks looked after by the Head of the Culture Office. He knows a lot of the people in the History Group personally, but they are seldom involved in or informed about the council work. In an interview, he made a special point of mentioning that the organizing of the Cultural Heritage Plan still is considered one of the main issues. Heritage work, however, is only one side of his job, and he has only one assistant helping him part-time. Local county council planning is taking place without any focus on preservation work

Nannestad represents the situation that usually takes place in local county council planning – there is no focus on preservation work. The way the planning process is organized is a precondition that concrete plans of various types are being presented for the regional cultural heritage department for verification. The local idealists characterize the county councils' heritage-profile as non-existing.

4.4 The formal planning process is in the hands of the regional responsible bodies

The formal planning process is held in the hands of the regional responsible bodies. This means that the responsibility for cultural heritage work in Nannestad primarily is taken care of on a formal basis – through the formal planning process in the hands of the regional responsible bodies. The regional heritage apparatus is mainly functioning as a controller. Societies' resources used to support these kinds of values, put into regional and local responsible bodies, are primarily acting on a minimum standard level – with anemic tendencies and without the commitment that enthusiasts bring into the work. Most of the activities that take place are solely on formal basis. There are very few preservation cases started on the sideline, and this gives the heritage work a bleak and rather invisible profile.

4.5 Expert-based systems

Planning within this category functions within what Anthony Giddens (1991) names as an abstract system, where focus lies on planning routines and procedures. An important trace in an abstract system is that what they do is emptying systems for local knowledge. Giddens calls this process 'to deskill', meaning emptying of skill. They replace local knowledge as common insight with another highly specialized type of knowledge. This process means that they necessarily are raising a partition between laymen and experts. At the same time more and more people in the modern society are forced into a situation where they are turned into lay-people. The highly specialized and expert-based systems claim that everyone has to have a fundamental trust to the functioning of the systems, without knowing themselves how. Giddens refers to this postmodern attitude as 'trust in abstract systems', and emphasizes that the nature of modern institutions is deeply tied up with this characteristic.

While the planning process can be defined as an abstract system, the wider preservation interest is based primarily in the personal, concrete space. People's feeling of belonging is closely tied to places they have their experiences from. They may have grown up in the area and have gained their knowledge from criss-crossing the county. Or they have their workplaces here and thereby gradually built up their local insight. These feelings are shaped through the years, through growing up in an area and gradually making and adjusting the stories about the local areas towards the person's own idea of the right story. This process stretches over time, but often takes place unnoticed when you are growing up in an area. People who have moved into a new locality, however, have to study actively to get to know the place by starting a process of active appropriation of an area. Through stories constructed over the years, time contributes to influence the form, and this effect is called 'narrative space' by Richard Sennett (1990). The time aspect will become an important element in a place where certain stories, narratives, are permitted to operate in everyday life. Sennett stresses that time starts to do the work of giving places character when the places are not used as they were meant to be, and he emphasizes the fact that space in this way gets encoded with time.

We may find a useful tool in the distinction made within social anthropology between experience-near and experience-distant positions. Social anthropologist Clifford Geertz (1983) describes the two viewpoints in the following way:

> "An experience-near concept is, roughly, one that someone – in our case an informant – might himself naturally and effortlessly use to define what he or his fellow see, feel, think, imagine, and so on, and which he would readily understand when similarly applied by others. An experience-distant concept is one that specialists of one sort or another ... employ to forward their scientific, philosophical, or practical aims. ... Clearly, the matter is one of degree, not *polar opposition*. ... And the difference is not ... a normative one, in the sense that one sort of concept is to be preferred as such over the other" (Geertz 1983).

He follows up by stressing the disadvantages of cultivating either one of these positions, as they both have their weaknesses: "Confinements to experience-near concepts leaves an ethnographer awash in immediacies, as well as entangled in vernacular. Confinement to experience-distant ones leaves him stranded in abstractions and smothered in jargon". In other words, Geertz stresses that these two viewpoints are not to be looked on as oppositions. The trick lies in combining them.

An important conclusion to draw from this discussion is that if good contact is not established between the two spaces or systems, they may go on existing relatively independent of each other, without much contact or influence upon each other. This is the situation that characterizes the minimum-standard of heritage work, that we can observe is being practiced in Nannestad today (Figure 5).

As this chapter shows, the preservation work tends to maintain an anemic character if it is not based on a strong local assistance and foundation. The history is based on local lives. It means that someone has to engage himself in telling the stories, to ensure a living memory about the history of the place.

5. THREE EXAMPLES FROM LOCAL PRESERVATION CASES

By presenting selected local examples, let us have a closer description of how the local preservation interests are formed. The first case is a study where description at close range is organized. With the help of planning documents, interviews and observations, we have formed a picture of which heritage questions the local cultural heritage interests have been concerned with.

Figure 15-5. Nannestad was up to recently considered a quiet, sparsely populated county with an economy primarily based on farming. Opening of the new international airport in 1998 brought a considerable amount of new activities to the county. Better communications towards Oslo have influenced people to move into the county while still working in Oslo, and several new housing estates are established. The landscape, however, still maintains a strong influence from farming activities, as this picture illustrates well. Photo: NIKU 2001

The first example is related to the local commercial center. It illustrates a modern conflict of interests – the need of modern welfare (shops, flats, schools) colliding within restricted fields with the care taking of old structures related to the local cultural history.

The second example is a modern housing estate. Despite the possibility of integrating parts of the old road and farms in modern structure, the actual development of the housing estate shows an insufficient ability to succeed in practical accomplishment. Private plans are often limited in their range and scale.

The third example presents a modern airport road. While the old roads were traditionally placed along the north-south structure of the valley, this new transport line cuts straight through the landscape to offer easy access to the airport. The landscape is being vastly transformed.

The chosen examples illustrate that greatly transformed developmental cases to a large extent lack a basis in solid local discussions in advance.

5.1 Example 1: a local commercial center

Low brick buildings dominate today's center. It is a combination of various shops, schools with stadiums and a civic town hall including a library. The buildings were mostly constructed in the 1960s-1970s. A former cotters' farm lies east of the town hall. It was regulated as a special preservation area in 1996 and is recently being rehabilitated. Among its neighboring settlement is another old farm, with an old grave close by.

The church catches one's eye first when approaching the center from the south. The oldest part is built in 1193, and it is known as an important Medieval church. On its eastern side we find a cluster of older buildings: a farm building called 'The Sheriffs House', the old civic town hall and the old bank. Placed as it is in a smooth hillside, it represents a characteristic porch to the center a short distance more north.

There are a series of potential conflicts in this area, which are closely connected to the fact that the center has planning status as a development area. The conflict focused on here has been chosen because it illustrates some of the built-in contradictions in the administration of cultural heritage work. The case chosen is 'The Wall', also by some ironically named 'Møkkakjelleren' (the Dung Heap). Long time ago, there was an outhouse building situated here, belonging to the main farm that gradually sold away its property. Over the years the farmhouses had to give way to modern buildings, but the wall was kept. Today, the wall functions partly as a meeting place and is suitable for local theatre performances etc. There is a certain resistance among local politicians against preserving the wall. On the other hand, it also has strong defenders, among others one of the shopkeepers who has given financial support. Recently the county council voted for its demolition. The decision is not finally signed, but the county council has given their point of view: "they want growth, and the wall is blocking it".

What sort of arguments can be used in preservation cases like this? It is not an archaeological remnant of high value, and not a building deserving its own preservation area. It is the local interest alone that turns into an important cultural heritage. It is a practical realization of the concept of identity. However, it often seems difficult to encourage the official cultural heritage management to defend this sort of heritage – it is considered not old enough, not rare enough, few interests attached from an architectural historical angle etc. Modern cultural heritage has a relatively weak position within today's legislation. Preservation orders according to *the Cultural Heritage Act* are mainly run by the central or the regional cultural heritage management. Using *the Plan and Building Act's* regulations about preservation areas is a tool on municipal level, but in practice the regional cultural heritage management is often included in the preparations. The local idealists have to fight on their own for local cultural heritage interests, since the resource situation seldom allows a versatile engagement. The local politicians claim that the wall is holding up important functions, and this argumentation seems to be supported more easily than diffuse arguments about symbolic value.

5.2 Example 2: a residential area, Maura

This neighborhood is situated in the northern part of the municipality. Along the road, houses of different ages have been built, and a new housing estate, Holkeby, has been planned during the last couple of years. It has been built in several stages, and one of the areas is called Lauvåslia (Figure 6). This section of the estate was introduced as a so-called *Reguleringsplan*, and according to the local planners in the municipality, it was called to life as a private development plan. By looking through the preliminary preparations, the plan strikes as a well-thought-through plan, where among other things the supply lines are well marked. An area on the outskirts of the estate is marked on the map in red, signalizing that this area is regulated as a preservation area according to *the PBL Act*. This brings forward an impression that cultural heritage interests are well looked after in the plan.

However, this positive first impression changes quickly when you walk around in the area. The properties have been sold one by one, and the choice of architectural building types and choice of entrepreneurs have been left to the owners themselves. It has led to a lot of competing expressions of bungalow ideals, which partly suppress each other. The building types are badly suited for the hilly terrain, and this negative impression gets intensified since the forest is cut down. East of the new estate, several farms are situated. A narrow, older road makes it possible to walk from the farms into the estate, and this road passes the farm now preserved. Instead of creating a functional line between the two areas, the old road is cut off at several places by new pedestrian paths. It strikes you that the plan seems to ignore the importance of integrating old and new elements. If an overall picture had been included in the plan, it would have been possible to adjust new elements to the existing old road, and it would then have been treated as an important resource.

Figure 15-6. Holkeby is a new housing estate that was planned in the recent years. It has been built in several stages, one of the areas being called Lauvåslia. This section of the estate was introduced as a so-called *Reguleringsplan*, and according to the local planners it was raised as a private development plan. The properties have been sold one by one, and the choice of architectural building types and of entrepreneur have been left to the owners. It has led to a lot of competing expressions of bungalow ideals. Photo: NIKU 2001

Interviews with the local planners supplement the picture of a case primarily run by private interests. The suggestion to label the farm as a preservation area was promoted by the owner himself, and it is this owner that sold the large area where the estate has been built. This case is one of the many development plans which private developers promote. It illustrates some of the insufficiencies attached to such plans: lack of an overall directing perspective for the total area and too much responsibility placed on the local owners. The inherent resources of landscape and cultural heritage values could have brought in special aspects, but have in this case been used insufficiently. It leaves an impression of positive intentions, but insufficient ability to be consequent in practical accomplishment.

5.3 Example 3: new road connecting the airport

Figure 15-7. Modern roads of today are breaking the traditional structure between farm and road. It represents new principles for transport planning. Even if landscape architects today play an important role in the physical planning of new roads, the consequences for the landscapes are tremendous. This road will connect the western part to the new airport. Photo: NIKU 2001

The new state highway (Figure 7) is being built to ensure a faster communication from the western areas to the new airport. Before starting the construction, estimations of the effects of pollution, nature resources, cultural heritage, aesthetic values and society have been made. In the planning process, led by the Regional Transport Department, different alternatives are being considered. It is done according to the demands for Environmental Impact Assessment surveys. The archaeological surveys have among other objects excavated a farm from the older Iron Age. The farm has probably been moved some time in the early Iron Age or early Medieval age.

The traditional farm structure suggests that the traffic either went through the farmyard or was connected by an avenue of trees. In other words there was established a close connection between farm and road.

While Environmental Impact Assessment is meant to function as a security measure by eliminating the most negative undertakings, an increasing amount of cultural heritage remnants are being sacrificed.

6. CONCLUSION

Pressure on the fringe of the cities is turning into an increasingly important issue within planning. On the local heritage level today the perception of landscape and the cultural history as a unity are gaining ground. The themes and perspectives attached to cultural history activities are widened and today include landscapes, nature conservation, places, processes, intentions and actions. Unfortunately the administration and implementation on the local and regional level are still tied up in a settled and fixed pattern, and parts of the cultural heritage work practiced in the local administration are still dominated by an object-orientated and museum-based profile. Despite the activities from several local idealists, the study shows that the cultural heritage work in our case study is primarily taken care of on a formal basis – through the formal planning process in the hands of the regional responsible bodies. Particular attention is paid to handling the controlling perspective that means handling the formal rejections when serious interests are at stake. But the heritage work is kept on this level – as a minimum-standard with anemic tendencies and without the commitment that enthusiasts bring into the work. This gives the heritage work a bleak and rather invisible profile, characterized by a minimum-standard of heritage work, as we have seen is being practiced in our case studies. Without strong local assistance and foundation, the preservation work will maintain an anemic character. The time aspect becomes an important element in a place when certain stories, narratives are permitted to operate in everyday life. A better understanding should evolve between the dependence of the mental perspectives, understood as the story-telling, and the physical structure, understood as the cultural heritage.

The cultural history is based on local lives. It means that someone has to get engaged in telling the stories, so they can stay on as living memories of the place.

REFERENCES

Antrop, M. (1997). The concept of traditional landscapes as a base for landscape evaluation and planning. The example of Flanders Region. *Landscape and Urban Planning 38,* 105-117.

Bedre kommunal og regional planlegging etter plan-og bygningsloven. Planlovutvalgets første delutredning (2001). NOU – Norges offentlige utredninger nr. 7. Miljøverndepartementet. Oslo.

De Certeau, M. (1984). *The Practice of Everyday Life.* Berkeley: University of California Press.

Gardermoen-prosjektet. Hovedplan. Ny vegforbindelse Lunner (Rv 4) – Nannestad (rv 174/120) Statens vegvesen Oppland ViaNova.

Forester, J. (1989). *Planning in the face of power.* Berkeley: University of California Press.

Forester, J. (1983). Critical theory and planning practice. In J. Blau, M. La Gory, & J. Pipkin, (Eds), *Professionals and urban form.* Albany.

Giddens, A. (1984). *The constitution of society: outline of the theory of structuration.* Cambridge: Polity Press.

Giddens, A. (1991). *The consequences of modernity.* Cambridge: Polity Press.

Geertz, C. (1983). *Local knowledge: further essays in interpretive anthropology.* New York: Basic.

Gregory, D. (1994). *Geographical imaginations.* Oxford: Blackwell Publishers.

Grimen, H. (1996). Tillit og rutinar. *Notater fra Norsk senter for forskning i ledelse, organisasjon og styring (LOS), 18.* Oslo.

Harvey, D. (1989). *The condition of postmodernity.* Oxford: Bail Blackwell.

Healey, P. (1997). *Collaborative Planning. Shaping places in fragmented societies.* Houndmills: Macmillan Press.

Muir, R. (1999). *Approaches to landscape.* Basingstoke: Macmillan.

Nesheim, O.N. (1999). *Mangfold, skjønn og formalisering i kulturminnevernet.* Doktoravhandling, Universitetet i Tromsø.

Relph, E. (1976). *Place and Placelessness.* London: Pion Limited.

Riksantikvaren (2001) Kulturminner og kulturmiljøer, plan og bygningsloven. Veileder. *Riksantikvarens rapporter nr. 29.* Oslo.

Røling, N.G, & Wagemakers, M.A.E. (1998). *Facilitating sustainable agriculture.* Cambridge: Cambridge University Press.

Sennett, R. (1990). *The conscience of the eye.* New York: W.W.Norton.

Schlereth, T.J. (1990). *Cultural History and Material Culture.* Ann Arbor, London: UMI Research.

Tunbridge, J.E., & Ashworth, G.J. (1996). *Dissonant Heritage. The management of the past as a resource in conflict.* Chichester: Wiley.

Urry, J. (1995). *Consuming places.* London: Routledge.

Chapter 16

'THE LONG CHAIN': ARCHAEOLOGY, HISTORICAL LANDSCAPE CHARACTERIZATION AND TIME DEPTH IN THE LANDSCAPE

Graham Fairclough
English Heritage, United Kingdom

...think for a moment of the long chain of iron or gold, of thorns or flowers, that would never have bound you, but for the formation of the first link ...
Charles Dickens, Great Expectations (1860-61)

1. WAYS OF SEEING LANDSCAPE: CHARACTERIZING THE HISTORIC DEPTH OF LANDSCAPE

Landscape Assessment in England in its modern sense has origins in the late 1980s, following unsuccessful attempts to produce objective, quantified methods (Countryside Commission 1987; 1993; Countryside Agency & Scottish Natural Heritage 2002). More broadly, the method can be carried back to the 1940s and 1950s and the creation of the first UK protected areas, known in UK legislation as National Parks and Areas of Outstanding Natural Beauty using criteria of 'specialness', perceived naturalness and aesthetic quality. Going back further, there is a long English tradition of landscape assessment based on the aesthetic values of landscape; this included interest in consciously designed high status ornamental landscape. For some heritage

H. Palang and G. Fry (eds.), Landscape Interfaces, 295–318.

managers and planners, 'landscape' still seems to mean only 'natural' or 'designed' landscape.

Scenic and visual criteria still stand close to the heart of English landscape assessment, with some added concern for geology and topography, and a few concessions to history. There is, however, limited recognition of the human processes by which environment has been modified. Most importantly, there is an underestimation of the importance of the interaction of people and nature over very long periods of history, and of the long-term sequence of change and continuity that creates so much of the landscape's character. It does not fully accept that historic human impact on the environment is fundamental in creating landscape, and that few landscapes can be classed as natural.

This chapter describes an attempt to redress some of these shortcomings in a heritage management context in England under the general banner of *Historic Landscape Characterization* (HLC), a technique developed to help people understand time depth in the landscape (Fairclough 2002e; English Heritage forthcoming a). As will become clear, HLC also addresses one of the shortcomings of heritage management itself, its limited engagement with heritage at landscape scale.

During the 1990s, archaeology and ecology both began to develop additional methods of understanding and assessing landscapes that properly reflected human cultural process (Macinnes & Wickham-Jones 1992). In particular they pursued the idea that landscape is the sum of all its parts – natural, cultural, ecological, archaeological, historic, aesthetic, social and mental – and that multidisciplinary approaches are necessary to do full justice to it.

Archaeologists understand the present-day landscape by means of long-term narratives and explanations rooted in social, political and economic historic processes. The main archaeological contribution to landscape is an understanding of time-depth, placing change at the center of landscape history and character, and indeed celebrating its effects (Fairclough et al. 1999), a perspective that encourages interest in the sustainable management of future change (Fairclough 2002d).

The basis of all today's cultural landscape, however, was created in the past, and is an equal concern for archaeologists. The cultural landscape can be regarded as one of our richest historical records, containing an archive of how human communities shaped, destroyed or created their environment over many hundreds or even thousands of years. It is an archive that exists both for prehistory long before documents illuminate the past, and even in historic periods it touches on areas of ordinary life rarely documented. It even enables us to know a little about our predecessors' cognitive landscapes. Historic landscape characterization, this chapter's main subject, was thus developed to use this physical archive to forge an archaeological perception of the present-day landscape.

HLC adopts the idea that *landscape* is not quite the same as *environment*. There is a need for archaeological and historical understanding of past environments, but the concept of landscape is seen to be something more than a mere description of the physical traces of the past. This additional factor, which changes environment into landscape, is the existence of an observer who constructs what we call landscape from the material environment. Because of such modern methods as air photographs, satellites and virtual-reality modelling, we no longer need to be constrained by the original definition of the word landscape as an area of land, a prospect, taken in by a viewer at a single viewpoint. We are also able to move away from seeing observation as merely a visual act, and to accept that landscape, whilst in the eye of the beholder, is also more importantly in the intellect, mind, heart and senses of its observer. All 'beholders' possess mixed as well as personal viewpoints, and thus an infinite multiplicity of responses and perceptions. This is what makes landscape such a powerful concept.

Landscape is in fact doubly cultural. Its components ('ingredients') within the environment are the product of hundreds, sometimes thousands, of years of human, cultural actions. At the same time, however, the landscape as a whole is cultural because it is created only in the present-day by our own cultural and social attitudes – it is not the same as environment but an intellectual construct.

HLC differs from landscape history in both method and sources. The two disciplines may as a result produce different 'stories', but in their different ways both are valid interpretations of their own data. History studies the past through old documents, and in the case of landscape, is essentially a documentary study of maps and documents related to landscape. Archaeology studies the past more directly, through material remains in the present. In the case of landscape, it is the landscape itself and the environment 'beneath' that are its main sources, even if sometimes they are studied through the proxy of maps and documents used in a supporting role. Historians discover meaning in documents, archaeologists attribute meaning to material culture as well as using it as 'document'.

HLC shares with landscape ecology a concern for historic processes in the landscape. It looks for example at the mechanisms that create the pattern of field and woodland, town and farmland, hedge and wall that sit at the heart of most people's perceptions of landscape. HLC probably recognizes a greater depth of time in these processes than landscape ecology, because whenever it can archaeology uses past material remains that survive in the modern world to look at longer time-frames. These often go beyond oral and even documentary history, and are taken beyond tradition and custom if possible.

Causal sequences that have led from one landscape pattern to another are equally relevant. They may not always appear directly to affect the present form of the land, but they are nevertheless important parts of the chain between then and now, a part of how we perceive and 'read' landscape in the way that we do. This is what Charles Dickens (*Great Expectations)* described in the slightly different context of human rather than landscape biography: "the long chain of iron or gold, of thorns or flowers, that would never have bound you, but for the formation of the first link".

2. DEEPER ROOTS AND A BROADER SPREAD – 'WHOLE LANDSCAPE' AND SUSTAINABILITY

In addition to its aim of helping people to understand the historic depth of landscape, HLC also had other origins and aims, largely growing out of the explosion of interest in sustainability in the 1990s (Fairclough 1995; English Heritage 1997). One aim was to provide an archaeological understanding of landscape that other landscape workers could easily absorb, understand and use. Greater integration between all these is needed, but all the separate strands need to be pursued first until they are sufficiently well understood. An integrative agenda is therefore at the core of HLC. Initially, this means using a spatial and landscape- based approach to create a common language so that ideas can transcend boundaries between disciplines. HLC therefore adopted landscape assessment approaches, but adapted them to an archaeological perspective in order to tell us about the past as well as the present, and to help shape a future.

HLC is also part of a trend to expanding archaeological and heritage resource management from individual sites to the whole environment. Landscape is an ideal vehicle for this. Historic significance can be read by archaeologists, not merely in those places where archaeological deposits can be excavated, but also in features whose simple existence and inter-relationship tells us why we have inherited the landscape that we have. HLC can embrace the patterns of trees, hedges and woodland, the distribution of land cover such as moor or heath, or the patterning of settlement and township boundaries. It has no chronological cut-off, being interested in remains of all periods, modern as well as ancient.

Working at landscape scale is also more accessible to public participation. People relate to landscape in more ways than they relate, for example, to wildlife, archaeological sites or fine buildings; as *the European Landscape Convention's* preamble insists, landscape is everyone's common heritage. Everyone 'owns' landscape, in memory and in daily life, everyone has roots in at least one landscape somewhere. Landscape connects people to their concerns about protecting sense of place and local character, and about the increased importance of local distinctiveness in the face of globalization. Almost any common measure of a place's distinctiveness mentions history, archaeology and the cultural processes that created and nurtured landscape.

A landscape viewpoint is also one of the most effective ways of delivering sustainability. Rare species cannot be preserved without keeping the wider landscapes – habitat, territory – that sustains them, and so too with cultural heritage. Archaeological sites without appropriate (not necessarily 'authentic') landscape contexts lose some significance, visibility and relevance to everyday life.

In Europe, all of these ideas, and more, have been brought together in a new aspirational instrument, *the European Landscape Convention* (Council of Europe 2000; Desjeants-Pons 2002; Naturopa 2002). This was opened for signature by its sponsor the Council of Europe at Florence in October 2000, and is likely to come into force (when ratified by ten states) by 2005. It is the first Convention in Europe, and perhaps the world, devoted exclusively to landscape. It covers not just special, attractive, or 'natural' landscapes but all areas of landscape, everywhere, no matter how cultural. It calls for equal integration of natural and cultural values, defining landscape as the perceived interaction of nature and culture. It sets these ideas into an explicit framework of sustainability, democratization, common popular heritage, and public participation.

In England, HLC is also part of a shift from protecting the past at special places ('designation') to managing change across the whole historic environment in broader, more socially embedded, ways that relate to all places, in all aspects. As such, is has been endorsed by recent UK Government policy for the historic environment, *The Historic Environment; A Force for Our Future* (DCMS & DETR 2001; for Scotland see HS 2002). The aim is to ensure that the entire historic environment is taken into account when changes are proposed, but this requires an acceptance that the goal is informed management of change rather than protection (Clark 2001; Bloemers 2002; Fairclough forthcoming).

Selective protection by designation of special places has been effective in the UK for over a century, but during the 1980s and 1990s opinion grew that additional approaches were needed to recognize that all aspects of the historic environment had significance. These ideas came from various quarters, such as the local distinctiveness movement, landscape assessment and archaeological resource management. Other conservation and environmental disciplines, notably ecologists, made the same transition from site-centered to area-based perceptions, recognizing that sustainable countryside management required more than the protection of nesting sites. Paradoxically, given their origins in the European Union's *Common Agricultural Policy* which is widely regarded as one of the most destructive agents of change to the cultural landscape, there emerged between 1986 and 1994 the beginnings of national, comprehensive incentive schemes to help farmers manage the land more sensitively. But it was sustainable development that gave this movement its greatest impetus, underlining the need to consider all parts of the historic environment when planning the future (English Heritage 1997).

3. INCLUSIVE, MODERN AND FLEXIBLE – CHARACTERIZATION AS A NEW HABITAT FOR HERITAGE AND LANDSCAPE MANAGERS

Characterization is one of the principal vehicles for putting these ideas into practice, with HLC, like landscape assessment, as a pioneer of the approach. It adopts a generalized understanding of large areas, noting their diversity and complexity but focusing on their essential overall character – in other words, a method for creating a comprehensive but generalized, neutral and interpretative description of the historic character of an area. It is a context for other information, and a framework for decision-making about change and management. Its main focus is on supporting conservation and historic environment resource management, but it also creates new understanding for a variety of uses, including research.

The task of creating contexts and frameworks for understanding and managing change is perhaps most important. It allows characterization to be enabling and capacity-building because it can open up more public debate about significance, rather than closing them with definite black and white answers from experts. It can encourage participation by informing anyone's decision-making without usurping decision making from communities of place or of interest. It is integrative, combining and synthesizing existing research and knowledge; and connecting different types of environmental management.

This approach supports a more diverse and flexible approach to the management of change than designation alone does. It recognizes that we preserve our past for our sake not for its sake, a recognition that can equally apply to protecting natural resources such as wildlife. These ideas accept that change should not be stopped, especially where landscape is concerned, but it should whenever possible (and wherever possible) be guided into sustainable and historically sensitive directions.

Alongside these ideas is a growing recognition of the need to manage change more broadly than protecting a sample of special sites or buildings. Traditional methods of monument preservation (essentially a 19th century approach) were until recently the only significant theory in historic conservation, but they worked well during the 20th century, albeit at the price of great selectivity and exclusion. Recent years have seen the approach more or less successfully expanded to other 'new' types of heritage, but such expansions cannot go on forever without changing approaches: not everything can be protected in that way, and recognition of everything is the goal, especially in landscape management.

Saying this is not meant to challenge the need for designation-based systems. These successful and effective methods have not suddenly failed, or become obsolete; rather it is society's ambitions for protecting and utilizing the historic environment that are changing, so that additional methods are now needed to work alongside the old tried and tested methods. These additional methods should ensure that all aspects of the historic environment (and other dimensions of the environment) are taken into account in planning change for the future, and that multiple ways of valuing are recognized. This is the aim that methods under the generic heading of 'characterization' are trying to achieve, particularly when dealing with the big picture of landscape character and processes.

There is, however, a trade-off at the heart of this concept as mentioned above. Characterization broadens the heritage canvas, so to speak, and gives a wider view that potentially includes everything, but it must paint a different picture on this new canvas. Characterization suggests that everything in the historic environment deserves to be taken into account in some way, however local or ordinary, rather than only ever fighting to keep a few places largely unchanged while often all around may change. There will be compromise and debate, with greater inclusiveness and participation, to seek the best achievable new development and the highest practical levels of historic preservation. Giving new weight to local context might mean for example that it is not always the nationally important things that matter most. Characterization aims at a different mentality to be applied everywhere, by planners and developers as well as by conservationists. It is a way of starting to ensure that the historic environment plays a fuller role - its proper role – in modern life.

To do this, the process of characterization has to be essentially neutral in the first instance. It does not in itself define relative significance, nor does it give easy answers to the question of what to keep from the past. Instead, it provides knowledge, interpretation and understanding to allow decisions to be made at the appropriate time, when heritage significance can be matched to the proposed change rather than judged in a social vacuum. It has multiple uses in the planning process, in agri-environmental and rural development policy, in urban regeneration and in conservation, and its users will come to include heritage and conservation agencies, government at local, regional and national levels, owners, developers and business, and of course among the wider public.

As an approach that seeks to use different ways of seeing and understanding in order to help manage change, characterization can be used at many different scales, from local through regional to national and beyond, as long as when used its perceptions are matched correctly to the level of decisions-taking. It works for townscape as well rural landscape, and for topics or thematically as well as geographically. It seems to work best at larger scales, however, being well suited to a landscape-scale. The most developed approach as far as the historic environment in England is concerned has been for the historic dimension of the rural (and now urban) landscape, through HLC.

3.1 Historic Landscape Characterization

HLC was invented and developed in England, and first used in Cornwall, during the 1990s. There was at that time wide consensus that landscape is the sum of all its parts and that assessment was needed (Brown & Berry 1995). But it is only possible to integrate ideas and information that exists, and we did not have genuinely landscape-led understanding of the historic environment to supplement landscape assessment. HLC was designed to fill this gap, with future integration in mind (Fairclough et al. 1999). English HLC now takes the form of a national programme that is over half-complete, promoted at county level by English Heritage (the government agency responsible for leading the protection, management and sustainable use of the historic environment, in both town and country) in partnership with local government archaeological services. It allows the time-depth of the landscape, its historic and archaeological dimension, to be read (Fairclough et al. 1999; Darlington 2002; Fairclough et al. 2002; Dyson-Bruce 2002; Fairclough 2002a; Hampshire County Council website). HLC techniques have begun to be adapted to urban as well as rural contexts, and is being used or tested in other countries in the UK countries, for example in Scotland (Dyson-Bruce et al. 1999; Macinnes 2002; Dixon & Hingley 2002), and further afield in other parts of Europe, such as Ireland (ERM 2000) and Sweden in the context of a Culture 2000 programme called *European Pathways to the Cultural Landscape* (www.pcl-eu.de; Nord Paulsson 2002; Nord Paulsson and Fairclough 2002).

The information and understanding that HLC creates can be used in very many different ways. First, most straightforwardly, it adds landscape knowledge to traditional *Site and Monument Records*, to strengthen their contribution to development planning, development control, Environmental Impact Assessment, designation, land management policy and agri-environmental programmes and hedgerow protection. It adds an understanding of time-depth and human actions to aesthetic-based visual landscape assessment and to the land management insights of landscape ecology.

In terms of research, HLC provides a new context for site-specific data, and it can be used as a predictive tool, or to identify gaps and weaknesses in existing knowledge, or to define research agendas. An HLC project carried out in 2003 can also be a benchmark for future change in the landscape and change in our knowledge and perception.

HLC also offers the potential for closer links to public awareness, encouraging 'democratization'. As suggested earlier, landscape is perhaps uniquely accessible to the public, particularly when assessed in ways that recognize that landscape is ubiquitous, everywhere and on everybody's doorstep, and that it can carry multiple not only specialist values. It has links to local identity and common heritage, and methods like this, therefore, as *the European Landscape Convention* recognizes, offer one way of enabling wider social participation in decision making, both about the landscape's significance and about its future management and sustainability.

3.2 The philosophy of HLC

All these considerations combine to create a series of principles that support HLC. They are closely connected with *the European Landscape Convention's* commendably straightforward definition of landscape as: "an area, as perceived by people, whose character is the result of the action and interaction of natural and/or human factors". For convenience, this chapter presents the summary version of these principles that was drawn up for the landscape network *European Pathways to the Cultural Landscape* (funded by the European Union Culture 2000 programme; www.pcl-eu.de). They summarise what HLC is setting out to achieve, and what distinguishes this approach from other methods of understanding and assessing landscape:

- a concentration on present-day landscape character, not past landscape; today's landscape as material culture – the main object of study and protection by HLC is the present-day landscape, as created by human action and perception;
- history not geography – the most important characteristic of landscape for the purposes of HLC is its time-depth, the way that earlier landscape and change can still be seen in the present-day landscape;

- area not point data – HLC-based research and understanding is concerned with landscape not sites; it is not a process of mapping find-spots and monument distributions;
- all areas and aspects of the landscape, no matter how modern or ordinary for example, are treated as part of landscape character; not just 'special' areas;
- landscape is a human construct – semi-natural and living features (woodland, land cover, hedges etc.) are as much a part of landscape character as archaeological features; biodiversity is a cultural phenomenon;
- landscape is different to environment – a characterization of landscape is a matter of interpretation not record, perception not facts; 'landscape' is an idea not a thing, constructed by our minds and emotions from the combination and inter-relationship of physical objects;
- peoples' views – an important aspect of landscape character in HLC will be collective and public perceptions laid alongside more expert views.

The most important of these, perhaps, are those which concern time and human agency (Fairclough 1994, 1999a). These concerns most define HLC's unique and special contribution to the wider integrative project of knowing the landscape by coordinating all the several ways of defining landscape character, including landscape ecology and landscape architecture as well as HLC.

The role of the passage of time is very important in HLC, which is most of all concerned to trace the imprint of the past on landscape, its time-depth. This can involve defining the survival of past environments at landscape scale, or looking for the deeper structures, such as long-established communications systems or deeply embedded field and settlement patterns that our landscape has inherited from the past. A particular concern is how these landscape components have either changed through time or survived through continuity.

Changes through time in this concept of landscape are caused mainly by human action, human 'agency'. HLC emphasizes the role of people in landscape creation, both in the past through physical activity and in the present through perception. Like landscape ecology, HLC recognizes the critical role of human land management practices on landscape character. It is particularly defined by its description of the interaction of humans and nature through the filter of social and historic processes (e.g. prehistoric woodland clearance, privatization of land in the late middle ages, fenland and coastal land reclamation). It works above the level of simple geological or climatic determinants to explore the more subtle nuances of human responses, some of the most significant of landscape's cultural attributes.

English HLC presently operates mainly at county level. This level allows generalization without being too distant from local distinctiveness; it also happens to match the level at which archaeological resource management operates. It is, however, only one of the levels at which HLC can work, and county HLC already fit into two national frameworks: *the Countryside Character Map* (Countryside Commission 1998; Countryside Agency 1999) and *English Heritage's Settlement Atlas* (Roberts and Wrathmell 2000, 2002). There have also been successful experiments in doing HLC more locally, and in producing higher-level regional views. The ultimate aim is for county HLC to be part of a 'ladder' of characterization rising from ultra-local small studies, through county and regional assessments, to national and international overviews. At each step, different aspects of the landscape can be considered, with greater or lesser generalization, to answer appropriate research and management questions and needs.

The HLC method takes an inclusive view of what are the 'building blocks' or ingredients of landscape. They must have a historic or an archaeological aspect, and they must operate to influence perception at landscape scale, but beyond that there are no limits. HLC does not restrict itself to 'monuments', conventional archaeological sites or historic buildings, which actually form only a very small part of HLC. It encompasses semi-natural components (hedgerow, woodland, heath), more abstract aspects such as the pattern of land use (arable, pasture, woodland), their inter-relationship to each other and to settlement patterns, roads and track-ways, or coastal movement.

Even the things often thought to be most natural have cultural depth. Wetlands in NW Europe have long-forgotten (except through archaeology) religious and ritual meaning, while rock outcrops anchored cognitive, mental maps to the real environment at many dates in deep prehistory. Less tangible aspects such as the historical processes that underlie the material record of the spatial and territorial patterns of estates, or town-lands are also part of the pattern of HLC. Importantly, it is not these elements in isolation that make HLC what it is, but the connection between them, both in space (geographical patterns) and in time (time-depth, successive layers, the rich palimpsest of the whole present-day landscape); most of all, HLC is about patterns.

Because HLC methodology concentrates on present-day historic landscape character before earlier layers and horizons, in England it tends to emphasize landscape character of the most recent few centuries. The focus on recent periods is also quite useful, representing part of a wider move away from site-based conservation, and from a single period focus, so it also reflects a wider growth of interest in 'new' heritage, even of the late 20th century.

Additionally, even the most 'shallow' HLC-based assessment of time-depth will spread backwards across several centuries, which is significantly deeper than most non-historical landscape characterization that sometimes seems to minimize human impact of any date in favor of contemporary or supposedly 'ageless' natural explanations. It is also worth commenting that the first historic landscape characterization of an area is also only a starting point to which greater time-depth can be added, later and in localized areas if need be, within an overarching framework that is anchored in the present-day. Indeed, a critical review of HLC would draw attention to the limitations inherent in current methods because of its choice of county scale. Further methodological development should focus on applying the principle of HLC both at more detailed local level, and at higher levels. Regional overviews are planned in England, which will produce different but scale-valid interpretations. In its present manifestation, HLC's planning and management outputs are mainly linked to strategic and spatial planning, and the setting of priorities for agri-environmental schemes. Applications for development control (e.g. using it to contextualize very local aspects of landscape character) still remain to be developed.

As practiced within English Heritage's national programme, HLC has also been tailored to types of landscape common in England. It reflects that country's dense and diverse, chronologically rich and morphologically varied patterns of hedge- and wall-bounded fields, closely linked to different types of settlement pattern and to a particularly fine-grained type of landscape. Where similar landscape exists – in NW France perhaps – the method could probably be used largely unchanged. It would be much less easy to use in more open unenclosed landscapes, such as parts of Portugal or in areas of Sweden heavily remodeled in the 18th or 19th century, which by and large lack visible marks of land division. Nor might it be easily used, for different reasons, in landscapes that on the surface appear to be overwhelmingly modern, such as those of the Netherlands. In such places, modification would be needed. The part of HLC that is most transferable is its general theoretical framework and its principles, notably those concerned with time-depth, the importance of change, the focus on human agency and the determination to characterize the present-day landscape historically rather than to study past landscapes.

3.3 How HLC is carried out – methods and uses

A detailed description of the HLC method is not necessary here, and several are published elsewhere (e.g. Herring 1998; Fairclough 1999; Fairclough et al. 1999; Dyson-Bruce 2002; Fairclough 2002a; Fairclough, Lambrick & Hopkins 2002; Darlington 2002; English Heritage 2002, forthcoming a). In summary, HLC uses area-based generalisations within a GIS environment to produce an overview of an area's broad landscape character. It is a desk-based vertical assessment of pattern, process and function in the landscape. It uses maps (modern but also historic), vertical air photographs and other datasets such as land cover, land use or ecological data, to define areas of land – created as polygons in the GIS – that have similar characteristics, described by attributes and data contained in related databases. These descriptive attributes allow complex modelling and classification to be carried out, and the result is a sophisticated spatial database that produces not just one simple HLC map but the ability for almost endless interrogation, analysis and thematic mapping.

The HLC method creates a continuous patchwork of small areas, each of which is a separate data-holding unit in the GIS (polygons). Polygons are defined subjectively, because of a general broad similarity within their boundaries, or because they share dominant characteristics (e.g. regular fields, fields cut out of woodland, ancient or new woodland, unenclosed heath, major industrial areas, or urban areas). The character of each area (or GIS polygon) is not normally homogenous, as that approach would become too detailed and tell us little other than that every piece of land is unique. What HLC seeks to do is to identify similarity and general patterns. It is not an environmental database, but a way of mapping interpretations.

The detailed attributes attached to each polygon provide the basic means to describe present-day historic landscape character. They allow identification of time-depth and other historic aspects, such as the shape and nature of field boundaries and the historic, social and economic processes that created them. They might also include indicators of past historic landscape character that explain present-day character, such as relicts of medieval open fields, the existence of diagnostic types of buildings, field ponds or indicative field or place-names. In urban landscape, HLC attributes reflect date, function (e.g. housing, industrial, civic space) or form (terrace versus semi-detached versus detached housing for example) of buildings, or variations in plan form of streets and spaces. HLC can also look below the modern town to characterise the pre-urban landscape, to show how this often still shapes the course of modern streets or the distribution of building types.

Many different classifications can be produced from these attributes, and maps can be made that reflect either the generality or some of the detail of the historic landscape. Themes can be addressed, previous past landscape character can be modelled; pie-charts and histograms can be used to analyse and compare the particular character of areas such as parishes and townships. Analysis of the distribution and particular types of historic landscape or areas that share similar combinations of attributes is also possible, as is the production of historically informed models of past or possible future landscape.

The use for which HLC was primarily designed (Herring 1998; Fairclough et al. 1999) is as a tool for conservation and historic environment management. Simply, it provides new information to guide decisions about proposed change to the environment. Understanding is a prerequisite for taking sensible conservation and development decisions, as part of a process known as *Informed Conservation or Applied Archaeology* (Clark 2001; Bloemers 2002). The most significant results of HLC are when it changes perceptions and understanding of the landscape. It can encourage people to review their assumptions (such as that a particular landscape is 'unchanged', 'natural' or 'age-old') by drawing attention to time-depth and human agency. Conversely it can demonstrate that even in very modern, or recently greatly changed, landscapes there are still traces of the more distant past that are important contributors to landscape character, sense of place or local distinctiveness.

Which fields in the countryside, for example, are the oldest or the rarest or provide particularly significant archaeological evidence for history? Which type and areas of field are likely to have been cleared and enclosed from medieval enclosure of woods and forests, and therefore what is the likely previous extent of woodland? Which fields, though perhaps modified, are the genetic descendants of prehistoric or early medieval patterns, which tell us most about the past, which contribute most to the landscape's historic character? Which types are most sensitive to change, or which can absorb change whilst keeping their essential historic character? Which areas have seen several successive layers of change, being dynamic zones where further change might seem most appropriate, and which are zones of longer term continuity that we might argue should remain largely unchanged? Why does woodland have particular locations, and a particular mix of trees? What are the human management systems that have made woodland, and woodland biodiversity? Such types of knowledge about the history of the landscape's historic character are essential for the sustainable planning of the future landscape.

In relation to English conservation and spatial planning, the types of new understanding that HLC provides are wide-ranging. Because HLC uses sophisticated GIS and spatial databases, it can produce an almost endless series of interrogations, analysis and thematic maps. Many different combinations of polygon attribute can be created, and many different maps, whether selective, synthetic or interpretative, can be made in response to different priorities and policies. Past landscape character can be modelled, change and activity can be charted, and benchmarks and baselines can be set for monitoring future change. Analysis of HLC in comparison with other archaeological, ecological, or architectural datasets can help to explain the patterns and biases of existing knowledge, highlighting gaps and illuminating factors of survival and condition. New research agendas can be created using HLC's insights, so that the process of understanding becomes iterative and ongoing.

4. REFLECTIONS – A WIDER LANDSCAPE OF LANDSCAPE MANAGEMENT

As discussed earlier, landscape can be regarded as 'merely' a construct of perception and interpretation. This means it is not the same as environment, even if is forged from the raw materials of the environment. Seen in this way, landscape is a product of cultural ideas, historic processes and change in many different ways. Of all heritage or conservation values and significances, those attached to 'landscape' are among the hardest to capture, and the most difficult to define. But for almost the same reasons, landscape is the most democratic, ubiquitous and socially-relevant aspect of heritage; at some level, it not only includes everything else, but it also allows everyone to be involved in appreciating and caring for it.

How we perceive landscape constantly changes, just as its component parts change through time. What people value in landscape, what they wish to preserve, why they wish to preserve it or manage it – all these change, all of the time. Our understanding evolves, and what we think is significant or worth keeping develops. All these changes are furthermore refracted through the autobiography, knowledge and aims of any observer. Finally, there is future change, the ever-present potential for more change and all the decisions and valuations that precede and accompany it.

Change takes many forms – slow long-term change or rapid change; recent change, with the immediate consequences of which we must deal; current change that can be mitigated or encouraged. In fact, if any one characteristic most deserves to be singled out as an essential defining feature of cultural landscape, it is surely its dynamism, the fact that it changes all the time, and the legibility or visibility of change within its fabric. The aim is surely to manage the physical and perceptual aspects of the cultural landscape, and at the same time to preserve the dynamism that characterises landscape and makes it live into the future. To do this, landscape heritage management will require more sophisticated and nuanced methods than the straightforward ones that are effective to manage sites and monuments.

For landscape, conservation or heritage management cannot be an activity at the edges of society, standing on the sidelines and intervening only reactively. It should be a constructive proactive process, concerned to shape the future, not merely to keep fragments of the past. It needs to be more of an activity at the heart of society and politics, a way of living not a pastime. This is the ultimate implication of accepting the need for sustainable development – a full integration of all three factors of sustainability – the environmental with the social and economic.

This chapter has tried to demonstrate that better understanding of the historic dimension of landscape – of what has created it and of what has survived from the past, time-depth in short – is needed if the future landscape is to continue meaningfully to reflect its past. Having roots in a landscape or place, visible everyday connections to history, tends to create better quality of life than when dislocated from any past, and an inclusive approach to understanding landscape and place is likely to be part of the way to achieve this. In an English framework, the programme of landscape characterization described above – HLC – attempts to begin to meet some of these needs by providing many different tools. Methods like HLC can measure long-term change and help to understand why the landscape has evolved to be like it is. They are therefore well placed to help with shaping the landscape of tomorrow, and many other European countries have research programmes on this theme. *The European Landscape Convention*, even though not yet in force, is already heightening this trend.

HLC's central aims are to increase understanding, and to use landscape as a vehicle to bring together disconnected information and different disciplines. It teaches a great deal about the historic depth of the present-day landscape, and equally important it puts this into formats that are more accessible to other landscape disciplines. In addition, it tells us a lot about the sites and monuments within the landscape, providing for them a wider meaning. It offers predictive tools that allow us to see where our knowledge is inadequate, where we may have looked with the wrong methods, where new archaeological sites might be found. More, it answers questions about landscape's historic character, helps to write new agenda for research, and asks new and better questions.

Looking forward in time, HLC will provide a benchmark for monitoring change, so in years to come we can see how the environment itself and landscape perceptions have both changed. The current search to define indicators to monitor and shape change (in the UK at least but more widely) is proving difficult, and it may be a mistake to measure them too quantitatively. Perhaps what we should be measuring as an indicator is the much more subjective idea of landscape character, what people think about landscape rather than what it is.

If landscape is truly personal, as argued above, it is also therefore ever changing, as people change. At a very crude level, in the UK, it is many years since most people worked on the land – a century since more than 10%

worked the land (Fairclough 2002), and current figures suggest perhaps only 1% in many parts of the country. This means of course that the land is being used differently and looks different, but more subtly, and equally inevitably, it means that most peoples' perception of landscape, and opinions about its value, are not the same as those of their grandparents. Almost everyone now sees the rural landscape as to some degree a conservation landscape, a leisure landscape, a backdrop for life, an ecological necessity perhaps, the raw material of the tourism industry, a place to have a second home or to build new better houses. For most people in the UK today, food production is not very high on the list of what landscape is actually *for*. So what matters most now in agri-politics – the view from the tractor or the view through the passing car windscreen?

The pace of change in peoples' perceptions is matched by the dynamism of the environment itself, which raises other issues of perception. We think the rate and type of change in the modern landscape is faster or worse than at any time in the past, but is this true? Comparison of present-day changes to false ideas of past 'timelessness' can say more about our values, and our response to change, indeed our dislike of change, than about the degree of change itself. Archaeology can highlight such issues by showing the scale of change in the past, and the degree to which the environment, and hence the landscape we create, is the product of change not continuity. It can also show how many times in the past people have damaged the ecosystem, and how they had done this

HLC rarely creates prescriptions or presumptions (e.g. in favour of keeping something unchanged) and does not necessarily dictate protection. Instead it provide better information to support proper discussion and sustainable decision taking between all those involved in deciding change (hopefully through a democratic inclusive approach). The overall aim is to ensure that historic character is not lost through ignorance, but only by conscious decisions that weigh all factors. Methods like HLC aim to extend consideration of the impact of change and development (e.g. urban regeneration, agricultural environmental policies) beyond a few special sites to everything within the environment, everywhere. The environment cannot be treated as a new empty page on which to write the future. We have to re-use the same page. Landscape is an old document, a palimpsest showing centuries of human remains, the human habitat we have engineered for ourselves. The past cannot be swept away without thought to make way for new development. If it were treated like that, as English experience in the 1950s and 1960s shows, the result is usually unsustainable and unsatisfactory, notably in quality of life terms; instead new development has to be fitted smoothly into old.

Studying and trying to characterise the historic dimension of the whole landscape through HLC shows what we have inherited in the landscape and the patterns of change and survival that lead to the present-day landscape, and it explains the human processes that make landscapes work. It therefore gives information to help us decide what could be preserved and passed on to the future as part of the next stage of landscape evolution. In the past, the drivers of landscape change have mainly been matters such as food production, shelter, political control, profits or industrial production. While these will doubtless continue to be important in future change to the landscape of the 21st century, there will also be newer drivers of change.

Public and political concerns about sustainability, and the pressures of increasing population and development, suggest that some of these newer factors will – and already do – fall under the heading of 'conservation', 'resource management' or 'local identity'. Landscape management like other areas of environmental management could become more proactive, with a larger share in deciding the sort of landscape we make for our descendants. The control of landscape change in the past was conscious, whether by farmer or by landscape architect, and to use a conservation and heritage ethos to take conscious decisions about the landscape is just a modern version of this. It does not make the resultant landscape change any less 'natural' or any more 'artificial', simply admits that conservation, or sustainability, is in itself a force for change not merely a constraint on change.

This is perhaps already to be seen in the concept of paying farmers to be custodians of the landscape as well as (and perhaps soon instead of) to produce food producers. It is also seen in the modern enthusiasm for 'recreating' lost habitats such as heath or woodland, even if this is not always in support of landscape character as opposed to creating new biodiversity, which are not the same thing of course. If these trends accelerate, it may be that landscape protection and management will become a mainstream part of the social processes that shape human interaction with use of the environment, and determine how people perceive the landscapes in which they live and work. In this sense, landscape has already changed. Most people in England for example now see landscape not as a place of work but as a sort of nature reserve, holiday location or pretty backdrop to everyday urban lifestyles. As the motivation for changing landscape becomes more conservation-led, and as traditional methods of land management fades from memory, we will need more and more knowledge and understanding of the landscape's historic aspects to help us take sustainable decisions.

It will remain difficult to come to terms with landscape change. Landscape has always changed, as all students of landscape accept, and studying that change is an important part of understanding our history: the landscape itself offers evidence of our past that is at least as important as historic documents. It can be argued that recent or current levels of change in relation to the *fabric* of the landscape are no greater or faster than they have been at other dates in the past. Without going back more than a few hundred years, the rapid re-structuring of the English landscape through Enclosure, or on Scania by the land reform shifts of the early 19th centuries were major upheavals greater than most witnessed in the past 50 years. The removal by human actions of all significant woodland in midland England by about 1000 AD was a remarkable change, as was the creation of upland and lowland heaths – now greatly valued for their 'wildness' – by over-intensive farming during prehistoric times. Such examples can be multiplied across the world.

What is perhaps different now is the accompanying loss of the *processes* that sustain particular types of landscape. Agricultural incentive schemes or consciously archaic methods of land management are having some success in replacing modern intensive agriculture with new, more environmentally sensitive methods that are not necessarily historic or traditional. On the whole, however, traditional land-management processes such as wood pasture land use, hay making, upland sheep grazing, cork production are vanishing, to be replaced by intensive monocultures, new energy crops, golf courses, new nature reserves or untended land. In areas of depopulation or demographic désertification, the landscape that they created will be difficult or impossible to maintain, but little of cultural (as opposed to natural, unmanaged) value will be put in its place.

It is particularly difficult, as mentioned above, to protect and manage landscape character where the activities that created it – notably traditional types of farming and land use – no longer take place. Is society prepared to create an artificial rural economy, for example continuing to subsidize hill-sheep farming in order to preserve the present appearance of upland landscapes, even though there is no economic demand for the lamb or wool? When do we accept that historic processes have stopped, and recognize that we need to create a new environment with new character? In some parts of Europe (southern France, Spain and Portugal, the Western Isles of Scotland, upland England and Wales) the problem of disappearing farming is, or threatens very soon, to change the character of cultural landscape very severely. Archaeologists have an option of simply accepting this as the continuance of the change that is archaeology's main subject matter, and welcoming it as the creation of 'new' historic landscapes for us to study and protect at a future date. Alternatively, we could offer an archaeological understanding of the present-day landscape to help society create tomorrow's landscape in ways that reflect or grow out of yesterday's, and allow future generations to enjoy the evidence of their long history in the way that we can.

A good starting point is to recognize that the protection of cultural landscapes is not first and foremost about preventing change but about managing it, living with it. Historic landscape character (perhaps more so than other types of landscape character) cannot survive without appropriate land use, and cannot be held in stasis. Because it so complex, it is perhaps not susceptible to the sort of artificial management that can maintain some aspects of biodiversity in isolation from cultural processes. Landscape has never been static during human history, despite the impression we sometimes think we have of timeless, unchanging aeons, especially in rural contexts when viewed with a post-industrial detachment from the land, a romanticised view of the past that has little justification. To try to keep the cultural landscape as a sort of unchanging fossil of itself or (worse) as a fossil of some mythical 'golden age' or 'rural idyll' (Fairclough in Grenville), will remove one of the things that create landscape character and makes it so valuable: the visibility of change and the evidence for past human actions.

Accepting change, however, need not lead to fatalistic acceptance of any or all change. Decisions are still there to be made,unless we are prepared not to interfere at all but simply to allow natural human processes take over, in the 21st century largely the workings of the global marketplace: change could just be 'allowed to happen'. Or we could try to steer it in various directions, preferably respecting historic patterns and traditional trajectories of change. There is a need for some clarity on why we make changes to the landscape. In the past it was usually clear enough – to grow more food, or to obtain timber, for example – but now? We need to explore more carefully our current motives for restoring field boundaries, replanting woodland or recreating wetlands. Such activities are sometimes said to be re-creating the past, or replacing lost biodiversity, but this is not really the case, not just because a replaced hedge, for example, cannot be truly authentic, but because more fundamentally the reasons and processes of creation are different. It is better to regard such landscape changes as attempts to produce new biodiversity (all biodiversity is cultural as well as natural) for people to enjoy looking at, or simply as the creation of new layers in the sequence of landscape history, not as a return to a better past. All landscape management is forward-looking; it can make new landscapes, but it cannot re-make old ones.

But we can take decisions about what new landscape should contain, based on understanding of what makes landscape, what has shaped it in the past, how it works and fits together, what is sustainable. A key aim of landscape management should be a balance between continuity and change.

Significant features of the landscape can be preserved, so that people may continue to identify and celebrate the work of their predecessors, in ways that allow a landscape feature to have a new use. We best protect what we value by giving it an appropriate (not necessarily its original) use, and this is true of the landscape as much as of an historic building. We should strive to pass on the ability to understand the past, by keeping its traces and imprints, by making time-depth and historic character.

Our goal, then, might not be to pass on to those who follow us the things that *we* think they will value, but rather to give them options by bequeathing a landscape that retains enough historic and archaeological depth for them to 'read' and study it afresh for themselves. This would be a living landscape in a new sense, that of being alive as a document that renews itself continuously, both because it is being changed and added to, and because age and experience normally makes re-reading a book worthwhile and productive of new meaning and understanding. In this way, the landscape can be read and re-read, each time surrendering up new meaning to new questions. This may ultimately be seen as more important than trying to pass on particular things that we value, in the hope that our descendants will agree.

REFERENCES

Bloemers, J.H.F. (2002). Past- and future-oriented archaeology: protecting and developing the archaeological-historical landscape of the Netherlands. In G.J. Fairclough and Rippon (Eds.) *Europe's Cultural Landscape: archaeologists and the management of change* (pp.89-96). EAC Occasional Paper no 2, Europae Archaeologiae Consilium and English Heritage, Brussels and London.

Brown, I.W. & Berry, A.R. (Eds.) (1995). *Managing Ancient Monuments: An Integrated Approach*, Association of County Archaeology Officers/Clwyd County Council, Mold.

Castro et al (2002). Archaeology in the south east of the Iberian Peninsula: a bridge between Past and future social spaces. In G.J. Fairclough and Rippon (Eds.) *Europe's Cultural Landscape: archaeologists and the management of change*, EAC Occasional Paper no 2 (pp. 133–142), Europae Archaeologiae Consilium and English Heritage, Brussels and London.

Clark, K. (2001). *Informed Conservation: Understanding historic buildings and their landscapes for conservation*, English Heritage, London.

Council of Europe (2000). *European Landscape Convention*, Florence, European Treaty Series - No. 176; www.coe.int/T/E/cultural-co-operation/Environment/Landscape.

Council of Europe (2002). *The European Landscape Convention, Naturopa issue 98* / 2002 European Landscape Convention, Strasbourg.

Countryside Agency (1999). Countryside Character volumes: vol 4 *East Midlands* (CA10), vol 5 *West Midlands* (CA11), vol 6 *The East* (CA12), vol 7 *South East* (CA13), vol 8 *South West* (CA14), Cheltenham. www.countryside.gov.uk/cci

Countryside Agency & Scottish Natural Heritage (2002). *Landscape Character Assessment: Guidance for England and Scotland*, CAX 84. www.countryside.gov.uk/cci/guidance

Countryside Commission (1987). *Landscape Assesssment – a Countryside Commission approach*, CCD18, Cheltenham.

Countryside Commission (1993). *Landscape Assessment Guidance*, CCP423, Cheltenham.

Countryside Commission, (1998). Countryside Character volumes: vol 1 *North East* (CCP535), vol 2 *North West* (CCP536), vol 3 *Yorkshire and the Humber* (CCP536), Cheltenham. www.countryside.gov.uk/cci.

Darlington, J. (2002). Mapping Lancashire's Historic Landscape: the Lancashire HLC programme (pp. 97-105). In Fairclough and Rippon (Eds.) *Europe's Cultural Landscape: archaeologists and the management of change*, EAC Occasional Paper no 2, Europae Archaeologiae Consilium and English Heritage, Brussels and London.

Déjeant-Pons, M. (2002). The European Landscape Convention, Florence. In Fairclough and Rippon (Eds.) *Europe's Cultural Landscape: archaeologists and the management of change* (pp.13-24), EAC Occasional Paper no 2, Europae Archaeologiae Consilium and English Heritage, Brussels and London.

Department of Culture, Media and Sport & DTLR (2002). The Historic Environment: A Force for Our Future, PP378, Department for Culture, Media and Sport & Department of Transport, Local Government and the Regions, London. www.culture.gov.uk/heritage.

Department of the Environment and Department of National Heritage, (1994). *Planning and the Historic Environment, Planning Policy Guidance (PPG) 15*, HMSO.

Dyson-Bruce, L., Dixon, P., Hingley, R & Stevenson, J., (1999). *Historic Landuse Assessment (HLA): Development and Potential of a Technique for Assessing Historic Landscape patterns, report of the pilot project 1996-98*, Historic Scotland, RCHMS, Edinburgh.

Dixon, P & Hingley, R. (2002). Historic land-use assessment in Scotland. In Fairclough and Rippon (Eds.) *Europe's Cultural Landscape: archaeologists and the management of change* (pp. 85-88), EAC Occasional Paper no 2, Europae Archaeologiae Consilium and English Heritage, Brussels and London.

Dyson-Bruce, L. (2002). Historic Landscape Assessment – the East of England experience, in G. Burenhult (Ed.) *Archaeological Informatics: Pushing the Envelope,* Computer Applications and Quantitative Methods in Archaeology, Proceedings of the 29th Conference, Gotland, April 2001, BAR International Series 1016, (pp. 35-42). Archaeopress.

Ede, J. & Darlington, J. (2002). *The Lancashire Historic Landscape Characterisation Programme*, Lancashire County Council/English Heritage, Preston.

English Heritage (1997). *Sustaining the historic environment,* London.

English Heritage (2002). *Historic Landscape Characterisation: Template Project Design for EH-funded county level projects*, English Heritage, London.

English Heritage forthcoming a) (2003). *Historic Landscape Characterisation: National Review of Methodology* by Oscar Aldred, English Heritage, London.

English Heritage forthcoming b) (2003). *Historic Landscape Characterisation: National Review of Applications and Uses by Jo Clark,* English Heritage, London.

ERM & ERA, (2000). *Pilot Study on Landscape Characterisation in County Clare*. Oxford, Dublin and Kilkenny: Environmental Resource Management and ERA-Maptec Ltd for the Heritage Council (Ireland).

Ermischer, G. (2002). Spessart goes Europe: the historic landscape characterisation of a German upland region. In Fairclough and Rippon (Eds.) *Europe's Cultural Landscape: archaeologists and the management of change* (pp. 157-168). EAC Occasional Paper no 2, Europae Archaeologiae Consilium and English Heritage, Brussels and London.

Fairclough, G.J. (1994). Landscapes From the Past - Only Human Nature. In Selman, P. *The Ecology and Management of Cultural Landscapes* (pp 64-72) (proceedings of an IALE UK conference at Cheltenham 1993 published - Landscape Issues vol 11 No 1).

Fairclough, G.J. (1995). *The Sum of All Its Parts: An Overview of the Politics of Integrated Management in England.* In Brown & Berry, (pp. 17 – 28).

Fairclough, G.J. (1999a). Protecting Time and Space: understanding historic landscape for conservation in England. In Ucko, P.J. and Layton, R. (Eds.) *The Archaeology and Anthropology of Landscape: Shaping your landscape.* One World Archaeology 30 (pp. 119-134). London: Routledge.

Fairclough, G.J. (1999b). Protecting the Cultural Landscape - national designations and local character (pp. 27-39). In Grenville, J. (Ed.) *Managing the Historic Rural Environment*, Routledge/English Heritage, London.

Fairclough, G.J. (2002a). Cultural landscape, computers and characterisation. In Burenhult, G. (Ed.) *Archaeological Informatics: Pushing the Envelope,* Computer Applications and Quantitative Methods in Archaeology, Proceedings of the 29th Conference, Gotland, April 2001, BAR International Series 1016, Archaeopress (pp. 277-294).

Fairclough, G.J. (2002b) Cultural landscape and spatial planning. In '*Historic Environments of the North Sea*', proceedings of an InterReg II conference, March 2001, South Shields.

Fairclough, G.J. (2002c). Europe's Landscape: archaeology, sustainability and agriculture. In G.J. Fairclough & Rippon (Eds.) *Europe's Cultural Landscape: archaeologists and the management of change* (pp. 1-12), EAC Occasional Paper no 2, Europae Archaeologiae Consilium and English Heritage, Brussels and London.

Fairclough, G.J. (2002d) Archaeologists and the European Landscape Convention. In G.J. Fairclough & Rippon (Eds.) *Europe's Cultural Landscape: archaeologists and the management of change* (pp.25-37), EAC Occasional Paper no 2, Europae Archaeologiae Consilium and English Heritage, Brussels and London.

Fairclough, G.J. (2002e). Steps towards the integrated management of a changing landscape: Historic Landscape Characterisation in England. In Swensen, G. (Ed.) *Cultural Heritage on the Urban Fringe, Nannetsad workshop report March 2002,* NIKU 126, (pp. 29-39). Oslo: Norsk Institut for kulturminneforskning.

Fairclough, G.J. (forthcoming). Cultural landscape, sustainability and living with change? In Teutonico J-M (Ed.) *Managing Change: Sustainable Approaches to the Conservation of the Built Environment.* The proceedings of the US/ICOMOS 4th International Symposium April 5-8, 2001), Philadelphia.

Fairclough, G.J. (Ed.) (1999). *Historic Landscape Characterisation* (Papers presented at an English Heritage seminar held at the Society of Antiquaries, 11 December 1998), English Heritage, London.

Fairclough, G.J., Lambrick, G. & McNab, A. (1999). *Yesterday's World, Tomorrow's* Landscape (The English Heritage Historic Landscape Project 1992-94), London: English Heritage.

Fairclough, G.J., Lambrick, G. & Hopkins, D. (2002). Historic Landscape Characterisation in England and a Hampshire case study. In G.J. Fairclough and Rippon (Eds.) *Europe's Cultural Landscape: archaeologists and the management of change* (69-83), EAC Occasional Paper no 2, Europae Archaeologiae Consilium and English Heritage, Brussels and London.

Fairclough, G.J. & Rippon, S.J. (Eds.) (2002). *Europe's Cultural Landscape: archaeologists and the management of change*, EAC Occasional Paper no 2, Europae Archaeologiae Consilium and English Heritage, Brussels and London.

Herring, P. (1998). *Cornwall's Historic Landscape. Presenting a method of historic landscape character assessment*, Cornwall Archaeology Unit and English Heritage, Cornwall County Council, Truro.

Historic Scotland (2002). *Past to the Future: Historic Scotland's policy for the sustainable management of the historic environment*, Historic Scotland, Edinburgh.

Macinnes, L. (2002). Examples of current national approaches – Scotland. In Fairclough & Rippon (Eds.) *Europe's Cultural Landscape: archaeologists and the management of change* (pp. 171-174). EAC Occasional Paper no 2, Europae Archaeologiae Consilium and English Heritage, Brussels and London.

Macinnes, L. & Wickham-Jones, C.R. (1992). *All Natural Things: archaeology and the Green debate*. Oxford, Oxbow Monograph 11.

Nord Paulsson, J. (2002). Raising awareness and managing change: the cultural landscape of the Bjäre peninsular, Sweden. In G.J. Fairclough and Rippon (Eds.) *Europe's Cultural Landscape: archaeologists and the management of change*(pp. 143-148). EAC Occasional Paper no 2, Europae Archaeologiae Consilium and English Heritage, Brussels and London.

Nord Paulsson, J. & Fairclough, G.J. (2002). Bjare and Bowland: Computer Applications (pp. 551-557). In *European Pathways to Cultural Landscapes,* a Culture 2000 programme. In Burenhult, G. (Ed.) *Archaeological Informatics: Pushing the Envelope,* Computer Applications and Quantitative Methods in Archaeology, Proceedings of the 29th Conference, Gotland, April 2001, BAR International Series 1016, Archaeopress.

Roberts, B.K. & Wrathmell, S. (2000). *An Atlas of Rural Settlement in England.* London: English Heritage.

Roberts, B.K., & Wrathmell, S. (2002). *Region and Place, Rural Settlement in England.* London: English Heritage.

www.hants.gov.uk/landscape: Hampshire County Council website for its Historic Landscape Characterisation

www.pcl-eu.de, the web-site of the European Pathways to the Cultural Landscape Culture 2000 network

Chapter 17

AUTHENTICITY IN LANDSCAPE CONSERVATION AND MANAGEMENT – THE IMPORTANCE OF THE LOCAL CONTEXT

Roland Gustavsson & Anna Peterson
Department of Landscape Planning, Swedish Agricultural University, Sweden

1. INTRODUCTION

1.1 Background, main goals, and main structure of the chapter

Authenticity is something we cannot escape taking into consideration, when dealing with landscape conservation and management, or with cultural heritage. Neither can we escape from dealing with authenticity, when it concerns landscape experience and important feelings linked to that, such as identity, responsibility and belongingness. One can try to get around it, inventing what could perhaps be claimed to be more objective and precise aspects, but authenticity will still be there, explicitly or hidden, but thrusting itself on us, forcing us to discuss it, and to evaluate its consequences. Maybe we should argue, right from the beginning, that authenticity should be emphasized more than today, and that it therefore also needs to be more fully understood. It should be emphasized more, because it also indirectly touches several other basic dimensions and aspects, such as the need for more integrative, overall evaluations and the respect for local territories as special landscapes. In a wider sense it comprises ethics, emotional feelings, perspectives held and initiatives taken in the future, and also creative aspects, involving both learning and aesthetics. It is a term which can be given an absolute definition, but more important: it is one of the key terms for communication in an action-oriented planning and management process,

H. Palang and G. Fry (eds.), Landscape Interfaces, 319–356.

linking the past with the future. With this chapter we hope to contribute to a better understanding of authenticity and its different dimensions, thus enabling us to handle it in very different conservation, landscape, and social contexts.

After a short discussion of conservation policies and ideologies as a background we present a theoretical framework focusing on authenticity, and its relations to a series of aspects and dimensions, such as historical correctness, perceived authenticity, dynamics, time flow, creativity, scale, object- or idea-based strategies, which are all important when practicing authenticity in the local context, but difficult to catch in systems which are primarily designed to function as top down systems.

This is later in the chapter followed by a contextual discussion, using three case studies in Sweden, allowing us to extend its exploration step by step based on our experiences of how to handle it in different types of situations, without losing the contextual dimensions. Furthermore, they help us to be concrete. The three case study areas have all been recognized as being of national or regional interest for nature conservation and cultural heritage. They are also essential as recreation, and tourist landscapes. Moreover, all three are small-scale landscapes with a mixture of small woodland areas, grazed half-open land and fields or former fields. On a closer look they also differ: in being complementary as types, if one considers cultural history, today's major functions, the conservation strategies chosen, management problems, and in what way authenticity is a relevant aspect. In the chapter the description of the cases is followed by a discussion more concerned with principles, in which we give special attention to 'authentic illusion' as a probable realistic concept for conservation and landscape management when applied in practice. Finally we have tried to pick out a few points which we consider as the main conclusions for practical application.

1.2 Some basic research approaches

Again and again administrators and politicians express their wish to protect old cultural landscapes and landscape elements. The art of preservation and conservation is regulated by law in probably all European countries, and there is a whole set of policies on European and national levels, which are supposed to interact, and serve as a help in practice to reach the many goals decided on. But if tested, how do these policy instruments become integrated in conservation and landscape management practice in the many local contexts? For example, when dealing with landscapes, which are most often dramatically changing, like in the rurban fringe areas, or when dealing with more distant landscapes, close to people but far from policy-makers; landscapes, which many times are particularly rich in old cultural elements and patterns, and consequently have been proposed for conservation for future generations? What will happen with them and in what way does it agree with how authenticity is perceived by the local people, who are still largely part of an agrarian culture, or by the policy-makers, stake-holders, and the increasing number of visitors, who belong to urban culture?

Figure 17-1. Excursion in Estonia during the conference in Tartu 2002. Researchers as tourists and outsiders. There are many hands to be shaken to make the necessary links between yesterday and today, between agrarian and urban cultures, between the old and the young generation and between the actors in the local context and the administrators jumping in now and then, demonstrating power and new ways to look at knowledge. Who are the researchers backing up and supporting this drama? Photo: R. Gustavsson

We hope the research this chapter is based on will give the chapter a special flavor. Over the years we have made an effort in our choice of studies to come as close as possible to understanding more of how and why transformation takes place in the landscape, how ideas, conservation policies and plans become a reality as it is experienced in the local situations and can be seen through the landscape character and its changes, the management, the landowners, the managers, and other actors involved, like administrators and advisers on regional and local levels. Even if it is true that researchers

(Figures 1, 2) will always be outsiders, we have aimed to get close to the concrete, practice and action, and stay there. Parallel with that we have made an effort to give space to conceptual thinking, which should be characteristic of planning and management research. Many of our approaches have come from the natural or social sciences, as it is common, but we have also got a great deal of inspiration from researchers who are primarily within the humanities.

Figure 17-2. The action performed by living and working people and the many time layers forming a piece of patchwork, which makes an environment alive and probably more authentic. Estonian farm. Photo: R. Gustavsson

Here we want to make clear the difference to us between research aiming to analyze the state of art, focusing on the collecting and testing of facts, and research which is directed towards the future, stressing solutions, the concepts and the understanding of knowledge (Toulmin 1972; Pedroli 2000; Woodhill & Röling 2000). We have primarily been aiming for the second type of research. This has led to a whole series of different approaches, such as deep reflections (Gustavsson 1999b), the use of a reference landscape approach (Fry & Gustavsson 1996; Gustavsson 2002), and methods chosen to create fruitful dialogues between the researchers and actors, often stressing the personal and the inter-subjective rather than the objective. In our research we have also tried to stress the understanding of knowledge as time-dependent, inspired by philosophers like Bergson (1928), Collingwood

(1946), and Ricoeur (1981) and the wish to search for methods for handling and keeping complexity and contextual knowledge (Gustavsson 1999a, b).

More precisely, this has brought us to research areas, in which we have been rediscovering old successful management traditions represented by 'identified masters' and their basic thinking concepts in a comparison between three different major management styles and knowledge cultures (Jönsson & Gustavsson 2002). In similar studies of local societies, we have picked out old farmers to find out their articulated understanding of 'traditional management', and how that might differ from the generally based, top-down concepts of today. It has led to the choice to often stay involved in the same case study areas during a long time, and thereby also combining researchers who have a deep local and professional insight with researchers who can act as outsiders, newcomers, and strangers. Thus the importance is stressed of going back and forth between the known and the unknown world, the local insight knowledge and the wide perspectives, the common and the daring innovative concepts. Here we would also like to mention our use of repeated photographing as a tool in a communication for extended interpretations rather than for illustrations or pure descriptions of changes.

2. CONSERVATION POLICIES AND IDEOLOGIES

2.1 Consequences of standardization, rationality and top-down policies

Technology and economy have during the 20th century caused great simplification in most rural landscapes in Europe. There are also other reasons than economy and technology that can easily be found supporting standardization and simplification. With the need to be effective it is easy for all actors, farmers as well as administrators, planners as well as management entrepreneurs, researchers as well as teachers, to be over-rational and contribute to an unwished standardization and simplification of the landscape. Moreover, to be effective it is tempting to use methods based on top-down rules, which are later forced on to the local landscape and social situations. Maybe the top-down approaches are also desired because they are meant to be more 'true' due to a preferable higher degree of generality, according to the dominating knowledge cultures and research disciplines (Ramírez 1995; Liedman 1997; 2001; Asplund 2002). No doubt, top-down systems also directly need to be simplified into categories that allow control. Furthermore, it is said that the top-down rules also obtain priority because it is more democratic if the rules are set up from a top-down perspective. No one should be able to take advantage. This might be questioned but is still a perspective that is very much alive (Röling & Wagemakers 2000).

Furthermore, through repeated studies the research society has wished to generalize knowledge to something much simpler than the single case in a local situation. It has also succeeded in doing so. An alternative would have been to stress complexity and adaptability; for each new case, the intention could be to try to improve the ability to work in situations with increasing degrees of complexity instead of the opposite, and adapt to such situations more and more successfully, while respecting the local contexts.

Maybe the simplification of the landscape and the view of management have gone far beyond what seems logical for just technological and economic reasons. Many claim today that it is also a question of style. It is well known that ideologies like modernism or functionalism applied the slogan 'less is more' in architecture, but this probably also happened very much within agriculture and forestry (Liedman 1997). The ideology in question also pushed for an international style and large production series with identical or similar products. The same ideology also claimed the functional, the one-functional rather than the multi-functional, and its utilization as something particularly positive. In the buildings it became popular to show the functions rather than to hide them or redesign them in a 'romantic' way. It could for example be the indoor design of water and lavatory pipes in a style that was called 'brutal architecture'. History was something passé, which one should not bother too much about. Building architecture, but also the modern agricultural and forestry landscapes of the last few decades in Europe acquired a style, which was meant to be brutal and hence considered authentic and positive.

2.2 Searching for a deeper understanding and future alternatives

Former defensive policies aimed at merely protecting what was considered as existing values. Later, especially since the 1970s, in most European countries policies have changed to become more offensive, partly because changes are unavoidable. Many changes are actually unpredictable over long periods, which mean they have to be dealt with in a flexible way. The changes in policies are reflected in the change of words from preservation to conservation, and in landscape restoration contexts from reconstructions to renovations, in which the latter is more than trying to copy form and get hold of old material. The need to take into consideration the often slow long-time changes, which are not always so easy to catch, is particularly important when considering the cultural countryside landscape. To conserve therefore means to be able to handle both strategies for existing qualities and their changes, and decide what to do with new qualities which are born or potential all the time.

The objectives of leading international principles for conservation of landscapes, nature, and natural resources have changed considerable also from 1980 till today. During the 1980s they were very much focused on physical resources and biodiversity. Since the 1990s the local society, people and cultural perspectives have received more attention (Green 1996). The degree of attention can vary very much between different European countries, and from national to local levels. In Britain and Norway the cultural heritage has received much more attention, in Sweden less. The new conservation instrument 'culture reserve', which was introduced in Sweden in 1999, with the aim to correspond to 'nature reserve' on regional and local levels primarily, has been one way to strengthen the cultural dimensions and the human science perspectives. Later we will refer to one of the few examples of culture reserves so far.

In the last few decades regional administrators have become more and more influential as policy-makers, affecting what happens in individual management situations. This should be seen as an improvement of the top-down system of today, but how does it help in the local context? Maybe the gaps in between the top and the bottom are attacked the wrong way? Maybe the local context has lost even more of power and status, and local actors have become even more reduced to doers, not being respected as intellectual, feeling, intention-led people, who are *also* doers. After interviewing the people advising and controlling the subsidies given to the farmers in one of our case studies, we observed the broad generalization as regards species, partly due to the very great time pressure. This is also confirmed by Stenseke (2001a). Furthermore, the advisers did not have a very articulated view of management, nor were they in most situations historically correct. But what is the most important issue concerning historical correctness? Is it that the local actor remains active, the action in itself and local life, the object itself or the idea of it being there? And do we really need site-specific accuracy – is that really possible in a dynamic landscape situation? If it is, is it even desirable? Is it not sufficient to have an old-fashioned look – which in landscape restoration and renovation contexts could be named an authentic illusion?

Even if today we have acquired new legal instruments and money to support conservation of particularly important rural landscapes, cultural elements and habitats, the conditions are still very often problematic, especially in a long-term perspective, and there is a need to improve these conditions. Many people, both politicians and researchers, would agree on this. They would probably also agree about the necessity to succeed better in activating local people and their creativity, which is needed but seldom used today as a driving force in the local system. We have to give much of the basic thinking and organizing of life and landscape a real shake to be able to improve and recreate.

More important than to continue a critical analysis like the one above is to try to answer what such approaches lead to in the local context, where ideas and knowledge have to be extensively transformed into reality, and how these approaches affect the wish to arrive at different aspects of authenticity. And what may be even more important: what are the alternatives, how can we better support the small-scale landscapes, so that they do not only survive, but also become perceived as something extra attractive, authentic and content-rich, with strong links to history and traditions even when they are changing.

2.3 Earlier conservation ideologies and their linkages to authenticity

Earlier conservation ideologies can, no doubt, be strongly linked to an overall authenticity evaluation and strategy. Today, overall considerations are more hidden, and nature conservation in particular has tried to identify more precise, objective indicators. But even if that would become successful, authenticity should perhaps still be a key term to be discussed and evaluated. However, authenticity does not mean one single strategy. To illustrate the spectrum of choices, which should be reflected, we are using the history regarding the conservation and restoration of buildings and other archaeological or cultural remnants over a 400-year period in Sweden (Hidemark 1997; Edman 1999). Probably there are many similarities with ideological trends if we compare with other European countries or with nature conservation from different periods. We do not attempt to go so far here. Rather, we want to explore possibilities of a more nuanced view, which we find relevant to the many different conservation situations of today.

- *the focus on a time of great esteem.* In the earliest days of restorations, in the 17th century, the focus was on age rather than aging. The reason for choosing at the time the more static view of 'age' was that it was important to try to highlight the memory of the past golden years for Sweden; the years when Sweden was a major power in Europe;
- *the focus on dynamics and aging, during the romantic period.* A century later the shift was very strong. The romantic period suddenly stressed aging, the symbols as part of experience, decay and death as important parts of authenticity.
- *historic accuracy, document-led, generalized ideal types.* The 19th century, again, meant ideology changes. Documents were used in trying to interpret on a general level what were the characteristic qualities. In

the specific case it was not so important whether the qualities had actually been there. More important was that they should have been there to be typical of the style of the time in question. This was part of the wish to find national and international styles within modernism;

- *technocratic interest, focusing on how to deal with aging elements. The image of the Sleeping Beauty.* Moreover, conservation became very much a technical concern. The main focus was directed on the state of the object in question, and how to deal with its aging process. Furthermore, it became more and more stressed, at least in the Scandinavian countries, that the historical object had the right to rest in peace;
- *perceived harmony and respect for the place and the local landscape when tourism increased.* Conservation should relate to the surroundings in a harmonic and respectful way, which often meant that the surroundings should have been mainly created in the same period. This explains why the landscape, too, suddenly became a relevant issue for the conservation of the cultural heritage. With increasing tourism human experience received greater attention, with greater importance suddenly given to subjective (inter-subjective) experience: to feel the presence of time, and the disappearing tracks of the passage of time. Nostalgia became important, restoration should be knowledge-based, but also performed with care;
- *modernism and progress as ideals.* In the beginning of the 20th century Curman was the leading authority in Sweden. For him restoration meant bridging between tradition and the modern concern with function. A perfect restoration for him was something that was invisible to the watcher. It should feel as if nothing radical had been done for a long time, and simultaneously, it should look well managed and purposeful. 'Historical continuity' was stressed as particularly important, and something that should look 'honest and authentic';
- *science-based as a prolongation within modernism.* Later, in the 1930s, a scientific perspective on history became more stressed in what was considered 'realism' within conservation and restoration. An object should be representative of its time and tell us about progress. However, behind the frame of the conserved outer forms the inner forms should live and become transformed again and again in a creative development. This concept of thinking was hence very directly influenced by Henri Bergson and his thoughts of a continuous flow of time (see below). To understand possible consequences in practice, an interesting comparison to make could be the one Van Buuren & Kerksta presented in the 1990s for Dutch landscapes (1993), which has a similar theoretical base, stressing the identification and development of a basic structure of forms and networks;

- *stressing the creativity and the local context.* A little later it should be noticed how much creativity was stressed in the conservation process. In addition to creativity as the key to a 'true development', Lundberg and Hidemark as the now leading authorities stressed the importance to make room for personal interpretations in the conservation of individual buildings as well as interpretations of their role in history. Our own time should be seen as a small passage in a long-term perspective, stressing the importance of our total understanding of the time flow. Practically, as a working model, this meant a focus on the combined reading of the history of the object and its technical condition today, searching for a harmony between old and new. Here Swedish authorities were critical of what they called 'the continental ideology' which favored fragmented conservation rather than wholeness, and a clinch between different times and styles rather than an expression of something fluent passing through time. This view remained a special Scandinavian concept till the very end of the 20th century. When culture reserve today is introduced as a new conservation instrument in Sweden with its emphasis on the larger landscape, also including surroundings in harmony with the preserved central part, this could be seen as very logical new step within the Scandinavian tradition. Also, what is considered as a 'place concept' very often belongs to similar traditions.

3. THEORETICAL FRAMEWORK

3.1 Authenticity and historical accuracy

Authenticity derives from the Greek term *authentikos,* with the meanings 'genuine, accurate in representation of the facts, reliable, undisputed origin or authorship, using period instruments, attempts to perform a piece as it would have been played at the time it was written, one who acts independently', the opposite of *plagiarism.* Compared with other terms used in landscape planning and management training, there is certainly an overlap between the terms *landscape identity*, *landscape character*, and *authenticity.* Here landscape identity corresponds to: "a characteristic arrangement of ecosystems and cultural elements, reflecting the connection between past and present, and between nature and society" (Pinto Correia 2000). To search for authenticity within a certain landscape character or identity gives a special dimension, but the search for authenticity also leads us in other directions. What we mean more exactly by authenticity will depend on the cultural and landscape context, and consequently to a large extent on the dominating conservation ideology during the time in which we think and act.

In an interpretation of the policies of today Goodchild (1995) states that a distinction should be made between *authenticity* and *historical accuracy*. Authenticity relates to the original fabric and its retention. The concept of historical accuracy refers to the new fabric, which is introduced to repair or reproduce the original: "Some think that new fabric should be distinguishable from the original. Others think the accuracy should support the historical integrity of the place in question".

> "Hay and cows for days on end, year in, year out. Eternal chores that never lead anywhere. Money that is never enough. That feeling both of guilt and impotent rage when the small coins for going to the cinema – something we had been nagging about – are slowly picked out of the well-worn wallet. While the summer visitors keep running around with a seemingly endless store of banknotes crumpled up in their pockets. But in another light, from the perspective of more than half a life, the picture contains something more. A feeling of authenticity. A feeling that there was something there that made life more real, gave it a dimension of genuineness. We can still feel it, as shivers down the spine, when we pass a tractor trailer shaking its way home at dusk with its load of hay bales, fence poles or strong feed. A feeling that it is us that should have been sitting there, nervously drumming the wheel, peering towards the sky, the thoughts of everything that has to be done before the next low pressure comes on. That is where we should have been. Part of a cycle of work, growth, life, death. More worn out perhaps. More tired. Not necessarily happier. But more real" (Smedberg 2003: 12).

When time passes and we leave one way of living for another, like going from an agrarian to an urban culture, increased possibilities to reflect suddenly open to us. We make it possible to watch ourselves by stepping out of ongoing situations or traditions. On the other hand, we thereby become more and more strangers to the past, making it more difficult to really understand earlier activities. In this context, no doubt, conservation ideologies are dependent on what has become rarer but also more problematic in the landscape and society. It is much more difficult to get the most ordinary life and landscape to shine (Ramírez 1995). When looking around we look especially for things that have become rare and that will become conservation issues in the near future. Moreover, we might find it difficult to notice all the trivial elements, but maybe we might react to at least something of that which is sticking out among the trivial. Taken all together, the question might be raised: what will be key elements in an authentic landscape experienced as strongly linked with the past? And the answer might be: traditional buildings, small-scale patterns, small roads. But maybe also something that symbolizes life itself, the atmosphere around the activities, and other things linked to local memories. Probably also elements of aging things, buildings which need to be handled, and new buildings on the way up, balancing the past with the coming. Maybe also the more and more rare milk cows, and maybe even working people who look agrarian and not urban.

The role activities play as part of authenticity, or the human life which belongs to it, or all the things 'in being' or 'on the way out' are probably not considered enough today in attempts to search for authenticity within conservation. But there are exceptions. At least it has influenced out-door museums in Europe, like the Danish Viking museums, in which life and actors play a central role. However, this should influence even more the choice of how to direct conservation more commonly in the countryside landscape. Alanen & Melnick (2000) mention how policies have changed recently in the USA regarding conservation in outdoor museums in order to make them feel more authentic. The old policy meant that all old buildings should look tip-top. However, on reflection, this standard was really not characteristic of old times. Today, many outdoor museums are, at the same time, constructing or destructing buildings, to look like ruins or seem so badly taken care of that they are just on the way to fall. By this they use a lot of resources to make it all more authentic. However, there is another side of the coin, which should be regarded as well. If we only focus on distant places, old buildings and old people we might also feel lost. We become tourists in reality, and we also make history be separated from reality. We may become more and more outsiders. In other ways, by focusing more on bridges, we could learn a lot from support and integration, not the least between the generations, and including the young coming generations by giving them leading roles. As the well-known Danish philosopher Kierkegaard has described a policy which should be successful when considering the meeting between the different times and generations: life should be lived forward-looking but be understood from a retrospective, reflecting from what has happened before.

3.2 The first generation of existentialists, creativity and local actors

Authenticity can be used to characterize a series of famous writers and philosophers, such as Kierkegaard, Nietzsche, Heidegger, Sartre and Camus. The first three of them are sometimes called the first generation of existentialists (Golomb 1995). In literature, art etc. authenticity was a kind of fashion to belong to, especially in the 1950s. Later it has rather fallen out of fashion. So, why bother to take it up again, and here? Because it probably highlights a lot of what can be seen as the missing forces and disadvantages of the dominant systems of today for conservation and landscape management. Golomb writes that the search for authenticity from the existentialists' point of view was a strong criticism of and revolt against: 'the blind, mechanical acceptance of an eternally imposed code of values'… 'the ethos of objectivity fostered by society and its institutions'… the anonymous… the too many words, gestures, and analysis… which had replaced life… and led to paralysis'... 'Authenticity calls for an ongoing life of significant actions'. Golomb continues: 'We create our authenticity; it is not delivered to us by higher authorities'.

Through authenticity the existentialists wanted to reach the genuine roots of ourselves, a sense of genuineness, real, actual, but still able to retain certain traditional motifs, to reach the sphere of human existence, to stress action rather than results or products, to imply a kind of essentialism, stressing that the relation between the self and the world has to be felt to be essential. They hoped to restore a personal mental power and sense of selfhood that modernity had diminished. Golomb includes a great number of quotations from different books of the authors. The following are a few with special relevance here. Kierkegaard holds that 'action requires passion'. Nietzsche asserts: 'Knowledge kills action; action requires the veils of illusion'. Sartre writes: To be is to fly out into the world… Authenticity is rooted in this flight, this refusal to exist as substance, this life-long attempt to fight facticity' (Golomb 1995).

Within the conservation and landscape management systems of today resources are handed down from the top to the actors in the local context, and the system is very much based on rational reasons such as to become effective, not to disturb democratic goals, and enable objective control. The existentialists rather stressed the local and the individual contexts, creativity and actions as driving forces, which were intention-led rather than just doing- and process-led. Reducing the actors, like the local farmers, to becoming just doers, would have been questioned. They would go for a support of the unique rather than try to repeat successful but normal traditional types, and rather try to enlighten the persons behind than trying to hide them, or replace them by neutral entrepreneurs who now and then came in to help. Furthermore, they would presumably stress the importance of basing an open system on emotional feelings like trust, passion, responsibility, and other feelings creating a sense of belongingness, which could be shared by many. Copies, standardizations, replications of conventional concepts, professional experts as outsiders making decisions by one-way communication they would revolt against.

3.3 The past, the back- and forward-directed forces

Taking into account authenticity as a quality within landscape character or landscape identity is something more than history. For us it very much presents basic features that will be a base for the future. Historical analysis can be part of it, showing links from the present to the past, from the customs to the traditions (Olwig 2001), and vice versa. Landscape character can thus, as an analysis of present and past, give important guidelines when deciding the way to the future. But that does not necessarily mean that the past will or should lead us forward. Authenticity can become part of the perceived also in other, more future-led ways. It is probably only in a few exceptional cases that 'a frozen landscape character' survives in a time to come, which means that most of what carries authenticity is actually moving. The forces or motives driving us forward can perhaps be found in the past, in our roots, but more probably in things, which lie ahead of us. This is stressed by Sir Carl Popper, the philosopher, in a reflective lecture given at the age of 84:

> "Future is open, remains to be resolved. Only the past is set. It has been implemented and is thereby passed. The present on the other hand is an ongoing process. In the view of the old world the past mechanically pushed us forward. But it is not the force or the kicks from behind, which make us move, it is rather the temptation of future that attracts us or drives us forward" (Popper 1996).

The search for future concepts and future landscape values should therefore not only be limited to involving conservation and reconstruction issues. The ties backwards lead not only to an understanding of the existent features and processes. They might also create impulses similar to what Popper describes as 'free thoughts' carried further into the future through an intuition of something coming. Maybe even conservation itself should have a high degree of creativity, and landscape character and authenticity should inspire us in a wider context, not just helping us to read the text backwards.

3.4 Expanding the now and the experience of stability and movements within the unavoidable dynamics

A big problem in dealing with landscape and conservation is time, 'the un-apprehensible past', which may make the talk about authenticity untenable, hopeless and even uninteresting because of the natural and cultural dynamics (Gustavsson 1999a). Olwig (2001) argues that we often consider 'the past' as a paradise, in other words, something better, and progress as something bad, with the exception of modernism, which stresses progress as particularly good. Olwig writes about 'the paradise – progress dialectic trap' and emphasizes the importance of finding ways of escaping this trap. However, we cannot avoid trying to handle the dynamics, the time flow, even if it is and always will be problematic. Dynamics in itself has one part, which can be perceived as stable, and one, which can be perceived as moving. In a fast changing society, the feeling of stability might be especially important to put forward to maintain balance – something that becomes essential when searching for authenticity.

In Sweden agriculture has existed for about 6000 years, but the disappearing elements, habitats or patterns of today do in most cases not have their roots so far back. Many of them are about 100-150 years old. For the moment, the Swedish cultural landscape is undergoing a dramatic change, but there have been changes before. The big land reforms in the 19th century can in many respects be regarded as more radical than what is happening now (Aronsson, personal communication), but just as far as we see now. We are just watching the very beginning of the consequences of modern and post-industrial agriculture, forestry, and other management activities in the landscape (Gustavsson & Ingelög 1994). There is a time delay that we should be aware of. This is shown in several landscape ecology studies, and as a phenomenon it is sometimes named 'the landscape memory' (Fry & Dramstad 1998). The time delay as regards the consequences of recent changes means that we really do not know which time is most radical in changes and effects. However, as pointed out in the Nara Conference on Authenticity, the concept of authenticity is dynamic in itself (Lowenthal 1994). Furthermore, according to Coeterier (2002), who writes that authenticity may permit several style periods together, it seems that an authentic rural landscape does not have to be a frozen time-specific reconstruction paradise. It can also be a multi-layer conservation paradise in progress.

Lambert writes: "Authenticity is not history, but a series of histories" (Lambert 2001). Dealing with landscape experiences and conservation our relation to the past and history will become an issue, largely based on our experience of time, which raises questions like how do we choose our time reference? How can we improve our ability to have a dynamic rather than a static or too reduced landscape and landscape planning approach concerning view of time? How can we train our ability to expand the now the way a professional musician uses an accordion, pushing and pulling it in a sensible way to create the tones and the semitones? Awakening a discussion of authenticity and creativity makes it impossible to avoid such reflections. In the early 20th century the French philosopher Bergson was highly influential in the search for knowledge, also within Scandinavia. Today he is probably rather unknown, but many people are rediscovering him. Bergson stressed the importance of a dynamic perspective:

> "The absent-mindedness of the qualitative differences is particularly a matter of distortion of the understanding of time. … Reality is a question of movement. Reality is the uninterrupted change of forms. The form is just an on-the-spot account, taken from a transition. … There is a clear relationship between motion and degree of consciousness" (Bergson 1928).

Studies of landscape changes, as well as conserving patterns, elements, and activities of the past, are all important in our relationship to the surrounding world, and to authenticity. To really understand the time movement becomes even more important considering that changes are also problematic to deal with mentally. Another French philosopher, Ricoeur, who has also studied our behavior towards time, talks about meaning-equipped time won through a dialectic use of chronological time and unchronological time (Ricoeur 1986). Ricoeur stresses the need to move knowledge back to time-consciousness to make it possible to understand, and to call all other abstract knowledge artificial.

The time delay could also be discussed in terms of preferences and attitudes, including those of planners (Turner 1996). This creates history, reaching right up to our time, and ways to make us improve our ability to reflect backwards and forward in a more fluent way. In this way we could argue about the importance of making historical knowledge include the present time. To go further we also need to *expand the time view within our own time* to be able to understand that the landscape keeps changing considerably the whole time. Husserl uses the terms *the immediate experience of what has just happened (retention)* and *the immediate experience of what is just going to happen (protention)*. By this extension of time we obtain the chance of having a time-flow experience expanding the static present to include a before and after (Husserl 1995). Here we call this the ability to *expand the now*.

3.5 Object-, process-, intention- or idea-led conservation?

Should the physical expression and the objects really be the most essential phenomena in focus of landscape conservation, or should it be the more anonymous processes? Many landscape architects as well as conservation administrators have been criticized in the last few decades because they are thinking too much in terms of objects. After Darwin had introduced process thinking, processes rather than objects became the right approach or the best perspective to use. More recently others have claimed that processes are far too passive, and result in too mechanical and generalized a view. They also lead to an abstract view of the world and thereby make things unnecessarily difficult to understand. Rather than focusing on the processes, we should focus on the actors and the intentions, following philosophers like Habermas and others. This has, again, created a reaction. If we choose to stress processes as well as intention-led perspectives, they actually both create abstract thinking, which is too difficult to understand and creates too few individual associations and engagements as an important prolongation. Schlereth (1990) then coined the expression 'material culture'. He is particularly skeptical of process-led perspectives and advises us to go back to objects, but now stressing the meaning linked to the objects much more than before, rather than just regarding objects as forms. Schlereth stresses the importance of our relationship with objects, and the meaning this gives in a cultural context, and in a time perspective, which makes it all understandable.

In almost all cases conservation is directed towards physical things – objects or habitats in the landscapes, instead of the idea – the thought – the intention behind why something was created in the first place. Process-led thinking is sometimes linked to physical objects, and sometimes not. This phenomenon can be recognized in all European countries. Today the object-habitat perspective is the entirely dominating perspective, and it is backed up by both professionals and amateurs. Two examples are given below to illustrate some possible consequences, which some of us might find severe:

– *the boarding house and the garden of a plant-lover.* A specific example concerns a house on the southern coast of Sweden. The house is owned by a university and run as a boarding house. The former owner was a professor of botany and he created a small arboretum in his garden. The people currently working with the garden are very enthusiastic and try to replace every dead tree with a new one of the same species on the same spot. Some years ago, a grateful regular guest at the boarding house planted a new tree in the garden to complement the wide diversity with a new species. The garden group immediately cut the tree down because a tree like that had never been in the garden before and according to them it was not authentic. Yet, would not the professor who created the arboretum have been happy to know that long after his death, trees of new interesting species were still being planted in his garden in the spirit of his work and original idea? A similar issue was discussed on a more general level at a garden history conference and somewhat later commented on by Keith Goodway (1995): "A creative way of conserving an arboretum would be to identify the historic trends in planting it in the past and to continue those trends today using new species";

– *the reconstruction of Tycho Brahe's Uranienborg.* Another illustrating example could be the reconstruction in 1992 of a quarter of the Uranienborg garden on the island of Hven, between Denmark and Sweden. This was the garden of the pioneer astronomer Tycho Brahe in the 16th century. Its design is known from an engraving, but there are no existing records of the original plants. One of the authorities on historical gardens, Professor Johan Lange, suggested that it was very likely that medical plants had been growing there, since Tycho Brahe experimented with plants for e.g. curing the plague and for alchemy purposes (Lange 1993). The planners made great efforts to find plants that might have been there in Tycho Brahe's time (Lundquist 1996) – thereby focusing on *the objects.* Since the place is the focus of attention because of the creator's Renaissance greatness and not because of the site in itself, perhaps Brahe's greatness would have been better represented by plants at the edge of today's science, rather than today picking out something retro-perspective, even if they would be representative of his time – thereby focusing on *the idea.*

3.6 Authenticity and scale

Authenticity might mean that our interest turns from isolated individual objects to larger sizes, which are perceived mentally as local landscapes or places. The central area has to be large enough to be understandable. Even the way the surroundings are visually and functionally linked in history and today this becomes strategically important. Even more so, if we join characteristic Scandinavian conservation ideologies, in which harmony plays an important role, compared to continental ideologies, which have been mainly based on contrasts. Here it should also be noted how 'landscape size' has been particularly stressed in the new conservation instrument 'culture reserve', which was introduced in 1999. Culture reserve is meant to become complementary to the usual 'nature reserve', but is meant to focus more on cultural aspects and values, and perspectives originating from the human rather than the natural sciences. Nature reserves are often small, concentrated around objects or a number of habitats. In the future such an area can become integrated in and still exist within a culture reserve, which reaches the level of a local landscape.

Thompson claims that the increasing interest today in larger sizes could also be explained in terms of ethics – the extension from biocentric theories, focusing on individual plants, animals or species to ecocentric theories extend the boundaries of moral significance to include life forms other than humans for whole ecosystems or the whole environment. This also has consequences for land size, which should be considered as a reliable concept of sustainable development. This position has been underpinned by a number of very influential philosophers like the Dutch Spinoza, the American Whitehead, and the Norwegian philosopher Arne Naess (Thompson 2000).

In both nature conservation and the cultural heritage of today, there is a trend to change the focus from a habitat, object or a stand level as a dominating perspective to a landscape scale. A trend, which might become even stronger in the future, making conserved areas like national parks, nature and culture reserves in the future reach sizes up to 1,000 km^2 or more. This is a size that many landscape ecologist researchers think is needed if the term landscape should be used. However, when getting down to how farmers think or how people as visitors perceive the surroundings we might need to choose another way to relate to a territory, and other scales. Otherwise we will create great problems if we want to stimulate a two-way communication. Our studies have clearly shown that farmers very much construct their views from an object perspective. It could be an individual field, but it could also be an individual tree. Comparable to the way a garden owner in a city thinks of what to do about his garden trees, the farmer views his tree in the pasture. Approaching management or conservation strategies based on a landscape view might most often mean to regard a series of scales rather than one, stressing a way of thinking spatially about how the phenomena are experienced, what they are part of biologically or functionally, and how they are linked.

4. THE CASE STUDIES

4.1 The Bråbygden case study

The case study area called Bråbygden is located in Kalmar County in the south-east part of Sweden. It is a forest-dominated area today with a history, which has to a large extent been based on grazing. The total grazed area is still very large. So is the number of cultural remnants, like traditional buildings, fences, and the trees and the shrubs in between the fields and meadows, and within the grazed areas (Aronsson & Gerdehag 1999). In-depth interviews with farmers were carried out, complemented with telephone interviews with advisers from six regional authorities in southern Sweden, on how they handle the nation's desire for local identity and biodiversity.

During the interviews with the farmers a walk was taken around each farm for 1-2.5 hours, using a small tape-recorder. It was surprising how well they knew their trees and the history of the trees, including all the periods back to a time when their grandfather was the actor. In a few cases there was an opportunity to talk to more than one person connected with the farm at the same time, for example father and son, or sister and brother. The administrators were asked questions like how trees and shrubs should be handled in their work as administrators for protecting cultural and biological values. They were also asked about how local identity was dealt with since that was especially pointed out as important from the authorities (Jordbruksverket 1998).

With the Bråbygden case study we had the chance to get closer to what authenticity might mean in one particular case. Part of the study was to see how top-down policies were transformed in the local situations. At the same time we hoped to be able to evaluate how much generalized knowledge would fit in with or have to be adapted at the local level. We also hoped to evaluate on what level one could talk about traditional management, depending on the variations between the different hamlets or farms or within one individual farm. The trees and the shrubs were focused on in particular as part of the ongoing farming, cultural heritage and nature conservation. Especially the treatment policies for trees and shrubs have been a very controversial subject in Sweden for a long time (Figure 3).

Today, administrators from the regional authorities advise and control farmers in order to make the landscape look traditional and to support richness in species. Subsidies, or rather economic compensations, are paid by the Swedish and European societies to what is considered as necessary management efforts (Jordbruksverket 2002). However, this ambition creates a whole bunch of difficult questions. The first question might be what did it really look like in previous times, what are considered as important periods, or periods we are able to study thanks to existing old documents?

In part of the case study we particularly stressed the homogeneity and heterogeneity at the local level among old farmers and their traditional management. We then found that the way trees and shrubs have been present and managed traditionally has varied much from one farm to another during a long period, which goes back at least to the beginning of the 20th century. Moreover, the thoughts of the farmers and the farm traditions could vary strongly not only between regions or villages, but also from farm to farm within a village. Such variations in management were far from just depending on the natural conditions of different villages or farms. Rather, it had to do with different human-related ways to act. This seriously questions the application of generalized knowledge to a particular farm, if local contexts of adapting the general are not taken into consideration.

Figure 17-3. Repeated photographing at two places with 30 years interval (1970 and 2000) from Bråbygden. One place is increasingly overgrowing, and the other place is illustrating that in some cases in the area even more articulated traditional management is taking place. This method was used to expand now by a retro-perspective. It was long enough to stress changes and dynamics, but at the same time it was short enough, so that we did not get the mental effect to go beyond from 'our own time', and a tempting possibility to go beyond our own responsibility. The study stressed the deeper interpretation of the photos as important, also when considering a place perspective, an articulated understanding of facts in a context-dependent situation, processes, intentions as well as the effect of policy instruments

Some of the overall results, relevant to questions of how to follow local traditions or aiming for authenticity, are presented below (Peterson, submitted):

- there have been large variations in tree and shrub species, considering both where they were placed and how they have been managed. This variation has not only existed between villages, but also between neighbors in the same village, and has had very little to do with differences in natural conditions;
- the vegetation in today's landscape has either been quite homogeneous corresponding to earlier situations, or the variation seems to be as great as before. In particular cases a sense of generalization was evident. Such cases concerned how the farmers handled situations when they were aiming for conservation; they knew what was 'the right thing to do and say'. They seemed to like to say the right thing, but on the other hand

they had problems with the advisers' recommendations when it came to which trees to take away;

- there was an over-representation of specifically cultural trees, i.e. trees that carry some sort of well-known cultural sign. There was an obvious increase today of pollarded trees, but only including lime and ash, and not the wide variation of species that used to be pollarded. Any shrub in a meadow was today almost certainly a clump of hazel. The birch, considered as beautiful and maybe more than anything else as a carrier of traditions, was protected everywhere. Due to what has been presented above, we also saw a loss in diversity in open as well as semi-open areas. There has been a reduction of common species like hawthorn, sallow, white beam, rowan, unpollarded lime and ash. Nobody bothers whether these latter species disappear or are removed. These species are not under threat of extermination but they are missed in the meadows, the pastures and around the houses where they were earlier of great importance;
- the telephone interviews with the advisers showed that they were well aware of the ambition to increase variation, but not down to farm or place level. More often they talked about what was common in the region. The advisers' working hours were too restricted for them to be able to talk properly with every farmer and even less to be able to do research in old maps and other documents;
- the ideal time to almost all of the advisers was the time just before mechanization, independent of farm type, and this pre-war time image was recommended to the farmers. Advisers made their recommendations to farmers in the name of history even though many of the recommendations arose from new knowledge of nature conservation, for example recommendations like leaving all cherries for the birds and saving high stumps for insect life.

The study confirmed that we have a history in which we find a wide variation of trees and shrubs from one farm to another and a growing generalization of conservation management. The view of farmers as like-minded actors with a rational economic way of handling their business still exists (Stenseke 2001b). However, the farmers have different ways of managing their farm and their trees, which a strategy could try to work with, and articulate, as much as to work against, trying to get them to adapt to the general knowledge of what is the best, based on facts. The farmers also tend to do what they get paid for and not start an argument about the subsidies given, whether these correspond to the farm's history or not. If this situation is allowed to continue, there will be homogenization within areas and we will get a false picture of the past that we set out to capture (Figure 4).

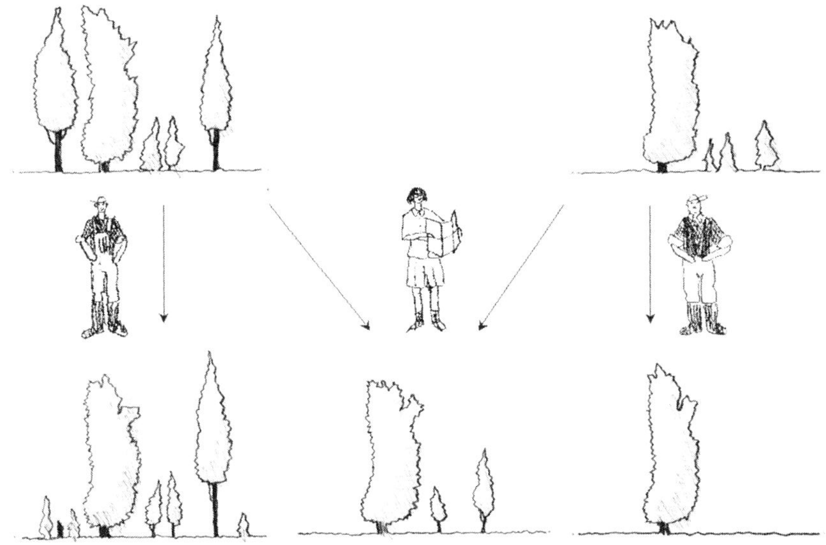

Figure 17-4. A concrete example of what is problematic about generalizations in dealing with variations in the local situations is the case of two informants who had recieved advice from the same adviser on how to deal with their juniper pasture. One farmer, who had a pasture, which was covered with junipers, was angry because he had to remove a lot of shrubs to get the economic surplus. The other farmer was angry because he was not allowed to remove as many juniper shrubs as he wanted to. They had always cleared it. We have not inspected these sites, but if the adviser got what he wanted, the two pastures would look quite the same today and neither of them would be recognizable from the past. But does it matter that the juniper pastures do not reflect their site-specific history? Apart from the farmer, who will know? Or is the articulation an attempt at historical place correctness important as much?

4.2 The Tjärö case study – creativity and landscape analysis in action

Tjärö is a 2.5-kilometer-long island of 70 ha in the archipelago of Blekinge County in the south-east corner of Sweden. It was made into a nature reserve in 1976 with the aim to conserve an old traditional agrarian landscape for mainly social-recreational reasons. The farming slowly became less and less intense during the beginning of the twentieth century, and the last farmer left farming in the 1950s. As early as the 1930s a tourist hotel and youth hostel was established on the island. The 1970s landscape was covered by shrubs and young trees, and it was logical to take restoration ideas into account to preserve the qualities for future generations. Today, instead of farmers, forestry management entrepreneurs do the practical landscape work together with sheep and highland cattle, and students from the Swedish Agricultural University (SLU) join for a week each year (Figure 5). Since the university course has been carried out for over 15 years now, it is possible to stress the time dimension in outdoor education, which is uncommon today. In several places, practical work and theoretical discussions have by now been carried out three times. The many visitors to Tjärö are not involved in the management – they are watchers. During the last decade the number of visitors has increased continuously, and they are today a real mixture of nature connoisseurs and 'ordinary' tourists from many countries (Skage 1992; Gustavsson 1997).

The dominating landscape management situations are either based on qualities on their way to disappear or lose their articulation, or on qualities on their way to become more fully expressed. Both extremes show a strong landscape dynamic, which needs to be taken into consideration, and therefore requires a high degree of creativity in the actor concerned. In such situations it is obvious that the central role that creativity should have, also in conservation, has been far from fully utilized. Tjärö illustrates the dynamic situation very well, but here the challenge is taken to highlight and train the creativity. To be more concrete, for us creativity and its importance has been clearly illustrated in the educational training course we have attended for a number of years, a course, which is based on close intimacy between theory and practice, landscape interpretations and management concepts. Till today this course has been taught for more than a decade as a special landscape management course within the landscape architecture and landscape engineer educational programs in Sweden. It has been experienced as a very intense one-week course on the island of Tjärö (Gustavsson 1997). Normally we separate educational periods in which we train practical items from theoretical-intellectual periods, but this course is a strong exception, and our chief experiences are – the more intense the physical efforts are, the better the students' thinking and the theoretical solutions they find. Furthermore, when having to deal with practical conservation particularly in what could be considered as 'places', the dynamics and the many qualities have to become integrated. This makes it much more a matter of creativity and design than of pure analysis and strategic analysis, which most landscape research primarily focuses on.

So, what about authenticity?

– At Tjärö the physiognomy of the landscape with its grazed trees and plenty of shrubs is still there and the students and the teachers try to use all smart advice and hard facts provided by the research world. To be true, the number of large trees gives the island another character than it had in the nineteenth and the early twentieth centuries. The priority today given to larger numbers is in most cases due to conscious decisions to stress the many new values that are linked to the presence of the large trees of today. Certainly the students and the teachers use a great deal of creativity, making it all very contextual and place-related, in a way which is meant to be experienced, understood and often also enjoyed by the visitors.

- Those who experience the landscape nearly all belong to an urban culture. They are really trying to deepen their view both of regional nature and of a traditional farming and grazing landscape. Most of the visitors probably experience it as an island giving them a good impression of an old traditional landscape, which is highly representative of the region. Connoisseurs like botanists, bird-watchers and mushroom-pickers are mostly satisfied because their interests seem to be taken care of, but a farmer from the old times would certainly miss the many small, cultivated fields as well as some of the mess around him and his activities. For him it would be too perfect to be true. Moreover, some of the visitors may have been wondering about the students as workers. In their eyes the students must seem to be too young and too stylish in their way of dressing to be true honest workers, and they do talk a lot. But are they not an important bridge over to now and the future in the understanding of time flow as well as of the past?

Figure 17-5. Photos from the educational course on the island of Tjärö. A course, which focuses on strongly integrated theory and practice, landscape interpretation and management.
I- The students have a show during the final day's presentation. Their grand idea was to let one huge tree stay alive as a time-teller, helping us to grasp how many years has regrowth been going on. All the other trees within the old field have been felled. The number of former trees, which have been felled, is indicated by the number of students, who are trying to cover the stumps by standing on them.
II- The older photo was taken in the 1940s or early 1950s from almost the same spot and shows a highly cultivated landscape in the immediate surroundings of the houses (Photo: the STF archives on Tjärö).
III- Long-term struggle to make a small-scale wooded meadow section survive. It is a struggle, partly because it so easily turns into a general, simplified, half-open grazed landscape; and partly because of ongoing management, which has to be taken further by a new management group as a never-ending story based on a deeper understanding of changes over time, from denser to more open, from younger to older individuals and groups. It also assumes an understanding of interactions between different tree layers and different generations of trees

4.3 The Ronneby Brunn case study

Ronneby Brunn opened as one of the first culture reserves in Sweden in 2003. It is located in the south-east part of Sweden, in Blekinge County. Our practical experiences from a project working with historical analysis and the implementation of a management plan gave us another colorful example of how the focus on objects or cultural remnants can be problematic and even questioned. This was for a time the largest spa in Sweden, and the historical remnants of today, the composition of the buildings, the park, and the woodland area which also belonged to the spa area, are still impressive, but easy to misunderstand. The original ideas of Henry August Flindt, deriving from 1873, emphasized nature and park as two complementary parts. The upper part was hilly and forested, and the lower part was used for a fairly formal park. Both parts were of crucial importance in the health and recreation program for the visitors. Parts of the area were carefully designed for the meeting, alone or in small groups, with sometimes picturesque and sometimes wild nature. Other parts were designed for socializing, providing opportunities to meet new people and to be watched. After more than a 100 years a number of the informal but very carefully designed elements and places in the nature parts have disappeared, in some cases altogether, while the formal ones are still easy to see.

If the management and renovation plan had aimed to be object-based only, the culture reserve and its history would have been very difficult to understand. A focus on just the formal part would have given us the beginnings of an entirely new story, without the balance between nature and park, informal and formal. Giving priority to freely growing, wild nature in the forest parts with their many walks, their picturesque nature and former cultural places, in order to contribute to nature conservation, would mean giving priority to a number of threatened and uncommon species rather than emphasizing the cultural base which had created the values associated with these species. This would certainly have been misleading. It is because of the cultural history that the species in question are still there. Even in a long-term perspective many of them need management to survive. Moreover, the woodland area continues to the south, an area which is also known to have been used in historical time for long walks and contact with nature in the spirit of Rousseau. However, there are today almost no remnants on the other side of the stonewall, with the exception of some impressively beautiful and very special passages through the stonewall, showing that the relation with the land behind was important. This raised the question if this distant land in the periphery should also be included in the cultural reserve. As an extra dimension, research about health issues today shows the importance of expanding time when you are out in the forest, something that could not be verified during the spa period. Following the arguments here for showing great respect for the original ideas, the cultural reserve should probably include the larger area. On the other hand, if the decision should only be based on old documents indicating that very few objects from old times remain in the southern part, it should probably not be included. Catching the ideas rather than keeping in order to present objects in old documents might involve many other aspects as well, considering larger landscape situations. It will increase the possibility to understand the background to the central area and the individual objects, the state of today, but also to make us see and invest in future potentials (Setterby et al. 2000).

The case related to Flindt becomes even more interesting when we take into consideration that he was the first Danish landscape architect who worked as a private consultant and he became involved in many of the most known park and landscape design projects of the period. Thus he is one of the most important pioneers in the Nordic countries. And his way of working makes him even more interesting today, because he used dialogue and a communication-led design, meeting the key people, making oral presentations, and writing letters. That means, on the other hand, that there are very few grand plans such as documents. Again, this makes it problematic, since we have here a person who works in a way that today would be considered politically correct, but because of this we have few documents helping us to be historically correct or authentic. However, with some effort it can still be done.

But these were not all the difficult decisions in our project. Suddenly we were standing in front of a ring of stones under a canopy of old pines and at the top of one of the most essential slopes, where we found a number of roughly cut stone steps. Old postcards revealed a café and a music pavilion, and after some time we had created an interest in this. It was agreed that there was nothing really to prevent reconstructing it. Why not, it was one of the major cultural places within the forest, but after a little more thinking we started to doubt it. Maybe it was even more important not to reconstruct it. Instead, use the imagination and let the visitor and the story teller create an even stronger impression or imagination, using the magic atmosphere of today to give credit to the time span between then and today. Through reviving the spirit of Flindt and others, the managers of today will be able to rediscover older cultural and natural patterns, the spirit of old cultural places and the visitors' possibilities to interpret and enjoy the area will probably increase.

5. DISCUSSION

5.1 Towards an authentic illusion as a future concept for conservation and management

What options do we have for future strategies concerning conservation issues? Are we trying to freeze moving landscapes? Do we wish to copy historical stages, and if so, why and how? Could it not be enough if the conserved landscape areas look old-fashioned and are managed so that the biological values can be preserved? Should we not rather aim to be satisfied with our conservation wishes if continuity is not dramatically broken and if the overall experience of authenticity is acceptable? If so, this gives us three main alternatives to choose (Figure 6):

1. To carry on like now, *generalizing* in the name of history, focusing on what was characteristic of the identified and categorized historical types, with a preference for the object. Sometimes remnants and historical documents make this concept more 'realistic' to use than other times. In many cases more recent research has found that earlier interpretations of 'historical correctness' have led us in the wrong direction or into an unacceptable simplification. This might make us more humble in future, or make us accept that we are just aiming for generalized historical types.
2. To stress the local context, its uniqueness in identity, and to be more place-related on the small scale, relating to the conservation site and the specific context. Within this, to look for the *historical authenticity and the local cultural heritage*. We could choose either the object or the idea, or a combination. The object-based way could be to choose a time period, in which material can be found to support the visual aspects of the chosen time. This is a way to preserve a farm as an old traditional museum, to show what it really looked like in the past. If the idea-based way of preserving was chosen, material can in many ways be much harder to get hold of and get to know. The result could lead to the same outcome as the object-based way, but it could also be the idea of something else, for example 'survival' – putting enough food on the table for one's children.
3. Moreover, we can choose to follow the *authentic illusion as a concept.* This is developed further from Sven-Ingvar Andersson, professor emeritus at the Royal School of Arts in Copenhagen: "The illusion has to be authentic. You should believe in it, you should be affected by it, even though you know you are being deceived. Like in a theatre, which both is an escape from reality and a concentration of it" (Andersson 1990:10) (Figure 6). This concept represents conscious generalization, where it is sufficient that the landscape is perceived as old-fashioned. Furthermore, we could gain a better understanding and feeling of belongingness. It

could be a pedagogic gain, if it achieves a good balance when connecting with history and the past. Moreover, this might be one answer to the difficulties of how to regard time, how to try to understand 'the past' in a correct way, how to try to consciously select what time should be regarded like, and then try to be precise about what consequences to stress, and what to highlight as characteristic. With the wish to acquire authentic illusion, you can put yourself above many of the problems concerning what historical correctness really means as long as you are aware of what you are doing, and it is based on the knowledge of history and historical documents. A farmer, who works consciously to visualize history in a way that simultaneously preserves all the biological values, could represent a type of agricultural development within frames. This is in fact a concept, which might come close to what was earlier described about Curman and Lundberg, and their modernistic view of restoration, inspired by Bergson and his time-flow theories. This is the idea-based way of conserving, the thought of making it seem old-fashioned as society wishes in a way that at the same time actively supports specific species and visitors' appreciation, but in a possible less resource-demanding way compared to the 'true historical way'. Today people have a preference for spending time in landscapes with a number of historically well-known elements (Hägerhäll 1999), and even if they may not go there in practice they feel comfortable to think that such landscapes still exist. This 'illusion-led way' of conservation could be stressed as important by arguments like farmers should produce what people want.

It can be dangerous to generalize here, when generalization is part of what we are criticizing. Despite this danger we believe alternative one is far too overused today in conservation and landscape management. There are many motives and research results to be found in this chapter confirming this worry of the authors. But, apart from the first alternative, neither alternative is better than the others in terms of practical realization. However, due to the enormous extent of practical and local contexts this vouches for a large variation and a use of all three. The adviser must have a site-specific awareness both of the place's history and its possibilities. The manager, whether it is the farmer or a state-paid landscape-worker, must have a similar ability, and, furthermore, must be able to adapt, communicate, and integrate the work with others. The place, intentions, creativity, and historical ideas must be in focus, and the work and money put into the place must be related to what is obtainable, not what is desirable. Bintcliff (1988) points out that the object-centered past requires guardians. He also points out that an understanding of societies that produce artifacts and ecofacts requires interpreters. In this Swedish context, we need to interpret what we have and guard it. However, guarding can be both protecting and taking care of in a creative, dynamic way.

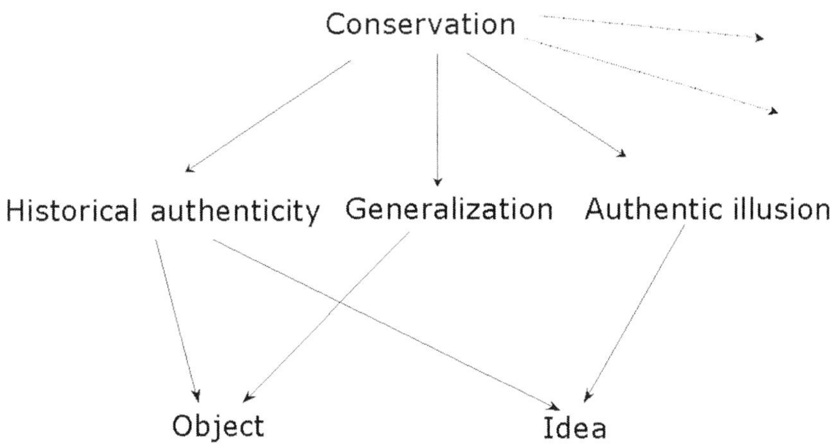

Figure 17-6. The three alternative conservation strategies of how to stress authenticity and its relation to generality, the local context, idea and/or object perspectives

5.2 The creators

In the past the farmer was the producer who dealt with the active management of the place, and the consumer was primarily passive and not present. Today, the active management is done either by the farmer or another, often a state-paid landscape manager. The consumer of the place is nowadays also present. An increasing number of people enjoy the countryside. Just a few percent of the people living or owning land in the countryside are still active as farmers and foresters. Because of this, a special kind of active consumer is on the increase, i.e. a group of people who are moving out to the countryside to improve their quality of life. These people are often those who seem particularly fond of old traditions and are willing to take part in work to maintain the special values of the rural landscape, such as mowing a meadow or keeping old kinds of farm animals. Another reason is that they are not economically dependent, since they obtain their incomes from elsewhere. The persons being active, whether it is the producers or the consumers, are the creators.

Considering the experience of landscape character and authenticity in the two cases of Tjärö and Bråbygden, the consumers have, no doubt, very different views of the two landscapes. Some people see Bråbygden as a remnant of something lost or as a pastiche of the same lost world. Others think it is a very beautiful place, full of traditions with strong symbolic values and harmony, full of authenticity. As authors and researchers we see it as a most authentic place, authentic in the sense that it is an area with a number of people belonging to the place or actively linked to the place. They are people of different ages who live there and obtain their incomes either from the land or from other jobs. The average age is low and families with small children are moving in. The land is authentic in other ways, too. Since the area is situated far away from urban regions, placed in 'nowhere', or 'in the darkest part of the region of Småland' many of the farmers have carried on 'like they have always done'. Now Bråbygden is getting rather famous, and visitors come here to see 'the lost agrarian world'. Therefore the inhabitants try to make it even more 'authentic'. As a consequence, 800 big trees have been pollarded this summer for the first time, rather than being re-pollarded, which is a remarkable fact these days. In one farm the farmer has pollarded more trees than there have ever been on that particular farm, but it gives him both money and attention. He produces what people and society ask for, as he and his ancestors have always done. An 'authentic' rural landscape is recreated, again and again.

On the island of Tjärö, 'nature' is created, nature which historically over many hundreds of years has been formed through strong cultural influences. It is particularly strongly linked to the grazed landscape rich in trees and shrubs, which is characteristic of the region, but the character also includes both old and new traditional buildings, fences, stonewalls etc. Many of the visitors who come to the island by themselves do not have a need to separate 'nature' and 'culture', and many see it as 'natural'. In the Bråbygden case visitors tend to regard the landscape more as a 'cultural landscape' than as 'nature'. The many cultivated fields, and the many well-kept fences are the major reasons for that (Hägerhäll 1999). But does this mean that 'authenticity' is stronger in one of the two examples? Do people have a chance to feel more deceived in one of the two cases? Partly this has to do with the expectations and the ideal landscape views of the visitors.

With the Ronneby Brunn case study we could take a prolonged step into urban culture, even if the countryside character here too is an important part of it. Early it was designed to attract visitors to the spa and much of what today has been considered as values from both social and biodiversity perspectives have rather been formed by this specific design than because of the earlier farming. This means that others have taken over as creators. Today the representatives of the cultural heritage together with the local community have made a shared agreement in the legal document declaring it a culture reserve, including how to achieve a balance between different values and functions, and in this communication authenticity has certainly been a key term both in the overall strategies and when deciding on the management of the many individual places.

We regard authenticity as a key dimension or one of the basic aspects and most important to the choices within conservation, to people and their experience of countryside landscapes, and to understanding time dimensions and the relations between now, future and past. Maybe it is most explicitly used as a key aspect within the conservation of the cultural heritage. Considering nature conservation, it is often one of many aspects taken into consideration, and the major issue is rather focused on the ground as a floor to keep the species alive in a sustainable way. Furthermore, nature conservation focuses on the biological and technical processes and the frames beyond the objects, and, to a minor degree on dimensions, which could be described as human experiences of landscape: the atmosphere of life itself, actions, myths, and stories, which belong to the places or the individual features. Even so, and even if it is very much invisible in documents, within nature conservation authenticity might also often be taken into consideration, because of its links to natural history and to agrarian traditions, which are much more often mentioned. Sometimes it is mentioned directly, or with similar terms like ‘true’, ‘historically correct’, or ‘traditionally honest’. Sometimes it is written between the lines of what is considered as an appreciated landscape. Authenticity as a dimension is living its life in probably a rather powerful way behind the whole battery of more measurable terms like the number of species, red-listed species, indicators, sustainability, and carrying capacity.

When using the term authenticity in landscape conservation it relates to how historically correct we can be depending on historical documents or on how much of the existing resources of today we are allowed to handle, and how far we can thereby get, but also on how ‘honest’ we today find a place character. Probably this latter meaning, touching our experiences of landscape, the overall character of a place, the time dimension, and activities which belong to the objects or the character is even more important to try to understand and should be given more emphasis in policy terms.

5.3 Conservation ideologies and creativity in the spirit of both development and respect for the past

To have the ambition to be historically correct is important for the future, but time-trueness is only possible in the spirit of enthusiasm, and maybe it is also possible in a museum context, with a story-teller as complementary to all the physical artifacts. Today the farmer produces landscape in the spirit of development. The artifacts are present as long as society needs them. Even though they are seldom there in their original function, they are needed in a symbolic sense, in which history, communication and society cooperate (Björkroth 2000).

Conservation itself depends on active actions, which again proves the old truth that conservation of historical environments often tells you more of the time when the conservation plan was decided than of the history it was intended to capture (De Olivier 1996). Rather than trying to find which alternative should be the best to use in practice when considering the large variety of ideologies present in the 20th century or earlier it would be wise to become more contextual in conservation ideology. However, this means that those involved should be aware of the choices. The discussion of conservation ideologies earlier in the chapter was given a great deal of space in order to make the reader understand what an important issue this is. Also to make the reader understand how much could be won if one is able to use a deepened insight rather than trying to act for the very best of all cases in a highly generalized way, which puts very little trust in the local actors. The earlier theoretical discussion has also provided many good arguments why to stress creativity more in the local context than it is common today.

6. CONCLUSIONS

In this chapter we particularly want to stress authenticity as a landscape management and conservation aspect, its deeper meaning and its key role in the conservation of landscapes, habitats, and landscape elements. Sometimes 'authenticity' is mentioned directly in policy documents, in management actions, or when experiencing a piece of landscape. Sometimes other words, which are said to be more objective or measurable, are used, but we believe that even on such occasions 'authenticity' is very much in the air, or should be so. Opening up to a deeper reflection of how to use 'authenticity' we have linked more general and theoretical discussions with a series of research studies, to be able to go deep, fly high, and hopefully also become concrete and understandable. In this way we open up for a number of perspectives aiming at administrators and farmers as well as other actors like creators, designers, managers or researchers. Below we have summarized what we particularly would like to stress:

- Authenticity means a search for the ethics, how a place or an element should be perceived overall, how original and true it is or can be. It approaches a sense of respect for the place, as well as for the local people, local history, but also how it can be linked to the future in a progressive way as part of a perceived time flow. A search for authenticity also stresses the meaningful elements and patterns, and furthermore actions, social life, the coming and the disappearing. It will often have strong links to earlier traditions.
- The search for authenticity catches a sense of essentiality, which is strongly linked to creativity. Conservation systems of today can be greatly improved if they succeed better in awakening creativity and pathos as driving forces in the local context. In such a case the systems should be based on trust rather than on mistrust and control. The role of the administrators and the experts should thus be changed, so that the distance to the farmers or the other actors would become less, admitting the administrators and the experts to work more as insiders than outsiders. Today the analytic logical tradition dominates the educational training of administrators and specialists as well as of other actors. This should be complemented, taking more impulses from design and poetic traditions, belonging to human science.
- Conservation is today very much object- or habitat-led. A search for authenticity rather directs us to use a place-concept and landscape perspective, in which the elements etc. can be discussed in relation to the surroundings.
- Stressing the understanding of knowledge as something basically different from a strict collecting, testing and presenting of facts should make us reflect about its possible consequences, from how to research, till how to support systems. For practice it involves how doing should not be separated from strategies and intentions, as is very much the case today. Furthermore, in landscape research it stresses the importance to search for authenticity, time-relations, integrative management, and a deep interest in the local context. Among other things it should also make us reflect about how to deal with generality, complexity, and contextuality — or how to avoid mechanical standardizations, and unnecessary simplifications. It also means that educational training should be better integrated into practice.
- Three alternative lines are suggested here as main strategies when searching for authenticity within conservation and landscape management. All three should be relevant for practice, but the one called 'authentic illusion' is here considered as the most interesting.

REFERENCES

Andersson, S-I., (1990). Den autentiska illusionen Gamla trädgårdar – restaureringsprinciper. *Kulturmiljövård 5-6*, 4-10.

Alanen, A. R. & Melnick, R. Z. (Eds.). (2000). *Preserving Cultural Landscapes in America*. Baltimore: The Johns Hopkins University Press.

Aronsson, M., & Gerdehag, P. (1999). *Bygden där Vinden Vände*. Jönköping: Prisma.

Asplund, J. (2002). *Avhandlingens språkdräkt*. Göteborg Bokförlaget: Korpen.

Bergson, H. (1928). *Den Skapande Utvecklingen*. Stockholm: Wahlström & Widstrand.

Bintcliff, J. (1988). A review of contemporary perspectives on the "meaning" of the past. In J. Bintcliff (Ed.) *Extracting Meaning from the Past*. Oxford: Oxbow.

Björkroth, M. (2000). *Hembygd i Samtid och Framtid 1890-1930*. Umeå: Umeå Universitet.

Coeterier, J.F. (2002). Lay people's evaluation of historic sites. *Landscape and Urban Planning 59*, 111-123.

Collingwood, R.G. (1946). *The Idea of History*. Oxford: Oxford University Press.

DeOliver, M. (1996). Historical preservation and identity: the Alamo and the production of a consumer landscape. *Antipode 28*, 1.

Edman, V. (1999). En restaureringstradition i svenskt 1900-tal. *Svensk restaureringstradition, Byggförlaget* (pp. 203-215).

Fry, G. & Dramstad, W.E. (1998). Biologisk mangfold på jordbruksmark – landskapsökologiske perspektive. In L. Eriksson (Ed.) *Landskapsökologi Slutrapport*. Trondheim: NINA.

Fry, G. & Gustavsson, R. (1996). Testing landscape design principles; The landscape laboratory. In R.H.G. Jongman (Ed.) *Ecological and landscape consequences of land use change in Europe* (pp. 153-164). Tilburg: European Centre for Nature Conservation.

Goodchild, P. (1995). Principles of protection for historic parks and gardens. In *The conservation of historic gardens in Europe*, Conference 1994, (pp. 3-6).

Goodway, K. (1995). Principles of protection for historic parks and gardens. In *The conservation of historic gardens in Europe*, Conference 1994, (pp. 42-43).

Golomb, J. (1995). *In Search of Authenticity. From Kierkegaard to Camus*. London: Routledge.

Green, B.H. (1996). *Countryside Conservation: Landscape Ecology, Planning and Management* (3rd ed.). London E. & F.N. Spon.

Gustavsson, R. (1997). Kunskap med känsla och vördnad. *Skog & Forskning 1*, 38-52.

Gustavsson, R. (1999a). Local Distinctiveness in Landscape Character. In M.B. Usher (Ed.) *Landscape Character. Perspectives on Management and Change* (pp. 41-45). Edinburgh: The Stationery Office.

Gustavsson, R. (1999b). Landscape management and planning concepts - reflections on the ongoing process, the scientific support, dominating aspects, and represented competences. In R. Gustavsson, & Å. Ode (Eds.) *Sustainable landscape management in Southern Scandinavia*. Rapport 99:1 (pp. 111-127). Alnarp: Department of Landscape Planning.

Gustavsson, R. (2002). Afforestation in and near urban areas. In T.B. Randrup, C.C. Konijnendijk et.al. (Eds.) *Urban Forests and Trees; Proceedings No 1*. Directorate – General for Research EUR 19861. Office for official publications of the European communities. (pp. 286-315). Luxembourg.

Gustavsson, R., & Ingelög, T. (1994). *Det Nya Landskapet*. Jönköping: Skogsstyrelsens förlag.

Hägerhäll, C. (1999). The Experience of Pastoral Landscapes. *Agraria* 182, Alnarp: SLU.

Hidemark, O. (1997). Den åldrande ruinen – en betraktelse. *Kulturmiljövård 4, 18-21.*

Husserl, E. (1995). *Fenomenologins Idé*. Daidalos, Göteborg.

Jordbruksverket (2002). http://jordbruk.regeringen.se/landsbygdsprgr/texter/uppdaterat.pdf

Jordbruksverket (1998). *Skötselhandbok för gårdens nature – och kulturmiljövärden.* Jordbruksverket, Jönköping.

Jönsson, A., & Gustavsson, R. (2002). Management styles and knowledge cultures; past, present and future, related to multiple-use and urban woodlands. *Urban Forestry & Urban Greening, 1*, 39-47.

Lambert, D. (2001). Present and future definitions of conservation success: the quest for authenticity. *National Trust Conference – Garden Culture, Bath.* http://www.gardenhistorysociety.org/archive/newsitems/gardenculturemay2001.shtml

Lange, J. (1993). Hortikulturelt plantmateriale i 1500-talets Danmark med specielt hensyn til Uranienborg-haven. In K. Lundquist (Ed). *Uranienborgs Renässansträdgård – Renässansens Växtmaterial, Stencil 93:7*, Alnarp.

Liedman, S.E. (1997). *I Skuggan av Framtiden. Modernitetens Idéhistoria.* Falkenberg: Albert Bonniers Förlag.

Liedman, S.E. (2001). Ett Oändligt Äventyr. Om Människans Kunskaper. Falun: Albert Bonniers Förlag.

Lowenthal, D. (1994). Changing Criteria of Authenticity. In K.E. Larsen (Ed.) *Nara Conference on Authenticity in relation to the World Heritage Convention: proceedings/Conference de Nara sur l'Authenticité dans le cadre de la Convention du Patrimoine Mondial: compte-rendu UNESCO*, World Heritage Centre, Tokyo.

Lundquist, K. (1996). Kulturväxter i Tycho Brahes trädgårdar. In Herre Erlandson, J. (Ed.) *Tycho Brahe Stjärnornas Landskrona kommun*, Landskrona.

Olwig, K.R. (2001). "Time Out of mind" – "Mind out of time": custom versus tradition in environmental heritage research and interpretation. *International Journal of Heritage Studies, 7* (4).

Pedroli, B. (Ed.). (2000). *Landscape - Our Home. Essays on the Culture of the European Landscape as a Task.* Stuttgart: Indigo, Zeist.

Pinto Correia, T. (2000). Landscape identity, a key for integration (pp. 145-151).. In B. Pedroli (Ed.), *Landscape - Our Home. Essays on the Culture of the European Landscape as a Task.* Stuttgart: Indigo, Zeist.

Popper, K.R. (1996). *En Värld av Benägenheter.* Stockholm/Stehag: Brutus Östlings Bokförlag Symposion.

Ramírez, J.L. (1995). *Skapande Mening. En begreppsgenealogisk undersökning om rationalitet, vetenskap och planering.* Nordiska Institutet för Samhällsplanering. Avhandling 13:2. Stockholm.

Ricoeur, P. (1981). *Hermeneutics and the Human Sciences.* Cambridge: Cambridge University Press.

Ricoeur, P. (1986). *Från Text till Handling.* Stehag: Brutus Östlings bokförlag.

Röling, N.G., & Wagemakers, M.A.E. (Eds.). (2000). *Facilitating sustainable agriculture. Participatory learning and adaptive management in times of environmental uncertainty.* Cambridge: Cambridge University Press.

Schlereth, T.J. (1990). *Cultural history and material culture.* London: UMI Research Press.

Setterby et. al. (2000). Ronneby Brunnspark. Vårdplan. Övergripande riktlinjer och detaljerade åtgärdsförslag för Ronneby Brunnsparks bevarande, vård och utveckling. *Rapport 2000:2.* Alnarp: Institutionen för landskapsplanering.

Skage, O.R. (1992). *Laga skifte och dess spår i Tjärölandskapet.* Särtryck ur Blekinges Natur. Karlskrona.

Smedberg Åke (2003). *Hässja; berättelser från en trakt* (Hay-drying Rack. Tales from a Neighbourhood). Stockholm: Albert Bonnier.

Stenseke, M. (2001a). Landskapets värden, lokala perspektiv och centrala utgångspunkter. *CHOROS 2001:1*. Göteborg: Handelshögskolan vid Göteborgs universitet.

Stenseke, M. (2001b). Lantbrukare – landskapets förvaltare. In Landskapet: Restprodukt eller medvetet skapat? *KSLA tidsskrift* 140:5.

Thompson, I. (2000). The Ethics of Sustainability. In J.F. Benson, & M.H. Roe (Eds.), *Landscape and Sustainability* (pp. 12-33). London: Spon Press.

Toulmin, S. (1972). *Human Understanding. The Collective Use and Evolution of Concepts*. Princeton: Princeton University Press.

Turner, T. (1996). *City as Landscape, a post-modern view of design and planning*. E & F.N. Spon. London: Chapman & Hall.

Van Buuren, M., & Kerkstra, K. (1993). The framework concept; a new perspective in the design of a multifunctional landscape. In C.C. Vos, & P. Opdam (Eds.) *Landscape Ecology of a stressed environment. Vol. 1. IALE-studies in landscape ecology*. (pp. 219-243). London: Chapman & Hall.

Woodhill, J. & Röling, G. (2000). The Second Wing of the Eagle: The human dimension in learning our way to more sustainable futures. In N.G. Röling, & M.A.E. Wagemakers (Eds.) *Facilitating Sustainable Agriculture. Participatory learning and adaptive management in times of environmental uncertainty* (pp. 46-69). Cambridge: Cambridge University Press.

Chapter 18

COMBINING APPROACHES IN LANDSCAPE RESEARCH

The Case of Saaremaa, Estonia

Helen Sooväli[1], Hannes Palang[1], Egle Kaur[1], Tiina Peil[2] & Irmine Vermandere[3]

[1] *Institute of Geography, University of Tartu, Estonia*

[2] *Stockholm Environmental Institute Tallinn Centre, Estonia*

[3] *Department of Geography, University of Ghent, Belgium*

1. SETTING THE SCENE

This chapter has a twofold purpose. First, to discuss research methods employed in interdisciplinary case study. Second, to demonstrate the role of such methods when undertaking a case study. The case in question is how people experience and value landscapes on the island of Saaremaa, Estonia, both in history and at present. We will argue that combining different methods allows for more insightful investigation than more conventional methods, such as spatial analysis, interviews or questionnaire surveys could achieve in isolation. A combined approach should provide a wider basis for argument and further discussion. The intention of this chapter is to illustrate through a case study conducted on the island of Saaremaa, how attitudes to, and representations of, landscapes are best understood and researched with a variety of methodological perspectives.

The political, social and cultural conditions have changed drastically in Estonia in the 20th century. This could not occur without leaving traces in the physical setting but even more extensive are the changes in the attitudes and perceptions of the Estonian people. The Estonian islands have recently been claimed as places for recreation and retirement. This development has led to a conflict in landscape conservation, since new factors forcing intense change in land use and attitudes are becoming more evident. Many conflicts

H. Palang and G. Fry (eds.), Landscape Interfaces, 357–374.

arise from the diversification of land use and often result in increasing pressure on land. We focus this study on a particular island, claiming that it is significant to analyze the image – mental picture that develops through cognition of memories, experiences and attitudes – of the island of Saaremaa and its influence in contemporary landscape evaluation to understand the ongoing changes in the landscape.

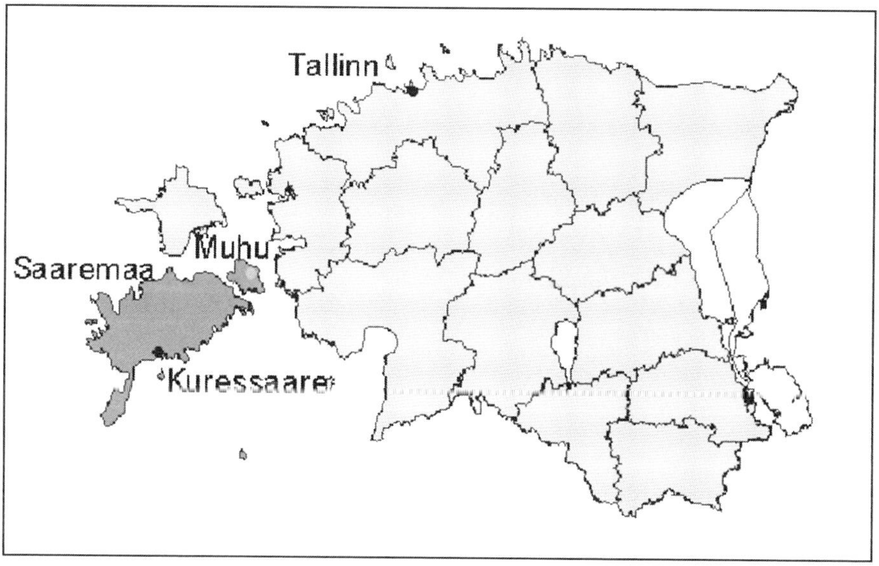

Figure 18-1. Location of Saaremaa, case study area

Our study area, the island of Saaremaa (see Figure 1), is one of the largest islands in the Baltic Sea with a current population of 37,000 and an area of 2,673 km^2. The coastline of Saaremaa is over 1,200 km long having both great natural and cultural value. The combination of sea and land is one of the greatest assets of the island, making this a favored destination for summer visitors.

Any island is marginalized when considering job opportunities and all-year-round activities. At the same time, the islands in Estonia are under increasing pressure due to attractiveness for second homes and several possibilities for leisure. Problems are caused in local valuable landscapes and habitats by increased property development which conflict with nature conservation objectives. Under Soviet rule, Saaremaa was a closed border area of the Soviet Union and visiting the island was strictly regulated, including the residents of Soviet Estonia. In some ways, the forced isolation helped to preserve the nature and lifestyle of the island freezing inevitable development and change.

The island landscape of wooded meadows, pristine coastline and dispersed farmsteads conveys an image perceived currently as typical of the Estonian isles. Domestic travel is increasing and foreign tourists have discovered the charm of the island. Children visit Saaremaa as part of their school curriculum; studies have shown that the island is among their topmost preferences (Palang 1993). Saaremaa is a favored place to spend summers with family and friends. Several old farms have been converted to second homes by Estonian town dwellers and increasingly by foreigners. Many tourists want to experience the rustic features of the island – Saaremaa is perceived as having maintained the character of the Scandinavian summer landscapes of childhood in the 1950s (Assmuth 2001).

Saaremaa is a hotspot for visitors; the landscapes of Saaremaa are under great pressure. Various social groups have expectations and interests in the landscapes: aesthetic, recreational and economic. The landscapes of Saaremaa are thus an arena for conflicting interests.

2. INTERDISCIPLINARITY IN LANDSCAPE RESEARCH

Landscape in this chapter is seen as the physical setting for human activities, as well as a source for human perception and a medium of changing attitudes to the life-world. The dialectics of people and land has its parallel when we study this dialectic: the dialectic of natural and human science. Social groups influence landscapes and are at the same time influenced by the landscapes (Tress et al. 2001). We argue that a precondition for examining and understanding landscapes in a holistic way is through interdisciplinary research. According to Fry (2001), seen from a landscape ecology perspective, interdisciplinary landscape research embraces attempts to go beyond the level of parallel studies and to understand the underlying relationships between subjects and theory across disciplinary boundaries. This is needed in order to gain a transdisciplinary understanding of landscape processes.

Landscape studies focus upon environmental, social, cultural, aesthetic, and economic issues simultaneously, and these work towards a common theoretical framework in order to understand each other as stated recently by Tress & Tress (2001).

One way forward in interdisciplinary landscape studies is simultaneous examination of the empirical reality of landscape together with their representations as shown in recent works of Daugstad & Grytli (1999), Peil (1999), Daugstad (2000), Setten (2001, 2002), Egoz et al. (2001), Akbar et al. (2003), Tress & Tress (2003) and Olwig (2003).

Our aim is to understand people living in their setting. Other researchers have also stressed the need for approaching inter- and transdisciplinarily. Widgren (2000) finds the key for understanding myths and representations of landscapes in the physical landscape itself. Fry (2001) stresses the importance of transdisciplinary landscape research, since this provides us with tools for the management of futurescapes, i.e. future landscapes that may be so far from our current landscape visions that they seem a fantasy today. Both cultural geographers and landscape ecologists thus agree that representations and attitudes towards landscape are understandable only in connecting with the actual physical setting. Massey (2001) indicates the increasing recognition of case studies, which provide a link between different disciplines. Combing methods and approaches in a case study allows us to bridge the gap between the disciplines, instead of conducting parallel research in landscape ecology and cultural geography.

2.1 Methods

In this chapter an example of conducting interdisciplinary research is presented. We start with the study of changes in the physical landscape and explore the reasons for these changes. We move on to multiple perspectives of how the landscape is represented and perceived by different social groups. We conclude by comparing the changes in physical landscape with that of the representations and perceptions.

Multiple methods for collecting both quantitative and qualitative data have been employed. The adopted methodological approach is historical/descriptive and include content analysis and discourse analysis of scientific texts, in-depth interviews, newspaper articles, tourist brochures and web sites for understanding the idea of past and current landscapes. With the help of questionnaires we can collect more data-based material for discussing perceptions and opinions on various subjects today. In-depth interviews help to focus on the issues in greater detail. Questionnaires and interviews provide rich material for comparison of media presentations and people's perceptions. Content analysis (Holsti 1969; McQuail 1991; Aitken 1997) is mainly used to analyze subjective viewpoints to process data derived from questionnaires and articles (Flick 1998). The content is transformed through objective and systematic application of categorization rules into data that can be summarized and compared (Rose 2001). Discourse analysis (Parker 1990; Fairclough 1995; van Dijk 1997; Flick 1998) gives the possibility of analyzing the texts more thoroughly than content analysis. Discourse analysis is a form of critical reading of meaning construction and a way of finding knowledge about how and why language is used the way it is used (Mattson 2001). According to Parker (1990), discourse analysis involves two preliminary steps: turning our objects into texts and locating those texts in discourses, in which material is interpreted and put into a linguistic form. In order to understand land use changes on Saaremaa we have studied series of old maps with the help of spatial analysis. This provides opportunities to trace both spatial and temporal changes.

2.2 Theory of social representations

We have chosen the theory of social representations to frame a study on perceptions and landscape representations. The theory of social representations seeks to understand individual psychological functioning by placing the individual in his or her social, cultural and collective environment. The social representations theory proceeds from the premise that any individual is a social being whose existence is rooted in a collectivity (Augoustinos & Walker 1995). According to Chaib (1996) and Moscovici (1996) the theory is based on people's representations about the world surrounding them. Moscovici defines:

> "A social representation is a system of values, ideas and practices with twofold function: first to establish an order which will enable individuals to orient themselves in their material and social world and to master it; and secondly to enable communication by providing a code for social exchange and a code for naming and classifying unambiguously the various aspects of their worlds and their individual and group history" (Moscovici 1963: xiii, edited in Wagner et al. 1999).

People develop representations according to their common understanding of what contributes to holding them together and orienting them in their social existence. Social representations, unlike stereotypes, are dynamic and they are in constant change.

3. LANDSCAPE DYNAMICS ON THE ISLAND OF SAAREMAA

3.1 The political system and its reflections in the landscape in Estonia

Landscapes are dynamic. Both natural and social processes bring along transformations in the physical landscape, depending on the processes in society. Cosgrove (1984) has argued that every social formation brings along new landscape patterns.

The appearance of Estonian landscapes has drastically changed in the 20th century, due to profound changes in the political system in 1919, 1940/1945 and 1991 (the changes and consequent results in the landscapes have been discussed in Palang et al. 1999; Sooväli et al. 2003). The Land Reform in 1919 ended the era of large Baltic-German estates and the land was allocated as small farmsteads to the Estonian farmers. The small independent farm was the dominant feature in the landscape till the Soviet occupation and the collectivization of agriculture in the late 1940s. The *collective landscapes* were dominated by new structures, particularly the *kolkhozes* with large fields and city-like village centers.

The land issue was one of the topmost priorities of the re-independent Republic of Estonia after the collapse of the Soviet Union. The Land Reform was introduced already in 1991 and included restitution of land to the descendants of the owners according to the 1939 cadastral list, and extensive diversification in land ownership. No one dominant philosophy can be found for interpreting the impact of this Reform on the landscape. Current landscapes, lacking both ideological and economic continuity may hence be gathered under the elusive umbrella of *postmodern landscapes*.

In summary, the Estonian landscapes, including those on Saaremaa, have been transformed from larger units (estates) with scattered smaller elements (farms) to extremely small units and diverse land use pattern, then back to large-scale production and settlement pattern (collective farms). The transformation proceeded with the land ownership diversification of the 1990s. This is hardly traceable in the physical landscapes, since despite the multiplication in landowners, the result has been landscapes becoming uniform (forestation and land left in fallow).

3.2 Population dynamics

The 20th century brought along significant changes in the population of the island. The population of Saaremaa declined somewhat due to emigration to the mainland between the two World Wars. The war, Soviet and German occupations, emigration, as well as deportations to Siberia caused the decline in the population from 55,000 to 40,000 between 1940 and 1945 (Assmuth 2001). The 1950s were characterized by extensive collectivization of rural people and industrialization. During this period the growth of urban and the decline of rural population were rapid. The population of Saaremaa was at its highest in the beginning of the 20th century when it amounted to 60,000. In the early decades of the 20th century, Saaremaa was one of the most rural regions in Estonia in terms of the share of rural population. Since then the urban population has increased; however, the island can still be regarded as relatively rural (Kään et al. 2002).

3.3 Land use dynamics

In accordance with the ownership and demographic situation, drastic land use changes have occurred in Saaremaa. The landscapes of Saaremaa have generally emerged similar to the general picture in Estonia, although several specific features may be pointed out. The number of manorial estates was extremely high on the island compared to the rest of Estonia, with a land ownership of 1,000-5,000 ha per estate (Oja 1994). The majority of the estates were public, rented to high officials of state and not privately owned with a few notable exceptions (Rennit 1986). The number of peasant farms was also high and the farms were small. About 10,000 farms existed on Saaremaa before World War II, with an average size of 23 ha. The first collective farms were established on Saaremaa in 1949. A year later, the arable land of 38,000 ha was divided between collective farms, on the average 567 ha in size. By the 1980s, the heyday of collective farming, these had been compacted into nine collective and state farms and one large state fishery (Palang & Kaur 2000).

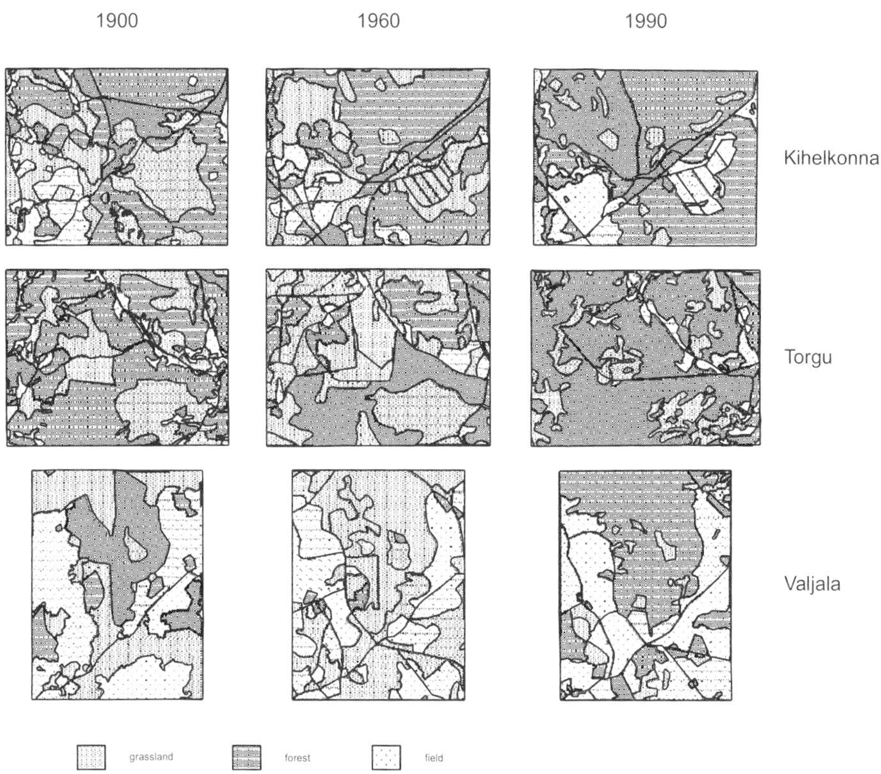

Figure 18-2. Land use dynamics in Kihelkonna, Torgu and Valjala sites, Saaremaa in 1900, 1960 and 1990

The island has been heavily influenced by the presence of the Soviet Army. A large part of the former agricultural land was used by the military: airfields, missile bases and border guard stations were built. Land amelioration has left its traces. These activities had already started in 1930 but peaked in the Soviet times. The main aim then was to remove the numerous stone fences and consolidate the former tiny fields into larger ones. By 1991, 731,000 ha of former grasslands had been improved (Palang & Kaur 2000).

Landscapes on Saaremaa feature limestone bedrock close to the surface and poor soils rich-in-lime. This has favored meadow vegetation and junipers. The land has mostly been used as pastures or hay meadows. Cultivation of crops has played a secondary role. The share of forest was low but is currently high, due to intensive forestation campaigns in the Soviet period – more than half of the island is forested today. According to official statistics (Palang et al. 1998), 88% of the island was in agricultural use (i.e. pastures, arable land, gardens) in 1918, while at the end of the 20th century this figure had declined to about 30%.

Table 18-1. Dynamics of the share of agricultural land on Saaremaa

1918	*1929*	*1939*	*1942*	*1945*	*1966*	*1975*	*1986*	*1992*
88.05	*69.22*	*69.6*	*73.16*	*70.69*	*37.5*	*32.7*	*30.5*	*30.53*

The changes in land use are illustratively shown in Table 1 and Figure 2. Land use changes on these test sites were followed on maps from four different time periods: Russian topographic map sheets (scale of 1:42,000) dating from 1896-1917, later referred to as '1900'; topographic map sheets (scale 1:50,000) published by the Topo-Hydrographic Department, Estonian Army, in 1935-1939, later referred to as '1935'; the 1960 land use map of Estonia (scale 1:50,000); topographic map sheets (scale 1:50,000) issued by GUGK in Riga in 1982-1990 and later referred to as '1989' (Palang et al. 1998).

4. OUTSIDERS' PERCEPTIONS IN THE PAST AND PRESENT

The maps and their analysis show a clear trend of diminishing agricultural land, which in the minds of the people still plays a dominant role. We argue that this trend was established in the 1930s and has been enforced by the outsiders. Hence we examined the perceptions of two groups of influential opinion-builders: the scientists of the 1930s and the tourism promoters today. The aim has been to trace the changes in the representation of Saaremaa landscapes.

4.1 Image of Saaremaa in scientific literature

A study of Saaremaa conducted by various scientists in the 1930s was used as a documentation of socio-economic conditions, as well as source material for context analysis of landscape change. The biggest achievement in the history of scientific research of Saaremaa was the publishing of the monograph *Saaremaa* (1934), which was edited by three academics from different disciplines: the historian Blumfeldt, the natural scientist Luha and the geographer Tammekann. People studying local lore also played an essential role in gathering the material for the monograph.

We are aware that the picture presented in the monograph is influenced by the prevailing scientific understanding of the world at the time. The majority of authors were, however, striving for objectivity and scientifically well-grounded argument, and their material can be relied upon as an adequate source for analysis.

Figure 18-3. Saaremaa landscape in the 1930s. Photo (ERM): Kalamees (1931)

Saaremaa (see Figure 3) was depicted in this monograph as the poorest region of Estonia. The soils of the island were unproductive and the cultivated area was small. Pastures and meadows were poor, covered either with junipers (*Juniperus communis*) or hazelnut shrubs (*Corylus avellana*). According to these researchers, the main reason why Saaremaa had fallen behind in its agricultural development compared to the rest of Estonia was because of not being enclosed (one large field could be divided between a hundred holders into narrow stripes or plots). Therefore it was hard to introduce agricultural innovations, which had to be accepted by everybody involved. Farming equipment was also old-fashioned. Both pastures and fields suffered from over-exploitation. Although the population density was high in Saaremaa, the farms were much smaller than those of the average Estonian farms. The share of forest compared to the rest of Estonia was four times less. This was explained by the clearing of land for tillage that had exhausted the soils. The educational and cultural basis was poor, which, the researchers argued, had its roots in harsh working conditions and extreme poverty. It is interesting to note that only women and young men were involved in agriculture because agricultural activities were regarded as not fitting for men. Men either worked at sea or moved to other counties to paid work and were thus not directly involved in shaping the agricultural landscape.

4.2 Images of Saaremaa in tourism material

We discuss the image of Saaremaa presented to foreign visitors today by noting the elements used to popularize Saaremaa and introduce its uniqueness. 23 articles were found in different newspapers published in English, German and French in the period from January 1999 to November 2001.

In the articles, mostly natural landscape elements were mentioned, followed by cultural landmarks and elements of the infrastructure such as the port or the ferry lines. Lake Kaali, a lake of meteoritic origin, was often followed by forests, the sea and the beaches. The Medieval castle of Kuressaare was mentioned most often of the cultural heritage sites, followed by the windmills.

Additionally, 15 English tourist brochures were reviewed. Not surprisingly, the sea and the coastline were depicted most frequently. Furthermore, the brochures wrote about Lake Kaali, junipers, meadows and pine forests of Saaremaa. The coastal cliff, local flora and the alvars were mentioned, followed in frequency by the historic town of Kuressaare, windmills, stone fences, manors and the medieval churches. Some of the manors are now functioning as guesthouses, which explain their place in the booklets.

5. LOCAL PERCEPTIONS OF LANDSCAPES ON SAAREMAA

In order to understand local perceptions of valuable landscapes on Saaremaa, we investigated the features, regarded as typical or distinct by inhabitants. An open question format questionnaire was composed and distributed to a group of people currently residing on the island all-year-round. The questionnaire included questions on peculiarity, favorite and unpopular places, need for protection and changes in the island landscape. The questionnaire aimed at identifying local values while defining value as a quality or qualities of an object that characterize its significance to a person, group or society. The survey was conducted between January 1999 and December 2000.

A sample of 320 people was composed of two respondent groups, the first consisting of secondary school pupils and the second involving randomly selected adult inhabitants of rural areas (Kaur et al. 2003). Previous studies have shown that it is important to study the attitudes of young people, since they are the ones who will be affected by, and will have to provide solutions to, the environmental problems arising from current actions (Lyons & Breakwell 1994). Secondly, the influential actors in landscapes are those actually living and working it. Thus people from randomly selected farmsteads were included. In addition, in-depth interviews were carried out with two groups: county administration specialists (planners, heads of communities, an architect, and specialists in environmental issues) and the people who are 'selling' the landscapes for tourism purposes, so called promoters. The latter included hotel managers, tourist information bureau people, farm accommodation providers and local travel agency representatives. The responses of the questionnaires and interviews are presented in the following discourses.

5.1 Percived landscape character

The most prominent aspects of the distinctiveness of the Saaremaa landscape among pupils and farmers were connected to the natural attributes of landscapes, such as juniper shrubberies and limestone outcrops. However, people, especially the adult respondents, conceived the character of the landscape as constructed and listed cultural features such as stone fences, windmills, historical farm and village patterns, as well as semi-natural elements of the landscape (e.g. wooded, coastal and alvar meadows). The opinions of the promoters and county administration specialists agreed with the farmers and pupils.

5.2 Perceived threats to landscapes

In general, the identified threats to the landscapes of Saaremaa Island were similar among all groups. The weight of danger factors was, however, perceived somewhat differently. While younger people were most concerned about the changes induced by economic practices (e.g. industrial developments and excessive and careless forestry), as well as about major infrastructure developments (e.g. constructing new roads, asphalt paving of roads, construction of a bridge between the neighboring Muhu Island and the mainland, and construction of a deep water harbor), adult respondents were most disturbed about the loss of a 'traditional' lifestyle brought about by general rural decline due to agricultural decline, tourism and newcomers. According to the planners, the threats to the Saaremaa landscapes were the real estate companies and agricultural decline. The promoters were most worried about general rural decline and new constructions particularly second home houses too close to the sea.

5.3 Need for landscape preservation

People were asked to name components of landscape worthy of management and protection on the island. The strongest need for conservation was put on the physical components of landscapes. The farmers, however, stressed the need for the protection of cultural heritage, since the human made elements were seen as an inseparable part of the scenery of the physical setting. The necessity to maintain semi-natural grasslands such as alvar meadows, coastal and wooded meadows and pastures that were seen as representing centuries-old local practice was mentioned by all respondents independent of their age and occupation. The planners added the need to protect the landscape from unsuitable development endeavors such as constructing cheap ready-made standardized summer cottages imported from Finland and Sweden. The promoters' greatest concern was nature. They expressed their concern for deteriorating sand dune and wetland areas, eroding coastline, loss of rare plant and animal species and ruined windmills.

5.4 Expectations regarding the future landscapes

In this discourse the most numerous answers both among young and adult respondents referred to their desire for well-maintained farmsteads with purposefully used fields and grasslands. Our data demonstrates that differences between perceiving a retrospective and a prospective situation are evident. The perception of the future landscape involved greater engagement of the people with the landscape, while in the perception of the existing landscape it was constructed as definite and comprising fixed objects and phenomena. The planners of Saaremaa pointed out two favorable trends: regulated tourism and small-scale rurality. Sustainable tourism was seen as producing local sustenance and maintaining the rural image of the island. The promoters saw small-scale tourism enterprise and 'sustainable farming' as the most desirable future for the Island of Saaremaa.

6. IMAGES OF SAAREMAA IN LOCAL AND NATIONAL MEDIA

In order to investigate whether and how local and national newspapers address landscape related issues, a local newspaper on Saaremaa as well as the major national daily and weekly newspapers were studied. The reviewed articles were chosen from the same period as the questionnaires were distributed and collected. The articles were selected and organized in accordance with landscape characteristics, threats and expectations regarding landscapes and the need for landscape preservation. The articles addressed the following issues connected to the landscape: construction of a deep water harbor; construction of a bridge between the Muhu island and the mainland; the issues of (county) planning, tourism, rural life, purchase of land by foreigners, waste management, valuable forest habitats and the restoration (or conservation) of heritage in the landscape. Conclusions on societal priorities can be drawn on the basis of the frequency of the occurrence of these themes in newspapers. In our study, more than two thirds of the articles discussed infrastructure developments. This indicates the main concern for the island future, and therefore in the following a closer look is taken at this issue.

Constructing a deepwater harbor at Undva, close to the Vilsandi National Park on the northwestern coast of the island, was the most discussed issue during the examined period. The occasionally extremely heated arguments about the pros and cons of the harbor construction took on much wider issues involving the future development of the whole island and its identity. Six societal groups clearly expressed their opinions: ornithologists, local inhabitants, local businessmen responsible for building the harbor, representatives of local authorities, experts, and nature conservationists. The prevailing positive attitudes towards constructing the harbor were expressed by local businessmen, local inhabitants, and representatives of local authorities. Arguments against the harbor were presented by ornithologists, nature conservationists and by local people. The experts remained neutral.

However, their arguments were used as a rhetorical tool both for and against building the port.

When comparing the results from the study on perceptions of different social groups and that of media there is a clear contradiction. On the one hand, local people fear large-scale infrastructure development on Saaremaa, which is, on the other hand, promoted by businessmen, and raised even to national priority in regional policy plans by politicians.

7. CONCLUSIONS

In this chapter we have focused on how to examine change in landscape representations and contradictory perceptions in the 20th century. By doing this we have explored the possibility of interdisciplinary landscape research and combining the material and methods of natural with those of social sciences. Landscape ecology and cultural geography, two of the central disciplines in landscape research, have to a large degree existed parallel to, and independent of, each other instead of working together and gaining insights into the landscape. We argue that the results therefore have been contradictory and seen as non-relevant or tendentious in every-day practice, remaining in the academic sphere and providing the researcher with a wide choice of arguments. We examined the landscapes of Saaremaa with a combination of methods – a questionnaire study, in-depth interviews, media analysis, as well as an analysis of land use change – to compare the landscape 'out there' and 'in here' or in the Estonian mind. The selection of methods is by no means exhaustive and further analysis of, for instance, maps and landscape painting and poetry could be useful.

Nevertheless, the results indicate that the image of Saaremaa differs greatly from that of the actual physical landscape today (see Figure 4). Saaremaa is represented and perceived as an open landscape contrary to the reality of a forested island where agriculture has a minimal role. Furthermore, we argue that the rural idyllic image of Saaremaa that the respondents and the media have, has its origins in the 1930s. However, the study of scientific texts and popular publications from the period makes it clear that survival in the island environment was a struggle at best and resignation and surrender (by leaving the island) at worst, in which all available resources were used to the highest degree. This resulted in a well-ordered, 'clean' landscape, which is perceived today as aesthetically pleasing. It follows that people today admire something that in reality was created and dominated by uttermost poverty and misery. This nostalgia driven imagery perceived as ideal has very little to do with the historic reality, and thus the 'real' social and economic conditions have only a peripheral connection to image construction and creation. Attitudes to various resources and living conditions as well as symbolism change over time. Indications, that image creation was well under way in the newly independent Republic of Estonia in the 1920s and 1930s, abound in contemporary press and tourism promotion material and Saaremaa gained a significant role in this campaign. Various improvement schemes – forestation, melioration and creating gardens – were embarked upon. The importance of past promotion in current image creation still needs further analysis. The aesthetization of landscapes, both imagined and real, has its roots in the 1930s.

Figure 18-4. Wilderness is taking over. Forested landscapes of Viidumäe, Saaremaa, 2002. Photo: H. Sooväli

A human characteristic is to desire what is not present, especially then people feel that they have been left out and forced into the role of passive observer or a chessman in a political game for decades. This explains why intensely managed farms as they are assumed to have been in the 1930s are the most desired form of existence and landscape organization for the inhabitants of Saaremaa. People, referring to the traditional landscapes and lifestyle of Saaremaa have in their minds that of small-scale farming and fishermen communities that to some extent persisted into the 1960s. These landscapes are often connected to childhood and remembered with nostalgia. The rural lifestyle of the small independent farmer played a significant role in Estonian myth creation. The image of Saaremaa as an idyllic and rustic 'traditional' part of Estonia was enforced in the Soviet period. Saaremaa became a closed and highly regulated part of the Soviet border zone and this contributed to the image of the island an exotic place among the Estonians, simultaneously being truly (untouched by degrading Soviet impact) Estonian and foreign (if you did not have any relatives, visiting Saaremaa was almost complicated). The case of Saaremaa illustrates that combined methods provide a new perspective in examining the particularity of the historical-geographical.

ACKNOWLEDGEMENTS

This paper was supported by the Estonian Science Foundation grant No 4589. Our sincere gratitude belongs to Gunhild Setten for her valuable comments on the manuscript and to David Sellar and Anu Printsmann for their kind help.

REFERENCES

Aitken, S.C. (1997). Analysis of texts: armchair theory and couch-potato geography. In R. Flowerdew & D. Martin (Eds.) *Methods in Human Geography* (pp. 197-213). Essex: Longman.

Akbar, K.F., Hale, W.H.G., & Headley, A.D. (2003). Assessment of scenic beauty of the roadside vegetation in northern England. *Landscape and Urban Planning*, in press.

Assmuth, L. (2001). On the margin, at the center: a view from a northern European periphery. In *The Baltic Sea Islands of Estonia: A Periphery in Transition* (pp. 9-18). Tallinn: Estonian Academy Publishers.

Augoustinos, M., & Walker, I. (1995). Social representations. In *Social Cognition. An Integrated Introduction* (pp. 134-162). London: Sage Publications.

Barnes, T.J., & Duncan, J.S. (1992). Introduction: Writing Worlds. In T.J. Barnes & J.S. Duncan (Eds.) *Writing Worlds* (pp. 1-18). London, New York: Routledge.

Blumfeldt, E., Luha, A., & Tammekann, A. (1934). *Saaremaa*. Tartu: Eesti Kirjanduse Selts.

Chaib, M. (1996). Sociala representationer. Om betydelsen av common sense i kommunikation. In M. Chaib (Ed.) *Begriplighet och förståelse. Texter om det kommunikativa samspelet* (pp. 37-57). Lund: Studentlitteratur.

Cosgrove, D. (1984). *Social Formation and the Symbolic Landscape*. Madison: University of Wisconsin Press.

Daugstad, K., & Grytli, E. (1999) How to study and manage a multihistoric landscape? *Norsk Geografisk Tidsskrift – Norwegian Journal of Geography 53*, 85-92.

Daugstad, K. (2000). *Mellom romantikk og realisme. Om seterlandskapet som ideal og realitet*. Dr.polit. avhandling. Trondheim: Geografisk Institutt, NTNU.

Egoz, S., Bowring, J., & Perkins, H. (2001). Tastes in tension: form, function, and meaning in New Zealand's farmed landscapes. *Landscape and Urban Planning 57*, (3-4), 177-196.

Fairclough, N. (1995). *Media Discourse*. London, New York, Sydney, Auckland: Arnold.

Flick, U. (1998). *An Introduction to Qualitative Research.* London, Thousand Oaks, New Delhi: Sage Publications.

Fry, G. (2001). Multifunctional landscapes – towards transdisciplinary research. *Landscape and Urban Planning 57,* (3-4), 159-168.

Holsti, O. (1969). *Content Analysis for the Social Sciences and Humanities.* Reading, MA: Addison-Wesley Publishing Company.

Kaur, E., Palang, H., & Sooväli, H. (2003). Landscapes in change – opposing attitudes in Saaremaa, Estonia. *Landscape and Urban Planning*, in press.

Kään, H., Mardiste, H., Nelis, R., & Pesti, O. (Eds.) (2002). *Saaremaa. Loodus. Aeg. Inimene.* Tallinn: Eesti Entsüklopeedia Kirjastus.

Lyons, E., & Breakwell, G.M. (1994). Factors predicting environmental concern and indifference in 13- to 16-years-olds. *Environment and Behavior 26*, 223-238.

Massey, D. (2001). Talking of space-time. *Transactions of the Institute of British Geographers New Series*, 257-261.

Mattson, K. (2001). (O)likheters geografier – Marknaden, forskningen och de Andra. *Doktorsavhandling. Geografiska regionstudier 45.* Uppsala: Uppsala Universitet.

Moscovici, S. (1963). Attitudes and opinions. In W. Wagner, G. Dureen, R. Farr, S. Jovchelovitch, F. Lorenzi-Cioldi, I. Marková, & D. Rose (Eds.) (1999). Theory and method of social representations. *Asian Journal of Social Psychology 2*, 95-125.

Moscovici, S. (1996). Förord till den skandinaviska introduktionen. In M. Chaib & B. Orfali (Eds.) *Sociala representationer. Om vardagsvetandets sociala fundament* (pp. 9-15). Göteborg: Daidalos.

McQuail, D. (1991). *Mass Communication Theory. An Introduction.* London: Sage.

Oja, T. (1994). *Eesti mõisad.* Tallinn: Olion.

Olwig, K. (2003). "This is not a ~~pipe~~ landscape". Circulating reference and land shaping. Paper presented at the 20th Session of the Permanent European Conference for the Study of the Rural Landscape, Tartu-Otepää, Estonia.

Palang, H. (1994). *Eesti maastike mitmekesisuse ja maakasutuse dünaamika XX sajandil* (Manuscript). MSc thesis. Tartu: Institute of Geography.

Palang, H., Mander, Ü., & Luud, A. (1998). Landscape diversity changes in Estonia. *Landscape and Urban planning 41,* (3-4), 163-169.

Palang, H., Kaur, E., Alumäe, H., & Jürimäe, K. (1999). Conflicts between landscape values as a driving force for landscape management. *Norsk Geografisk Tidsskrift – Norwegian Journal of Geography 53*, 153-159.

Palang, H., & Kaur, E. (2000). *Saaremaa kultuurmaastike hindamine.* (Manuscript). Tartu: Institute of Geography.

Parker, J. (1990). Discourse: definitions and contradictions. *Philosophical Psychology 3,* (2-3), 189-203.

Peil, T. (1999). Islescapes. Estonian small islands and islanders through three centuries. *Acta Universitatis Stockholmiensis 8.* Doctoral dissertation. Stockholm: Almqvist & Wiksell International.

Peil, T. (2001). Maastike keskel. In H. Palang & H. Sooväli (Eds.) *Maastik: loodus ja kultuur. Maastike käsitlusi Eestis. Publicationes Instituti Geographici Universitatis Tartuensis 91* (pp. 57-66).Tartu: Tartu University Press.

Rennit, M. (1986). *Saaremaa kroonumõisad ja kroonutalupoegade sotsiaalmajanduslik olukord 1840. aastatel.* (Manuscript). Ba thesis. Tartu: Department of Estonian History.

Rose, G. (2001). *Visual methodologies. An introduction to interpreting visual objects.* London: Sage Publications.

Setten, G. (2001). Farmers, planners and the moral message of landscape and nature. *Ethics, Place and Environment 4,* (3), 220-225.

Setten, G. (2002). *Bonden og landskapet. Historier om natursyn, praksis og moral i det jærske landskapet.* Dr. polit.-avhandling. Trondheim: Geografisk Institutt, NTNU.

Sooväli, H., Palang, H., & Külvik, M. (2003). The role of rural landscapes in shaping Estonian national identity. *Proceedings The 19th session of the Permanent European Conference for the Study of the Rural Landscape, Egham and Aberystwyth, UK, September 2000,* in print.

Tress, B., Tress, G., Décamps, H. & d'Hauteserre, A-M. (2001). Bridging human and natural sciences in landscape research. *Landscape and Urban Planning 57,* (3-4), 137-141.

Tress, G., & Tress, B. (2001). Capitalising the multiplicity: a transdisciplinary systems approach to landscape research. *Landscape and Urban Planning 57*, (3-4), 143-157.

Tress, B., & Tress, G. (2003). Scenario visualisation for participatory landscape planning - a study from Denmark. *Landscape and Urban Planning*, in press.

van Dijk, T.A. (1997). The Study of Discourse. In T.A. van Dijk (Ed.) *Discourse Studies: A Multidisciplinary Introduction. Vol. 1.* London, Thousand Oaks, New Delhi: Sage Publications.

Widgren, M. (2000). Landskapsforskning som kulturforskning. *Presentation in a seminar Kultur och hållbar utveckling*, KVHAA, RBJ, Stejersund, September, 2000.

Chapter 19

LANDSCAPE – A MATTER OF IDENTITY AND INTEGRITY

Towards Sound Knowledge, Awareness and Involvement

Jan Diek van Mansvelt[1] & Bas Pedroli[2]
[1] *PETRARCA, the Netherlands*
[2] *Alterra Green World Research, the Netherlands*

1. INTRODUCTION

Landscape has of old been perceived as a complex whole, studied by a wide range of disciplines. As a result, in our understanding, it has a wide range of aspects coming together, ranging from those covered by the natural sciences to those covered by the social sciences and the human sciences (Van Mansvelt & Van der Lubbe 1999), or, in other words, covered both by the geo-bio-sphere as well as the noosphere (Tress 2000). Early in the 20th century Carl O. Sauer formulated it as follows: "The cultural landscape is fashioned from a natural landscape by a culture group. Culture is the agent, the natural area is the medium and the cultural landscape is the result" (Sauer 1925). People thus produce landscape.

More and more landscape scientists and landscape managers allow themselves to be aware that all the facts and figures of a landscape do not make the actual landscape that people perceive and experience when living in, working in, passing through and spending leisure time in it. It is the overall experience of the landscape, with its range of forms and colours, structures and smells, its dynamics over time and its links to the observer's reminiscences that make the landscape what it is to each and everyone.

In this paper we first discuss subsequent levels of observation that build up from the level of sensorial observation (collection of facts/appearance), through the levels of imagining ('seeing'/'regarding') the facts as found in

H. Palang and G. Fry (eds.), Landscape Interfaces, 375–394.

their spatial and temporal coherences and feeling the qualitative character up to sensing the landscape's uniqueness, its identity.

After that we move away for a moment from the landscape itself towards the landscape-education. This to address the human-development aspect of the landscape, as the way people experience, appreciate and get involved in the landscape is seen here as crucial for its development and management in the future. Finally some prerequisites for an improved landscape management will be indicated.

1.1 Healthy landscapes for healthy people

Already in as early as the 19th century, Alexander von Humboldt stressed also the inverse effect, stating: "the closest reciprocal relationship between the earth and its inhabitants: Land affects the inhabitants and the inhabitants affect the land". In that same time already, working the land(scape) and experiencing it were recognized and applied in their therapeutic dimensions, for example in the Canada Lakeshore Psychiatric Hospital's garden (Paine 1997), but also in various places on the continent. Interestingly, both creating and elaborating the landscape on the one hand and experiencing/enjoying it on the other were seen in their therapeutic dimensions. Eating from that same landscape, as yet another way of healing interaction was so self-understanding that it was hardly ever mentioned.

As insanity was seen as a brain disease largely brought about by psychological stress, the 'healthy' design of the architectural and landscape-environment of therapeutic centres was regarded as a crucial tool for the patients' recovery. Recent studies showed that for the recovery of patients in hospital, a view on the park importantly contributed to the recovery of patients as compared to those viewing the wall of an industrial building (Larsen 1991; Mooney & Hoover 1996). Actively meeting landscape in various ways, ranging form survival trips to farming-practice weeks and walking-weekends, becomes more and more seen as an important tool to re-connect to the real-world qualities and thus to their own essence. This holds especially for all the people that become more and more surrounded by the large range of 'virtualities' that characterize today's urban life.

The same holds within the landscape, for example for agriculture, where consciously including the production of landscape as an issue in farming makes quite a difference for the landscape experienced by the visitors as well as by the farmers themselves (Tress 2000; Bohnet 2002).

1.2 Working, enjoying and living in the landscape

Here, the awareness of the regional quality of agricultural products is also at stake. Quality wines and cheeses, fruits and meats are among the products well known for their exclusive regional qualities.

Key-notion of this first point is that beside the well known effects of man on the landscape, there is as well a wide range of sensorial landscape qualities that have quite an important impact on the people living in and visiting that landscape. A rough scheme may illustrate what is meant is presented in Table 1

Table 19-1. Some interactions between people and their landscape

Landscape **People**	People affecting landscape *Projecting themselves into*	Landscape affecting people *Internalising into themselves*
Intellectual/Mental	Recognising, understanding, Aiming toward the truth.	Frame of reference, Preferably intelligible.
Emotional/Affectio-nal	Appreciating, validating, Aiming toward the beauty	Feeling home, Preferably beautiful.
Voluntary/Physical	Working, changing, managing, Aiming toward the good.	Eating from and raising in, Preferably creating well-being.

Obviously, both people and landscape are far from uniform, and so are the ways in which they actually affect each other and also the degree of awareness that people have about the reciprocity of that relationship. This holds for everyday use of the landscape in people's practical life, but also for academia and politics. For practical life raising awareness on the mutual interactions in their material and as well as immaterial (qualitative) dimensions is crucial. For academia the challenge is to find ways to study the individual, unique qualities instead of mainly focusing on the general, average quantities (Bockemühl 1997; Van Elsen 2000). For politics the challenge is to find the balance between regional and international interests (Lapka et al. 2001) and to assess the landscape quality appreciated by the landscape's inhabitants (Cassios 1994; Bosshard 2000).

2. STEPS IN OBSERVING A LANDSCAPE: PORTOFINO, ITALY

2.1 A first impression – appearance (observations/facts)

When you get off the train at the small station of Santa Margherita Ligure on the Italian Riviera, and descend the steep stairs between the houses, you suddenly find yourself on a pebble beach along a boulevard on the Mediterranean Sea. Between the palms you notice green forested hillsides above ochre, yellow, pale orange and sienna house fronts. The Monte di Portofino is a steep outlier of the Apennines in the Ligurian Sea. Away from the sailboats, motorinos, ice cream booths and gesticulating tourists with their cell phones, already at few minutes from the promenade you can find the quietness of walled gardens. Narrow streets lead further up. The villas with view on the distant sea over neatly shaven lawns under manicured olive and apricot trees, gradually make way for terraced olive groves. Only small tractors can alleviate the hard work of old farmers on these steep slopes. Many terracettes on these slopes have been abandoned. The path crosses the road that leads to the splendidly located restaurant further uphill, and arrives at the church of a small village leaning to the slope. Steeply uphill behind the church, some vegetable gardens have remained between the encroaching forests and soon you walk between stems of old sweet chestnut trees. Even here the slope has been terraced, you can notice that the wild boars have laid out their tracks across them. Further uphill the forest is less maintained, large chestnut trees lay scattered through the hornbeam forest and everywhere are the wild boar tracks.

Figure 19-1. View on Sta Margherita with the Apennines in the background

Resting at the crest you stand between tall trees, tree-heather and pistachio shrubs on rock soil. An age-old paved road, eight cobbles wide, runs along the crest. At the other side of the road the shadow of the forest gives way to the warmth of the Mediterranean sun. Looking back, you can still see the town down on the wide bay beneath the faint blue-green

Apennines (Figure 1). Towards the sun, the difference between sky, horizon and sea at the other side of the peninsula can hardly be distinguished through the dense maquis. A few minutes later along a winding path downhill you can suddenly discover the breathtaking view between light Acacia woods on the shore of an azure blue sea deeply beneath (Figure 2). The nearly unvegetated rocks dive straight down into the sea, with a rim of bright white foam (Figure 3). In a small bay you can see an old building and a solid tower, closely together between holm oaks and olive groves. Descending, you pass a spring level. You notice that the olive yards on these slopes have now completely been neglected, and largely overgrown with bramble and bushes. Only close to the former monastery, some olive groves are still maintained (Figure 4). Down there, a surprise is waiting: instead of serenity, around the age-old cloister of San Fruttuoso you only find bars and souvenir shops. The small beach is full with noisy children. Several small ferryboats bring the tourists here from Portofino (Figure 5) and other small harbors; there is no need to make the effort of the 1.5-hour hike.

Figure 19-2. Maquis-slope

Figure 19-3. Bay of San Fruttuoso

Figure 19-4. San Fruttuoso between olive groves

Figure 19-5. The former fisherman's village of Portofino

2.2 Spatial coherence

Interestingly, a landscape cannot be described from one point of view only. It becomes an image as soon as the observer has combined in his mind the observations of many sites belonging together. The old chestnut trees are inseparable from the eroded terracettes discernible around and from the village back in the valley. The pines in the maquis belong to the same system as the lizards on the bare rocks. Some parts may have deep, moist soils, others dry and shallow, some parts of the slope are steeper than others, the smell and the noises may vary, but this all together characterizes the same landscape component. These are the phenomena as they appear physically, and we have to accomplish the spatial coherence ourselves. Just like a specific tree produces a richer image in us when having observed it from several sides, the landscape's image is richer, multifaceted, when filled with diverse observations in spatial coherence. The spatial coherence also tells us about the indicator value of plant or animal species for specific soil or habitat conditions, or about the secrets behind a natural lake and the background of buildings and constructions.

2.3 Temporal coherence

Another dimension is the coherence in time. The flowers in the olive trees promise fruits next winter. From the branching of the old chestnut trees it can be deduced that they have been taken care for during hundreds of years until a few decades ago. The abandoned terracettes tell of periods of intensive agriculture. And the black scars on the pine trees show that fire has occurred. Following the landscape during the seasons, or even during one day, enriches strongly the experience of the landscape. At this stage, also the question arises whether the current landscape is a result of the past or whether it just presents the potential for the future. This dimension in the arrangement of observations yields an image that is constantly in motion. The same landscape component exhibits many different faces during the day, the seasons, the years. The observed phenomena are continually in transition like the water flowing from a spring. It requires an active thinking effort to build up a conscious image of this unsteady but none the less characteristic picture.

2.4 Character of a place: the message from appearance and succession

The character of a landscape can be seen as the combination of appearance aspects and features of succession, brought together in one's mind. For every landscape component this character is different, resulting in different processes, plants and animals present. Upper slope, middle slope and lower slope can be differentiated. This is reflected in plants and animals, in the presence and absence of water and in the soils. At the same time, together they are 'the slope' as a whole, which can be characterized as such. Altogether, the slopes of a mountain ridge can be characterized as a whole on a still higher level, with the slopes and slope-parts as subsystems (organs or organelles). The character is what everyone knows about a landscape, when knowing it well. The inhabitants of the region know what is the difference between the northern slopes of the Monte di Portofino and the southern ones (see Figure 6). It can even be communicated between us, without being required to sum up it's quantitative characteristics like solar radiation, plant species composition, etc.

At this stage, it can help to mark the character of landscape components using summarizing conceptions. In general, for the northern slopes we would speak of 'an earthy blanket of trees', whereas for the southern slopes 'a sunny and fruity openness' is at stake. The character of the side slopes, where water-rich valleys invited people to build water mills can be generalized as 'flowing meeting places'. Of course, these conceptions are not exclusive and they are – depending on time of the day of the year and on scale of detail – relevant in all landscape components, but they may inspire the composition of a target image of specific landscape components as a whole.

Figure 19-6. Overview over the Monte di Portofino peninsula

2.5 Identity of a place – *genius loci*

Why is the Portofino landscape different from the Cinque Terre landscape, a comparable landscape some 40 km further down the Riviera? In both landscapes very comparable physical phenomena can be observed, comparable processes play a role, and a comparable – though not identical – character may be attributed to the identified landscape. But still these landscapes differ completely from each other. Like no river is identical to another river, every landscape has its own identity. In the specific composition of the characters of the landscape components, the identity of a landscape is defined. The Monte di Portofino biography is characterized by the presence of the small harbor and the monastery, both inaccessible by road transport (Portofino only got a road in the 1950s, San Fruttuoso still is only accessible by boat or foot path). However, they allowed for strong connections over sea, contributing strongly to their unique *genius loci*. The Cinque Terre landscape is characterized by active agricultural villages leaning against the hill slopes that are just as inaccessible as those of Portofino. This combination gives that landscape a completely different identity.

Moreover, it is also the cultural appreciation of the landscape that determines to a large extent the identity of a landscape. Whether the landscape has this influence on society, or society on the landscape, is an unsolved question (Schama 1995). At any rate landscape identity is the combined result both from the physical appearance and from the human perception. Man is associated inseparably with landscape. Thus, to find target images for landscape rehabilitation, we should find those images that are realistic and complete, referring to natural physical processes, as well as to the variation of those processes in time, but also explicitly to the changes society has brought about, and which in most instances are irreversible. Even if reversed into the former 'ideal' situations, completely different situations would result, because of the changed boundary conditions. Landscape integrity is a dynamic concept, with remnants of the past and images from the future interacting.

The described observations together give a firm, yet imprecise, personal impression of the Portofino area. This is the way it works for every researcher what makes it worthwhile to make use of. To put the observations in order, we propose to exercise a gradual approach to the landscape's identity, leading through appearance, succession and character as described below (Figure 7).

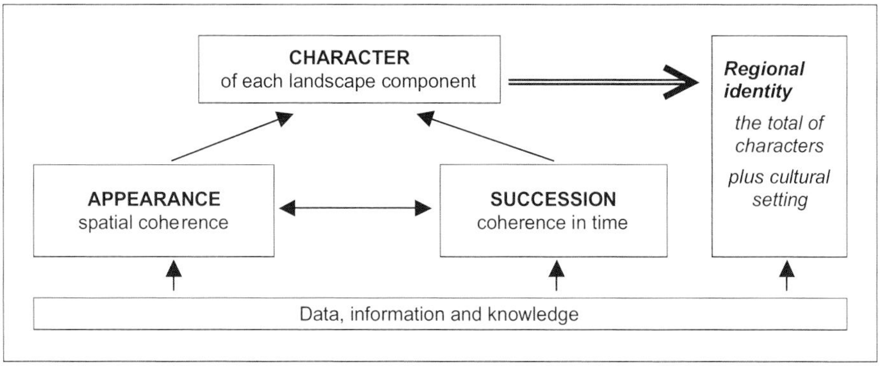

Figure 19-7. Data, appearance, succession and character of a landscape as stages in its identification

3. FROM SELF-EVIDENCE TO LANDSCAPE AWARENESS – HOW TO APPROACH IDENTITY IN LANDSCAPE?

The example of the Portofino mountains landscape can illustrate how important it is to develop a conscious image of the landscape, as a possible target for landscape development. The development of the concept of landscape in history starts at a relatively unconscious but fully involved awareness of the landscape as a whole, e.g. in the old estate times. Since the Renaissance, a detailed analysis of a wider range of disciplines based on an outsider/onlooker position lead to a renewed involvement in landscapes as a whole, now perceived as interacting subsystems contributing to the overall (landscape) system. A specific single issue arising from that approach was production per surface unit. However, a wide range of other functions, such as tourism, nature conservation, cultural inheritance, traffic, watershed management, came to compete in the landscape as an arena without a value of its own. In fact, landscape at the identity level was lost (Arnesen 1998).

Now the landscape as such starts to become an issue of conscious awareness. This means that a wide range of disciplines is challenged to contribute to the landscape on a higher, more general level than the object of their particular discipline and its sub-disciplines. Here, environmental soundness, diversity of species and ecosystems, sustainability of the management, aesthetics of the landscape are at stake as aspects of the landscape's identity in its historical development (van Mansvelt & van der Lubbe 1999). This requires an explicit move from expert's analysis to task-group's synthesis, as well as a change from a scientist's objective outsider position to that of the insider.

Here, the issue of training and education is at stake, as working in interdisciplinary task groups requires awareness of the own discipline's position amidst the other ones. It also demands the capacity to translate one's data and values into generally understandable language, some modesty and much appreciation of 'the others'. That is why the next chapter is dedicated to the awareness of levels in education, starting with an awareness of the parts and moving toward a more holistic, self-completing view of the whole.

3.1 Filling the bucket or lighting the fire

When education and training are at stake, one approach is to 'change the format', that is to provide the pupils and trainees with updated and revised information, referring to the newest findings of research and the latest positions of policy. Using the language of Heraclite, a philosopher of ancient Greece, this would be a kind of filling the bucket. Heraclite himself, however, recommended that teaching should be more like lighting a fire than filling buckets, meaning that teaching should be the art of helping the student finding, sharing and evaluating ways to solve problems, practical ones as well as theoretical ones. This in contrast to the so-called knowledge transfer from 'high' (research) down to 'low' (practice and training). In the fire-lighting approach, learning about oneself as learning goes hand in hand with learning about the topic of interest. It is about exploring the self as the ultimate research instrument to learn about the world around, in this case the landscape (Bockemühl 1997). Here, experts of education and training like Bawden & Valentine (1984), Forsythe (1984), Rushby (1985) and MacRae (1989) agree with earlier experts like Bloom (1956) and psychologists like Maslow (1970) arguing that the most humane education contributes to the students' free and autonomous self-development. For the European Landscape Convention, a 'fire-lighting' approach of education and training would be appropriate, that can also be indicated as Human Resource Development in its true sense (Van Mansvelt & Kólster 1990; van Mansvelt 1990). Emancipation and empowerment are leading criteria in this approach that addresses the intellectual education (knowledge oriented – cognition) as well as the emotional education (finding out about the feelings and values – affection) and also the motorative education (knowing about doing, how to practice – conation). In the mentioned literature these levels are referred to as cognitive, affective and conative (Bawden & Valentine 1984).

3.2 The cognitive domain of education

In the cognitive domain of education, Bloom (1956) differentiated some 6 steps, from a passive memorization of facts ('Factual knowledge') to full understanding of the facts in their methodological context ('Evaluation').

Table 19-2. Steps in the cognitive domain

From passive memorization of facts ('Knowledge') to full understanding in context ('Evaluation')	
Factual knowledge	facts/data to be memorized and reproduced on demand;
Comprehension	simple 'if...then...' connections between the data, direct associations;
Application	useful application of abstract regulations and prescriptions in a well-known context;
Analysis	explicit determination of different structural elements in publications/situations, recognize intentions/manipulations etc.;
Synthesis	reconstructing or reorganizing all kind of given situations, trials, explanations; reviewing, planning and explaining clearly/convincingly; formulating laws of nature etc.;
Evaluation	getting explicit hold of essentiality, realistic judgment of essential values/ultimate quality requirements; summarizing the essence of a paper / case / situation.

Here the first steps or levels of cognition require relatively superficial individual involvement in the landscape, as compared with the later ones. They are in general sufficient for those studying for credits, but do not satisfy those studying for interest in the landscape. Therefore, the presented sequence is also a sequence of increasing involvement of the student into the landscape, which is, at the same time, an increasing internalization of the whole landscape in the student. In this same process of internalization-by-involvement, the need for an outer authority, to give the necessary help or orders to make things work out, decreases, as emancipation, and the scale of own practicable responsibility, increases.

Whereas with cognition of the type of the lower levels many actions can be taken in an 'automatic'/'habitual' way, the higher levels of cognition demand an increasingly constant awareness of the situation and an increasing self-reflection. The division of the scale into six steps is basically comparative and qualitative. Studying any of the landscape's features on all these levels or steps helps to discover and communicate these levels. It is interesting to notice that such discussions exceed the purely technical dimensions of the landscape, leading the participants of the discussion to express themselves more personally, even individually on their concepts and perceptions of their landscape and the research methods used to explore it.

3.3 The affective domain of education

In the affective domain of education, five steps have been differentiated by Bloom (1956), starting with a relatively passive/neutral 'Reception' or noticing of the information to a state of involvement that he called 'Characterization' where one has become a representative of a chosen paradigm by identifying oneself with it. These steps can be described in Table 3.

Table 19-3. Steps in the affective domain

From relatively passive / neutral 'Reception' or noticing of the information to a state of involvement that here is called 'Characterization':	
Receiving	from 'untouched awareness' to 'controlled attention' or 'passive looking for repetition(s) of the event';
Responding	from 'goody-goody compliance' to 'satisfaction in joining';
Valuing	from 'tentative acceptance of a value in a passive way' (OK, call me a student of landscape science) to 'effective commitment to a value in an active way' (let me tell you how wonderful it is to be a landscape ecologist);
Organizing	from 'ideal conceptualization of a chosen system of values' to 'harmonization or integration of different complex value systems within one value system';
Characterizing	becoming a prototypical representative of a chosen philosophy of life, as a result of its complete internalization.

Here, as in the case of the different steps in cognition, the first ones require the least personnel commitment, and, going from receiving to characterization, the commitment with landscape increases, together with the incorporation or internalization of the relevant landscape values. Thereby, the third level marks the important transition from being a more or less passive onlooker/outsider, to becoming a convinced participant/insider, through a process of gradual identification with the landscape. But, as this field or domain of affection touches the human being much deeper than the field or domain of cognition does, touching it in a less conscious and therefore more vulnerable way, its elaboration is a much more sensitive matter than that of the cognitive domain. To be explored in a fruitful way, this domain requires mutual respect among those involved in practical engagement in the landscape.

However, often concern is expressed on the growing disengagement, the lack of interest and, in general, the increasing alienation of "modern" people with regard to landscape. Many point to the fact that all children and students are somehow influenced in their affective field by their tutoring staff. So there is no sense in denying, underestimating or suppressing this aspect of education. On the contrary, by giving it appropriate attention, this field of affection becomes more and more open for self-conscious and self-responsible management. It can be argued that an increasing clearness about one's own attitude towards landscape implies something like emancipation in regard of one's own affections and emotions. This emancipation, decreasing the dependency on uncontrolled emotions, does not at all mean a complete abstinence from all empathy (sym- or antipathy), but rather an increasingly clear awareness if its indispensable signaling function.

Here, it can be realized that, it is precisely the outsider/onlooker situation that leaves a person much more captive of, and dependent on, his own/subjective feelings, which thus tend to fail in supporting a clear, communicative relation with the 'others'. The implicit, non-communicable socio/emotional dependence of scientists in general was discussed earlier (van Mansvelt & Kólster 1990) as a counterproductive, irrational barrier against the acceptance and introduction of innovation in landscape (Miller 1984; MacRae et al. 1989; van Mansvelt & van der Lubbe 1999; Pedroli 2000).

3.4 The conative domain of education

Now we come to the conative domain of education (Bawden & Valentine 1984), which refers to the implementation of the understanding, in a certain affective state, through handling, into the practical living landscape where doing is essential. Referring to literature (see van Mansvelt 1990, 1992). Several levels of autonomy in handling can be distinguished, ranging from the initial imitative acting, to acting out of a free fully self-conscious dedication to the landscape.

In the sequence presented in Table 4, the level of pure motoration gradually transits into conation, as the amount of consciousness and effective responsibility on all actions, gradually increases with the years of individual, autonomous experience. Where the discussion on education of the affective domain was already obviously at stake in recent decades, the explicit education of the conative domain is essential for the century we now live in. The famous French philosopher (also minister of cultural affairs) François Mauriac stated in the mid-1980s that the 21st century will be the century of ethics or it will not be at all. Just like before, here again it must be argued that by not explicitly including this realm into educational objectives, and be it only provisionally, can no more be justified once one recognizes that implicit ethics are incompatible with human emancipation. The success of this type of education will be reflected in the landscapes of the 21st century.

Table 19-4. Steps in the conative domain

From the initial imitative acting according to the masters example, to acting out of a free fully self-conscious dedication to the landscape	
Imitation	Learning practice from practice is where apprenticeship starts: "don't talk (so much), just look and do like I do". The choice for a master whom to imitate, and in what aspects, is the student's;
Handling	This ranges from 'dutifully adjusting complex manipulation' to 'personal concern for flawless performance'. The master's example is present in the student's mind, imaginary but efficient. This level complies with the 'mate/journeyman' in the old guilds, or the traineeship in landscape design and management offices;
Mastering	Here the craft or art is mastered: the student is ready for independent self-employment as the outer authority has become sufficiently incorporated. Here a transition from freedom of outer control and interference into freedom for individual motivation may develop, based on increasing experience and a widening world-view;
Engagement	Here the challenge is to perform increasingly creative, perfect and outstanding in the profession itself. This demands an ever-increasing engagement with and into the relevant landscape(s) including its socio-cultural conditions (and constraints). Ongoing self education in fine-tuning of the individual capacities and performance to those of the colleagues, i.e. in professional organizations;
Dedication	Appropriate identification with the essence of the chosen landscape leads to convergence of the own interests with those of the partners at stake: human and natural, individuals and entities. The responsible and self-conscious individual gradually replaces the self-centered ego.

Using the degree of emancipation and internalization of the cognitive, affective and motorative-conative capacities as a key to their comparison, Table 5 gives an attempt to integrate them.

Table 19-5. Overview of steps in the three psychological domains of human education

Domains → Levels↓	Cognitive domain	Affective domain	Conative domain
Pre-emancipatorian levels: emphasis on Staff-initiated education ↓	Knowledge Comprehension	Receiving Responding	Imitation Handling
Increasing internalization of learning into self-development ↓	Application Analysis	Valuating	Mastering
Emphasis on student-initiated learning on the post-emancipatory levels	Synthesis Evaluation	Organizing Characterizing	Engagement Dedication

It must be stressed that a scheme like this should in no way be taken as a strict, one-way, exclusive approach. It is meant as a tool to increase the awareness of gradients in the process of learning, as a tool for understanding various levels in learning and in human capacities. This could structure human resource development education and training in such a way that it includes teaching both *theory* and *practice* in a way that includes the training of the *affective* domain. Especially the latter domain is important in landscape awareness, while concerning the *real* landscape. It is only meant as a tool to implement the demands made in section 1 – for the full conscious relationship with the landscape at stake, for research, development and – most important – living, working and enjoying in.

4. PREREQUISITES FOR ENGAGEMENT

In the light of the previous elaboration of steps in landscape research and the elaboration of landscape awareness in education, some key aspects of ways and degrees in engagement to the landscape should here be referred to.

First there is the process of gradually becoming aware of the landscape one lives and works in. For most people, the landscape they are born and raised in is the 'normal' one. Mountain people wonder about flatlands, forest people about arable fields, urban people about the countryside, mainland people about the coastal areas, and vice versa. Living 'elsewhere' can be exiting, challenging and pleasing. 'Coming home' gives often provides two kinds of feeling: one of relaxation and familiarity and another of strangeness. People see, hear, smell, in short; experience the old familiar surroundings in a new way. Emancipation from the homeland apparently went along with some alienation, bringing about a higher level of conscious awareness of the homeland's peculiarities in contrast to those of the new landscape.

In the meeting of a new landscape, several steps can be discerned. They follow one another in a never-ending sequence of steps that can be described in the following way: 1. Sensorial observation/collection of data; 2. Conceptualization of the observations made, leading to some kind of understanding/making a model of system of the collected data in the light of a certain theory (1, 2); 3. Validation of the landscape as understood, leading to some kind of Appreciation through the use of a chosen system of values (1, 2, 3); 4. Action on viz. into the landscape, implementing the appreciation in the light of the understanding (1, 2, 3, 4). Interestingly, in the process leading from step 1 to step 4, the objective outsiders' position (1) is stepwise transformed into an insider's subjective engagements position (4). In step 1 reflection is required whereas in 4 full involvement is at stake.

Reflection on this sequence, by the way, rather clearly shows how objective knowledge and understanding inevitably are a product of making objective the as such inevitably subjective research path one has followed (Feyerabend 1975). Reporting on all cognitive, affective and conative steps in the pathway that has lead one to come to a certain awareness or conclusion, making it sharable with 'the others', that is making it objective.

From this point of view, looking that way, under these conditions, 'you can see what I mean' could be a phrasing of that process of 'objectivation'.

Quite often these 4 steps are made by different partners in a landscape development process: the data-provision (1) is done by another group of people then the use of the data for planning and drawing of proposals (2). Deciding on landscape plans (3) once again is done by other people then those collecting the data or those providing the proposals. And finally, working the landscape (4) according to the plans (2) as decided for (3) is most often done by other people then those previously mentioned. No wonder that alienation, lack of interest, lack of care and repeating fights on rights for whatever subgroup of land users interested in a specific aspect of the landscape can easily come about. Based on the experiences that lead to these considerations, the proposal is made to make sure that in landscape planning all four categories of experts and functionaries are sufficiently involved in all four steps (Table 6).

Table 19-6. Steps as usually made and steps proposed to be made in regional landscape planning

Steps	Groups usually involved	Groups to be involved in ***overlapping*** steps
1. Sensorial observation/collection of data	Field data monitoring groups	Monitoring groups and Planners Decision makers and Implementing Practitioners (iterative actions in overlapping sessions) and Representatives of various interest groups relevant for and active in the particular region/landscape to be sufficiently engaged in steps 1 to 4
2. Conceptualization of the observations/making a model	Landscape planners	
3. Validation of the landscape/Appreciation through a chosen value system	Policy and Decision makers	
4. Action on viz. into the landscape/implementing the appreciation in the light of the understanding	Implementing practitioners	

This means that appropriate involvement of all those affected by the landscape: that is by its management, its planning and its perception, is crucial to warrant that management to fit the actual demands of those living, working and spending leisure time in it. Local involvement of those living in the landscape and general expertise in various options for the land use in a national context may be mixed in the mentioned process to warrant the ongoing development of the landscape's identity in a regional and subsequently in a national and supra-regional context. At the same time, all landscape discipline's experts should be asked to contribute their expertise into the landscape development as a whole, without forcing their disciplinary logic and values on the landscape, subduing the others. Whereas the latter would lead to monotonous, single purpose landscapes that lack appreciation, the co-operative approach as mentioned could bring about a highly diverse, multifunctional landscape. It would allow the individual landscapes to function as ecologically and economically sound socio-cultural organs in the national organism.

REFERENCES

Arnesen, T. (1998). Landscapes lost. *Landscape Research, 23 (1),* 39-50.

Bawden, R., & Valentine, I. (1984). Learning to be a capable systems agriculturalist. *PLET, 21 (4),* 273-287.

Bloom, B.S. (1956). *Taxonomy of educational objectives. Handbook I: Cognitive Domain.* New York: David Mckay Company.

Bockemühl, J. (1997). Aspekte der Selbsterfahrung im phänomenologischen Zugang zur Natur. In G. Böhme & G. Schiemann (Eds.) *Phänomenologie der Natur.* DTV. Stuttgart.

Bohnet, I. (2002). *Exploring landscape character: A socio-ecological analysis in the High Weald area of outstanding natural beauty.* Imperial College, University of London.

Bosshard, A. (2000). Landschaft zwischen Objekt und Subjekt. In B. Pedroli (Ed.) *Landscape – Our Home/Lebensraum Landschaft* (pp. 65-73).

Cassios, C.A. (1994). Landscape values and bio-aesthetics. In Cassios (Ed.) *Proceedings for the 2 Day Conference on EIA; Methodology and Research on EIA Studies.* European Centre of Delphi.

Feyerabend, P.K. (1975). *Against method: Outline of an anarchistic theory of knowledge.* London: New Left Press.

Forsythe, K. (1984). The human interface: Teachers in the New Age. *PLET, 20 (3),* 161-166.

Lapka, M., Cudlinova, E., Rikkon, S. & Maxa, J. (2001). Intergrating nature, culture and society: The concept of landscape field. *Ecology Bratislava, 20 (1),* 125-138.

Larsen, L. (1991). *Nature as therapy: An assessment of schizophrenic patients' visual preferences for institutional outdoor environments.* Master of Landscape Architecture thesis. University of Guelph: School of Landscape Architecture.

MacRae, R.J., Hill, S.B., Henning, J. & Mehuys, G.R. (1989). Agricultural science and sustainable agriculture: A review of the existing barriers to sustainable food production and potential solutions. *Biological Agriculture and Horticulture, 6,* 173-219.

Maslow, A.H. (1970). Motivation and Personality. New York: Harper and Row.

Miller, A. (1984). Professional dissent and environmental management. *The Environmentalist, 4,* 143-152.

Mooney, P. & Hoover R. (1996). The design of restorative landscapes for Alzheimer's patients (pp. 50-55). In C. Wagner (Ed.) *Proceedings 1996 Annual Meeting, American Society of Landscape Architects Washington DC.* American Society of Landscape Architects.

Paine, C. (1997). Creating and re-creating landscape for therapy and recreation. In *Proceedings, 34th IFLA World Congress.* Buenos Aires: International Federation of Landscape Architects.

Pedroli, B. (Ed.). (2000). *Landscape – Our Home/Lebensraum Landschaft. Essays on the culture of the European landscape as a task.* Zeist: Indigo.

Rushby, N. (1985). All our colleagues have PhDs: Objections to Educational Technology. *PLET, 22 (1),* 81-84.

Sauer, C.O. (1925). The Morphology of Landscape. In Ingerson, A.E. (2000). Changing approaches to landscapes. Institute for Cultural Landscape Studies, Harvard.

Schama, S. 1995. Landscape and Memory. London: HarperCollins.

Tress, B. (2000). Landwirt schafft Landschaft – Umstellungspotential und landschaftliche Konsequenzen der ökologischen Landwirtschaft in Dänemark. PhD thesis. *Research report no. 121, Dept. of Geography and International Development Studies*. University of Roskilde.

Van Elsen, T. (2000). Naturentwicklung – Zukunftsaufgabe der Landwirtschaft? In B. Pedroli (Ed.) *Landscape – Our Home/Lebensraum Landschaft* (pp. 65-73).

Van Mansvelt, J.D. (1990). Education and training in organic agriculture. Agriculture II: Methodological aspects of appropriate Human Resource Development and indications of their implementation. In J.M. Besson (Ed.) *Biological farming in Europe: Challenges and opportunities*. Swiss Federal Research Station/FAO, Bern.

Van Mansvelt, J.D. (1992). Human Resource Management in organic types of agriculture. In U. Koepke & D.G. Schulz (Eds.) *Proceedings of the 9th International Scientific Conference of IFOAM.* Wendel.

Van Mansvelt, J.D. & P. Kólster (1990). *Education and training in organic agriculture I: Present situation and polar aspects of educational content.* FAO expert consultation. Bern.

Van Mansvelt, J.D. & Van der Lubbe, M.J. (1999). *Checklist for sustainable landscape management.* Amsterdam-Lausanne: Elsevier.

Chapter 20

LEARNING FROM TARTU – TOWARDS POST-POSTMODERN LANDSCAPES

Jussi S. Jauhiainen
Institute of Geography, University of Tartu, Estonia & Department of Geography, University of Oulu, Finland

1. INTRODUCTION

Landscape is a text that can be written and read in various ways. The current debates in social and cultural theories underline the importance of situated knowledge and the partial perspective of the researcher. The myth of universally valid concepts and theories is being dismantled in the postmodern era. Knowledge, theories and concepts are not floating over the landscape but always in relation to it. Knowledge is always situated within some contexts.

The study of landscapes is one of the most interesting and challenging possibilities to approach debates of situated knowledge and partial perspective. Landscapes are at the same time local and non-local. They are specifically place-bounded in their local embeddedness but necessarily open for interpretations that derive from different landscapes and contexts. Today, the task of a landscape researcher is to reflect critically upon these different conceptual and empirical contexts. This means that all interpretations of landscapes are always enriched by the experiences and sensitivity of the researcher – thus the partial perspective.

Landscape researchers do not share the same idea of what is a landscape and how it should and could be studied. Swedish landscape researcher Mats Widgren (2002) has categorized landscape studies in three perspectives that are:

– landscape as a way of using (resource: land use, production, capital, etc.);

H. Palang and G. Fry (eds.), Landscape Interfaces, 395–406.

- landscape as a way of communicating (institution: customary law, social order, etc.);
- landscape as a way of seeing (scenery: representation, etc.).

There are several possible approaches landscape: from empirical description to positivist analysis, from humanist interpretation to structuralist categorization and from post-structuralist interpretation to artistic insights, praxis and activism (see Cosgrove & Daniels 1988; Duncan 1990; Barnes & Duncan 1992; Rose 1993; Harvey 1996; Claval 2002; Olwig 2002; Widgren 2002).

In principle, cooperation between these three landscape perspectives would be useful for understanding the complexity of landscape. Now, cooperation is significantly more difficult when one considers the more profound questions that are necessarily embedded in the study of landscapes. These are, for example, the nature of reality, the ways to categorise reality and the modes to speak, write or approach reality through scientific, social and artistic ways.

In this article, my concern is to reflect upon possibilities to study landscapes from post-postmodern viewpoint. One aim of post-postmodern landscape studies is to overcome and abandon dichotomies. This is illustrated in the 'learning experiences' from London, Las Vegas, Los Angeles – and from Tartu. Another task in this article is to discuss the methodological possibilities for post-postmodern landscape studies based on partial perspective and situated knowledge.

Rurban landscape is the empirical dimension I discuss here. Rurban can be seen as a location somewhere in the urban fringe outside the densely built and intensive urban environment but before remote areas with sparse population density and extensive land uses. Nevertheless, this rurbanity exists as also as 'pockets' of rurban mentality in the minds of people who might live and act either in dense and intensive urban environments or in remote rural areas. Rurban areas are located materially and immaterially in-between what used to be defined as urban or rural. A rurban landscape cannot be precisely located despite visual characteristics that can be found more or less between the urban core and the extended rural areas.

I use 'rurban', first, as a specific example to blur the dichotomy between urban and rural and, second, more generally to emphasize the need to pay attention to categorization. I mean that in the contemporary globalizing world there are no 'pure' rural or urban landscapes (see Antrop 2000; Hidding & Teunissen 2002). The dichotomy between urban and rural has been blurring already for decades. The difficulty to progress today in landscape studies is partly due to the 'territorial trap' of fixed (often binary) categories – here exemplified through urban/rural. However, I do not mean that we should be satisfied by claiming that the categorizations are fuzzy today. Neither is enough that we argue for new more precise typologies or triads instead of binaries. The challenge is to overcome or even abandon the binaries and categories.

2. METHODOLOGICAL CHALLENGES

Traditionally, most landscape research has paid attention to commonalities in physical landscape. The task of the researcher is to depict (as much as possible) homogeneous landscape through standardized methodology leading to clear definable landscape regions. One common method has been landscape morphology that concentrates on the physical appearance of landscape. Earlier landscape scholars conducted this exploratory method through direct field exercises. Recently, the modern technologies such as remote sensing and geographical information systems have been used and often replaced these field practices, especially in landscape ecology focused on visual landscape. Landscape is seen as bounded space and a container. The task of the researcher is to trace enough detailed information from the landscape. Scholars are interested in the extent of homogeneity needed to define the unity of landscape region by using, for example, classic land use categories. Categories for measuring landscape are taken for granted and the position of the researcher is considered 'objective' towards landscape. If the level of precision is adequate then the community of researchers following the morphological, taxonomic and categorical approaches are satisfied with the outcomes. Landscape is thus depicted through universally acceptable categories. Landscape is a resource or more rarely an institution.

Many questions can be posed on this approach. What about the scale – are the results different if one measures landscape in square centimeters or in square kilometers? Do we get exactly those results we want by changing the scale and categories? Does this mean that the more precise we are in measuring the less we are able to understand landscape? The recent definition of 'multifunctional landscape' seems to be a naïve empirical statement depending on the categories used for measuring and monitoring the diversity of multifunctionality. One tries to 'mummify' the contemporary landscape (or land use) without paying attention to its ideological background and meanings. Maybe above all else the meaning of landscape is about continuous change of the interplay between individuals, society and nature. Also interpretations change continuously, so needed is much more conceptual and theoretical reflection about the landscape than the classic morphological approach can provide.

In the contemporary fragmented world of multiple meanings it may not be possible to share and to depict the same landscape at all. We co-habit different landscapes in the same place at the same time. However, our memories about our past landscapes are different and they remain different – thus the co-habitation in the same place cannot ever really be the same, neither the landscape. Landscape is no more one unity but a multiplicity of meanings, readings and misunderstandings that are open to many interpretations. This means much more than the perceptual differences between individuals grasping the landscape through different senses. Instead of paying attention to similarities and commonalities and bounding space as much as possible into homogeneous regions, the unexplainable aspects of diversity and differences are developing into interesting possibilities for "borderland" landscape studies.

In broader and more explicit terms the changes in the understanding of the world require new methodologies for landscape studies. Modern theory was seriously attacked already in the late 1960s when deconstructionists proclaimed the death of the author. This means that there can no longer be any grand theories, solid methodologies, or universal interpretations. The reader of a landscape (as a text) is part of the process to claim what is a landscape. If all readers make the landscape (individually, as a text, despite the intentions of the landscape interpreter, i.e. the landscape researcher) we enter into a situation in which the interpretation can never be fixed. It is impossible to trace the various *genii locorum* of landscape. The wholeness of landscape cannot be defined. The changing memories, mentality and attachment of the multiplicity of readers to different landscapes are too much to be understood. On the other hand, without readers there is no sense to study landscapes. For the matter of convenience many scholars do not problematize these ontological and epistemological aspects of landscape studies. They make significant methodological compromises, especially those interested in normative aspects of landscape, such as landscape policies. Nevertheless, the more there are normative applied goals in landscape research the more one should be interested in these methodological issues. If not, then what landscape we are defining, preserving and developing and for whom?

In general, geographers were late to consider postmodernism. It was in the mid-1980s when geographers and planners came out of the 'modern closet' by reflecting upon the postmodern urban development, planning and landscape (see *Environment and Planning D: Society and Space* 4:3). Critical social and cultural theories helped to deconstruct the modern view of urban landscapes. Later scholars in landscape studies reflected upon the symbolic values and meanings in landscape (Cosgrove 1984; Cosgrove & Daniels 1988). There has been a sea-change in landscape studies; from physical morphology towards representational approaches in which landscape is understood as a way of seeing. I still dare to claim that landscape scholars have not taken seriously enough the challenges proposed by postmodern and poststructuralist theories. Few if any ideas of post-postmodern approaches have been presented.

In the early 21st century it seems that the gap in landscape studies between traditional ('modern morphologists') and contemporaneous scholars ('representational postmodernists') has grown to the extent that there is no common landscape to be discussed or debated upon. The classic systematic morphological approach for landscape depiction was based on a universal and commonly shared method to study landscapes everywhere. From a postmodern perspective the morphological approach is one method among many and not more 'true' than free associative-writing or action-painting that would be considered as nonsense from a modern perspective.

Following the demise in faith in a universal truth many landscape researchers agree that there can be different representations regarding a particular bounded physical landscape. Many scholars have abandoned the aim to find 'an average' homogeneous landscape regions based on universal comparative research methodology. As indicated above, this 'representational turn' has become a mainstream approach among 'younger' landscape scholars. Nevertheless, often these postmodern approaches miss the critical methodological reflection. Many so-called postmodern landscape scholars still believe in common concepts that can be used everywhere. Thus the difference between the traditional and contemporary approaches in landscape studies is not so big and ontologically it often withers away. Also many engaged in the emancipated ('third way') approach on landscape – situated between modernism and postmodernism – have failed to be sensitive enough. Important themes in landscape studies such as social justice and gender are often based on universalism, for example, about the role of capitalism in transforming the physical landscape and its meaning (Rose 1993; Harvey 1996). Also social justice seems to be a concept that has not been opened and deconstructed enough in the emancipating praxis.

One additional issue is language. Language is one key instrument to discuss landscape and the language (with words) prevails over the way how we as scholars discuss landscape. I do not enter here the discussion whether we can communicate about complex issues without language, what is the relationship between matter, meaning and language, and how linguistics, geography and landscape might be related (see Claval 2002). In landscape studies already the use of different languages is a challenge. The structure of different languages is constructed differently and the notions related to landscape are not the same (of the origin of the word *landscape*, see Olwig 2002). We cannot speak exactly the same way about the same landscape through different languages. The very place-bounded empirical reality influences on the concepts (or simply – words). Ultimately, we cannot transfer concepts from one language to another because the concepts are place and situation sensitive and always in relation to some particular landscapes. The traveling of the concepts into different contexts is an issue mostly still uncovered in comparative and non-comparative empirical landscape research – except the 'mother' of all concepts – *landscape*. The traveling of concepts goes with individuals. Also the body of the researcher has to be reflected in landscape studies.

3. FOUR CASES – LEARNING FROM LONDON, LAS VEGAS, LOS ANGELES AND TARTU

To illustrate the above discussed methodological challenges I briefly turn to four empirical cases that reflect on the concept of rurban landscapes. A significant transformation in landscape has been caused by industrialism from the 18th century onwards and post-industrialism from the latter part of the 20th century onwards. Recently, suburbanization, counterurbanization and the development of modern information and communication technologies have enlarged the territory of 'rurban' landscapes. Nevertheless, it should be remembered that there has always been partnership between urban and rural from ancient Mesopotamia to contemporary cities and towns. There are no 'pure' rural or urban landscapes. Rurban in the urban fringe is physically a particular 'fusing' landscape but also a significant political landscape.

Here I turn briefly to three cases of rurbanity that cover the 20th century: London with its Garden Cities, Las Vegas with its commercial streets and Los Angeles with its multilayered linguistic landscapes. To my mind these are all rurban landscapes despite the population density or built infrastructure. The cases illustrate how the concept of rurban landscape is particular, political and context-sensitive.

3.1 Learning from London

The Garden City was proposed by Ebenezer Howard in his book *To-morrow – or the Peaceful Path to a Real Reform* published in 1898. Howard proposed his scheme by reflecting on what he experienced in the North America and Britain of the late 19th century both empirically (see, for example, urban development of larger cities and industrial towns in 19th century Britain and Friedrich Engels' *The Condition of the Working-Class*) and conceptually (see, for example, early socialist writing in Henry George's

Progress and Poverty and Edward Bellamy's *Looking Backward 2000 – 1887*). The aim was to combine the virtues of urban and rural activities in a rurban landscape (Figure 1). Several localities were developed around London and elsewhere in Britain (Welwyn, Letchwork, Hampstead etc.). The idea of Garden Cities gained popularity also in Europe and in North America. In post-war Europe this idea of rurbanity and rurban landscape continued with the policy of new towns.

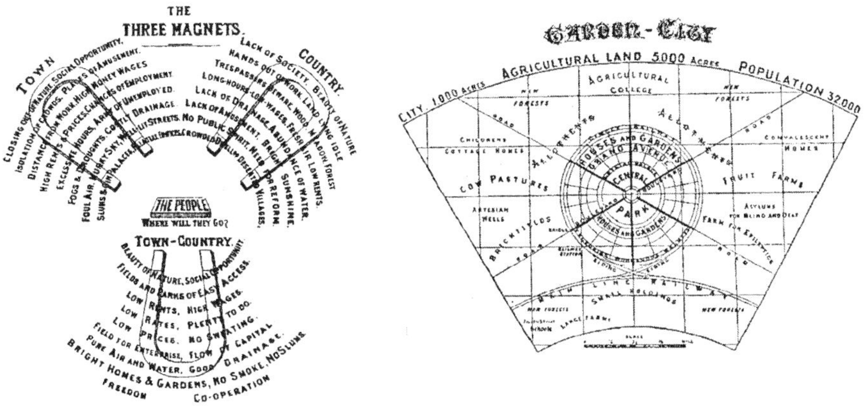

Figure 20-1. Learning from London. Source: Howard (1898)

What we can learn from the case of London is the relationship between the conceptual and empirical world. The concepts and theories of Howard were meant to be universal. This failed and concepts became significantly modified when they were contextualized in Britain and other countries. The morphology, meaning, and political significance of rurban Garden Cities became different in different localities. The next generation of landscape studies should reflect more upon the concept formation (relationship between the individual and the place-bounded experiences) and upon what the traveling of concepts means for their contextualization (modifying of landscapes in particular contexts).

3.2 Learning from Las Vegas

Robert Venturi, Denise Brown and Herbert Gans wrote in the 1960s a book, *Learning from Las Vegas: the Forgotten Symbolism of Architectural Form* that facilitated the deconstruction of the myth of simple modern (urban) landscape. In this book they consciously paid attention to popular architecture. Out of their observation came the notion of how the main commercial strip, then and today so common in the USA, was almost 'all right' as a landscape. The 'primitive' vernacular and popular-trivial architecture opened up possibilities for free use of historical ornaments in the construction of urban landscape – or rurban landscapes. Here urban received a new meaning by fusing it with rural and in-between urban and rural mentalities in and around Las Vegas (Figure 2).

Figure 20-2. Learning from Las Vegas. Source: Venturi et al. (1966)

What we can learn from Las Vegas is the possible target for the next generation of landscape studies. What we need in the post-postmodern world is to pay more attention to the trivial, popular and vernacular in landscapes and in the landscape studies. The use and organization of common or 'trivial' rurban landscapes is significant in the understanding of which ways landscapes of the past and present are understood. This does not mean that the proposed 'learning' would be trivial or naïve falling into the territorial trap of local particularities. Furthermore, our interpretation of past landscapes is always modified by the contemporarity. In this sense the past landscape does not remain 'still in the history' but it is continuously changed by new interpretations.

3.3 Learning from Los Angeles

The paradigmatic case of the late 20th century rurban landscape is Los Angeles. The case of *LA* has been studied *en vivo* for the past two decades by several scholars living in and around Los Angeles – Ed Soja, Michael

Dear, Steven Flusty, Frederick Jameson and Mike Davis to name but a few. This means that both theories and practices are being developed at the same time and place. This escape from modern to postmodern approaches of rurban landscape means that a completely new vocabulary has to be invented (Figure 3). Language, linguistic turn and new symbolism are significant. The concepts are reflections of the empirical and vice versa.

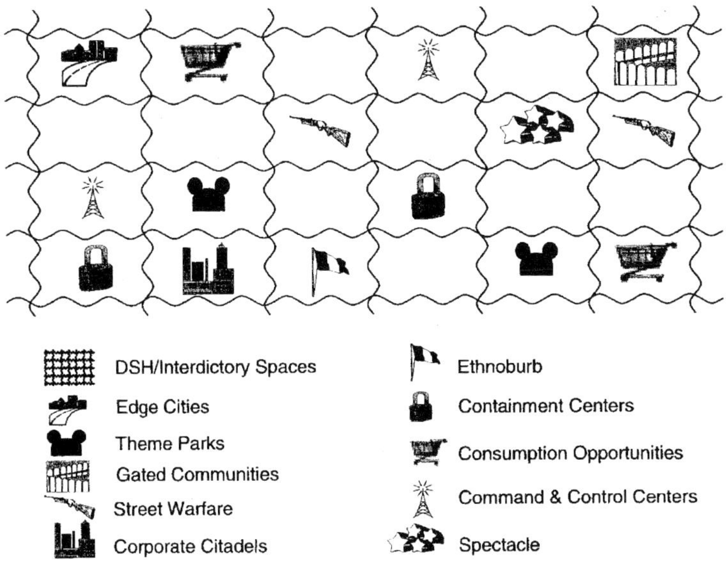

Figure 20-3. Learning from Los Angeles. Source: Dear (with Flusty) (2000)

The new concepts (*LA jargon of the LA school of thought*) are related to local particularities. Many proponents of 'the school' claim that there is no school and they do not necessarily agree with all the details of this new rurban grammar and vocabulary. The scholar-based contextual intuitions mean again the impossibility of transferring these concepts outside Los Angeles. Despite this many scholars have noticed that somehow similar phenomena are taking place in urban fringes in different countries and continents. Nevertheless, the differences between rurban landscapes are evident when studied in detail. Furthermore, the case of Los Angeles illustrates the impossibility of translating the terminology of rurban landscape into different languages. In most cases it is non-understandable for the native speakers and not recognized by the scholars of different schools of thought.

What we can learn from the case of Los Angeles is that in the post-postmodern landscape research language has to be taken more seriously. It might result in landscape studies that we will not share any more the same concepts or that their meaning is consciously modified when used in local contexts. Also, words will not remain any more the main instrument through which the findings and outcomes of the landscape research are opened, presented and criticized. More attention is paid to the different modes of art as modes of communication as well as onto non-visual representation and un-written communication (speech, discussion, legends, everydayness).

3.4 Learning from Tartu

To finish this article about the possibilities of post-postmodern landscape studies I come to Tartu, a town of 100,000 inhabitants in South-Eastern Estonia. In landscape studies Tartu is nothing special and it has not become a global case. Can we use 'global' theories and concepts to study rurban landscapes in Tartu – or anywhere? At first glance, we find seemingly similar rurban landscapes in Tartu that were mentioned in the cases of London, Las Vegas and Los Angeles. The idea of Garden Cities found its way into Tartu in the early 20th century, for example, in the district of Tähtvere that was designed for the new upwardly moving society. The commercial strip of signs indicating the banal, trivial and commercial is visible, for example, in the main Turu Street from the town center outwards. What is physically trendy and conceptually new in postmodern rurban landscapes of LA exist in some details, for example, in the new housing district of Uus-Ihaste (Figure 4).

Figure 20-4. Postmodern rurban landscapes in Tartu. Photo: V. Puik

Is it possible to learn something from Tartu in landscape studies? The rurban landscape as a text appears similar in Tartu and in many cities around the world (though obviously the scale and details matter) compared to the case cities mentioned. However, the readers of the landscape are not the same, the meaning is not the same, nor is the language. Knowledge, experience and language remain place-enriched; thus also the physical outcome, use and meaning of Tartu's rurban landscape. It is not possible to transfer the conceptual and contextual particularities of Tartu elsewhere. This signifies that the analysis and interpretation of rurban landscapes cannot be the same everywhere. The separation of theories and concepts from their contexts and their flexible implementation around the globe may result as an unsuccessful mixture.

4. TOWARDS POST-POSTMODERN LANDSCAPE STUDIES

Learning is a continuous process in which one has to reflect critically upon the past, the contemporary and the future. To my experience – and that is not much – in landscape studies an emphasis on traditional morphological approach concerning pastoral rural landscapes of the past has prevailed despite the world is intensively urban. Following the generational change the postmodern representational approach for urban landscape studies has become increasingly popular among scholars.

What possibilities there are then in the 21st century for comparative approach or collective discussion about (rurban) landscapes? Does virtuality of landscapes in information and communication technology-based world direct our attention from the material to the imagined? I think that the answers cannot be found in traditional modern or current postmodern approaches Modern and postmodern approaches have to be abandoned but the extreme relativist 'anything goes in landscape' or solipsist 'this landscape is true for me' cannot be the answers for the 21st century landscape studies. However, the issue here do not concern only the methodology or methods how to study particular landscapes. From my perspective, in the post-postmodern landscape studies there will be much more sensitive analysis and interpretation of landscapes due to enriched contextualism. Some scholars have indicated this possibility concentrating on biographical landscapes, particularities of individuals and home, political significance of landscape, and landscape as a geobiography (Karjalainen 2000; Olwig 2002; Tuan 2002).

For post-postmodern approaches in landscape studies – contextually and conceptually sensitive, multiple, anti-universal, mono-linguistic and multi-interpretable – there are not clear examples or answers yet. Characteristic for the methodologies is that we do not translate, we do not represent, and we do not bind. Instead we listen, we feel, and we open landscapes in multiple ways. In situational grounded landscape studies the methodology is contextually enriched. Embedded conceptualizations give new insights that arise from the partial perspectives of the researcher.

In the sensitive landscape approach there are no universal theories, concepts or meanings. The point is that in the end there is not a closing synthesis with the landscape of general commonalities but the landscape with its appearance and meanings remain necessarily open. Sometimes this means a wordless, uncategorized and endogenous interpretation – embedded knowledge and situatedness for continuously open and alternative interpretations of landscapes.

REFERENCES

Antrop, M. (2000). Changing patterns in the urbanized countryside of Western Europe. *Landscape Ecology, 15* (3), 257-270.

Barnes, T., & Duncan, J. (Eds.). (1992). *Writing worlds: discourse, text, and metaphor in the representation of landscape*. London: Routledge.

Claval, P. (2002). *The languages of rural landscapes*. Paper presented at 20th Session of The Permanent European Conference On The Study Of Rural Landscapes, Tartu-Otepää, Estonia.

Cosgrove, D. (1984). *Social Formation and the Symbolic Landscape*. London: Croom Helm.

Cosgrove, D., & Daniels, S. (Eds.) (1988). *The iconography of landscape: Essays on the symbolic representation, design, and use of past environments*. Cambridge: Cambridge University Press.

Dear, M., & Flusty S. (2000). *Postmodern urbanism*. London: Routledge.

Environment and Planning D (1986). Society and Space 4:3.

Duncan, J. (1990). *The city as text: the politics of landscape interpretation in the Kandyan Kingdom*. Cambridge: Cambridge University Press.

Harvey, D. (1996). *Justice, nature and the geography of difference*. Blackwell, Oxford.

Hidding, M.C., & Teunissen, A.T.J. (2002). Beyond fragmentation: new concepts for urban-rural development. *Landscape and Urban Planning, 59* (1), 29-42.

Howard, E. (1898). *To-morrow – or the peaceful path for real reform*. London: Sonnenschein.

Karjalainen, P.T. (2000). Earth writing as a humane art. In K. Lehari, & V. Sarapik (Eds.) *Place and Location II*. Tallinn: Estonian Academy of Arts.

Olwig, K. (2002). *Landscape, nature and the body politic. From Britain's Renaissance to America's New World*. Madison: University of Wisconsin Press.

Rose, G. (1993). *Feminism and geography: the limits of geographical knowledge*. Cambridge: Polity.

Tuan, Y.-F. (2002). Foreword. In K. Olwig, *Landscape, nature and the body politic. From Britain's Renaissance to America's New World*, xi-xx. Madison: University of Wisconsin Press.

Venturi, R., Brown, D., & Gans, H. (1966). *Learning from Las Vegas: the forgotten symbolism of architectural form*. Los Angeles.

Widgren, M. (2002). *Three perspectives on landscapes*. Paper presented at 20th Session of The Permanent European Conference for the Study of the Rural Landscape, Tartu-Otepää.